Mesenchymal Stromal Cells

Mesenchymal Stromal Cells
Translational Pathways to Clinical Adoption

Edited by

Sowmya Viswanathan
Institute of Biomaterials and Biomedical Engineering,
University of Toronto;
Krembil Research Institute,
University Health Network, Toronto, ON, Canada

Peiman Hematti
Department of Medicine,
University of Wisconsin-Madison,
School of Medicine and Public Health;
University of Wisconsin Carbone Cancer Center,
Madison, WI, United States

AMSTERDAM • BOSTON • HEIDELBERG • LONDON
NEW YORK • OXFORD • PARIS • SAN DIEGO
SAN FRANCISCO • SINGAPORE • SYDNEY • TOKYO

Academic Press is an imprint of Elsevier

Academic Press is an imprint of Elsevier
125 London Wall, London EC2Y 5AS, United Kingdom
525 B Street, Suite 1800, San Diego, CA 92101-4495, United States
50 Hampshire Street, 5th Floor, Cambridge, MA 02139, United States
The Boulevard, Langford Lane, Kidlington, Oxford OX5 1GB, United Kingdom

Library of Congress Cataloging-in-Publication Data
A catalog record for this book is available from the Library of Congress

British Library Cataloguing-in-Publication Data
A catalogue record for this book is available from the British Library

ISBN: 978-0-12-802826-1

For information on all Academic Press publications
visit our website at https://www.elsevier.com/

Publisher: Mica Haley
Acquisition Editor: Mica Haley
Editorial Project Manager: Lisa Eppich
Production Project Manager: Chris Wortley
Designer: Victoria Pearson

Typeset by Thomson Digital

Dedication

To Paddi, Thambi, and Shreya: this, and everything else is only possible because of you.

Sowmya Viswanathan

To my parents, Parvin and Manouchehr, for all their sacrifices,
To my wife, Shirin, for her endless love and
To my daughters Neeloufar and Faranak, for all the joys they bring to our lives.

Peiman Hematti

Contents

Contributors

Eytan Abraham
Lonza Walkersville Inc., Walkersville, MD, United States

Amin Adibi
The University of British Columbia, Vancouver, BC, Canada

Gillian Armstrong
The EMMES Corporation, Rockville, MD, United States

A. John Barrett
Hematology Branch, National Heart, Lung, and Blood Institute, National Institutes of Health, Bethesda, MD, United States

Jacqueline Barry
Cell and Gene Therapy Catapult, London, United Kingdom

Minoo Battiwalla
Hematology Branch, National Heart, Lung, and Blood Institute, National Institutes of Health, Bethesda, MD, United States

Shashank Bhatt
Krembil Research Institute, University Health Network, Toronto; Cell Therapy Program, University Health Network, Toronto, ON, Canada

Tania Bubela
Public Health, Alberta School of Business, Alberta, Canada

John M. Centanni
Regulatory Affairs Scientist, Institute for Clinical and Translational Research, University of Wisconsin School of Medicine and Public Health, University of Wisconsin, Madison, WI, United States

Ian Copland
Department of Pediatrics, Emory University, Atlanta, Georgia; Aflac Cancer and Blood Disorders Center, Children's Healthcare of Atlanta, Atlanta, Georgia; Department of Hematology and Oncology, Winship Cancer Institute, Emory University, Atlanta, Georgia, United States

Giulia Detela
Cell and Gene Therapy Catapult, London, United Kingdom

Neil Dunavin
Hematology Branch, National Heart, Lung, and Blood Institute, National Institutes of Health, Bethesda, MD, United States

Ashraf Fiky
The EMMES Corporation, Rockville, MD, United States

Jacques Galipeau
Department of Pediatrics, Emory University, Atlanta, Georgia; Aflac Cancer and Blood Disorders Center, Children's Healthcare of Atlanta, Atlanta, Georgia; Department of Hematology and Oncology, Winship Cancer Institute, Emory University, Atlanta, Georgia, United States

Adrian P. Gee
Professor of Medicine and Pediatrics, Center for Cell and Gene Therapy, Baylor College of Medicine, Houston, TX, United States

Patrick Ginty
Cell and Gene Therapy Catapult, London, United Kingdom

Siddharth Gupta
Lonza Walkersville Inc., Walkersville, MD, United States

Peiman Hematti
Department of Medicine, University of Wisconsin-Madison, School of Medicine and Public Health; University of Wisconsin Carbone Cancer Center, Madison, WI, United States

Sunghoon Jung
Lonza Walkersville Inc., Walkersville, MD, United States

Robert Lindblad
The EMMES Corporation, Rockville, MD, United States

Anthony Lodge
Cell and Gene Therapy Catapult, London, United Kingdom

Erika McAfee
Lonza Walkersville Inc., Walkersville, MD, United States

Christopher McCabe
Department of Emergency Medicine Research, Department of Economics, School of Public Health, University of Alberta, Alberta, Canada; Academic Unit of Health Economics, University of Leeds, Leeds, United Kingdom

David H. McKenna
Department of Laboratory Medicine and Pathology, Division of Transfusion Medicine, University of Minnesota Medical School, Minneapolis, Minnesota, United States

Natalie Mount
Cell and Gene Therapy Catapult, London, United Kingdom

Muna Qayed
Department of Pediatrics, Emory University, Atlanta, Georgia; Aflac Cancer and Blood Disorders Center, Children's Healthcare of Atlanta, Atlanta, Georgia, United States

Amish N. Raval
Division of Cardiovascular Medicine, Department of Medicine, University of Wisconsin School of Medicine and Public Health, Madison; Department of Biomedical Engineering, University of Wisconsin, Madison, WI, United States

Eric G. Schmuck
Division of Cardiovascular Medicine, Department of Medicine, University of Wisconsin School of Medicine and Public Health, Madison, WI, United States

Sowmya Viswanathan
Institute of Biomaterials and Biomedical Engineering, University of Toronto; Krembil Research Institute, University Health Network, Toronto, ON, Canada

Deborah Wood
The EMMES Corporation, Rockville, MD, United States

Foreword

It has been fascinating to observe the study of mesenchymal stromal cell (MSC) biology burgeon dramatically over the past 20 years. The impetus to investigate this extraordinary cell population does not seem to be abating—there are at least five new publications daily on MSCs, as reported by PubMed. Much of the work involves cultured, rather than primary in situ cells, and focuses on laboratory research. Part of the excitement concerns the observation that similar cells can be cultured from different tissues and is not merely confined to bone marrow as a source.

Over the past decade however, the focus has shifted toward constructing preclinical models that explore the role of MSCs in ameliorating tissue injury, mitigating inflammatory responses and more recently, in establishing effects on tumor growth or suppression. These studies, in turn, provided a rationale for conducting early phase clinical trials. Many of the preclinical models and the phase I/II clinical trials showed efficacy and held promise for MSCs becoming part of our armamentarium in regenerative medicine.

Outcomes of prospective randomized trials however, have been less salubrious—indeed a number of studies have shown that MSCs are no better than standard of care. These developments suggest that the field is at crossroads and that thoughtful reassessment is now required. This book by Drs Viswanathan and Hematti is therefore extremely timely.

The book examines the critical interface between laboratory observation and clinical implementation. It is entirely possible that many of the topics covered here can help address the limitations we have recently encountered in the clinical translation of MSCs. Donor variability in functional activity, clinical trial design, and cell manufacturing methodologies are among the many issues that spring to mind. Additional topics that provide key insights in translating MSC research to the clinic include the important regulatory aspects of this form of cell therapy.

A. Keating, MD
University of Toronto

Chapter 1

MSCs Translational Process

S. Viswanathan*,**, A. Adibi[††] and S. Bhatt**,[†]

*Institute of Biomaterials and Biomedical Engineering, University of Toronto, Toronto, ON, Canada; **Krembil Research Institute, University Health Network, Toronto, ON, Canada; [†]Cell Therapy Program, University Health Network, Toronto, ON, Canada; [††]The University of British Columbia, Vancouver, BC, Canada

CHAPTER OUTLINE

Mesenchymal Stromal Cells. http://dx.doi.org/10.1016/B978-0-12-802826-1.00001-5

1.1 MSCs AND GLOBAL CLINICAL INVESTIGATIONS

Mesenchymal stromal cells (MSCs) can be isolated from various tissues including bone marrow, adipose, umbilical cord, synovial fluid, periosteum, and others [1], and because of their multifunctional properties [2] are of therapeutic interest for an array of indications. A search of global clinical registration websites, including clinicaltrials.gov, European Union (EU) trials registries, Australian New Zealand Clinical Trials Registry (ANZCTR), and Chinese Clinical Trial Register (CHiCTR), reveals over 490 MSC trials (Table 1.1) (Fig. 1.1). In most trials, the sponsor, the individual, or the institution responsible for the proper conduct, management, and financing of the clinical trial is an academic investigator or center. Given that a majority of early phase trials are still being led by academic investigators with limited personnel and funding resources [3,4], a review of translational requirements and challenges is timely, and is the main focus of this chapter.

1.1.1 Translation Challenges for MSCs

The timeline to translate small molecules and other pharmaceuticals through a developmental pathway that requires rigorous assessment of safety and efficacy parameters in a number of animal and clinical studies is usually over

Table 1.1 Global Overview of Clinical Registries With MSC Clinical Trials Information

Clinical Registry	Number of Trials	Indications	Total Patients
ANZCTR	11	Musculoskeletal, cardiovascular, HSC transplants	239
ChiCTR	34	Cancer, neurology, musculoskeletal, cardiovascular, liver disease, eye disease, diabetes	1,850
ClinicalTrials.gov	344	Cancer, neurology, musculoskeletal, cardiovascular, liver disease, eye disease, diabetes, skin disease/defects, infertility, pulmonary, viral infection, and others	5,369
CRIS	6	Musculoskeletal, cardiovascular, liver diseases	124
CTRI	17	Neurology, musculoskeletal, cardiovascular, liver diseases, pulmonary	351
EU Clinical Trials Register	44	Cancer, neurology, musculoskeletal, cardiovascular, liver disease, eye disease, diabetes, skin disease/defects, infertility, pulmonary, viral infection, and others	997
German Clinical Trials Register	2	MDR/XDR tuberculosis, cancer, pulmonary	120
IRCT	7	Diabetes, neurology, musculoskeletal, eye disease	120
ISRCTN	3	Musculoskeletal	180
JPRN	19	Musculoskeletal, cardiovascular, liver diseases, skin disease/defects, others	291
Netherlands Trial Register	3	Musculoskeletal, neurology, cardiovascular	73

Information gathered from WHO website (http://apps.who.int/trialsearch/) searched for "MSC," "mesenchymal stromal cells," and "mesenchymal stem cells"; searched on June 2015.

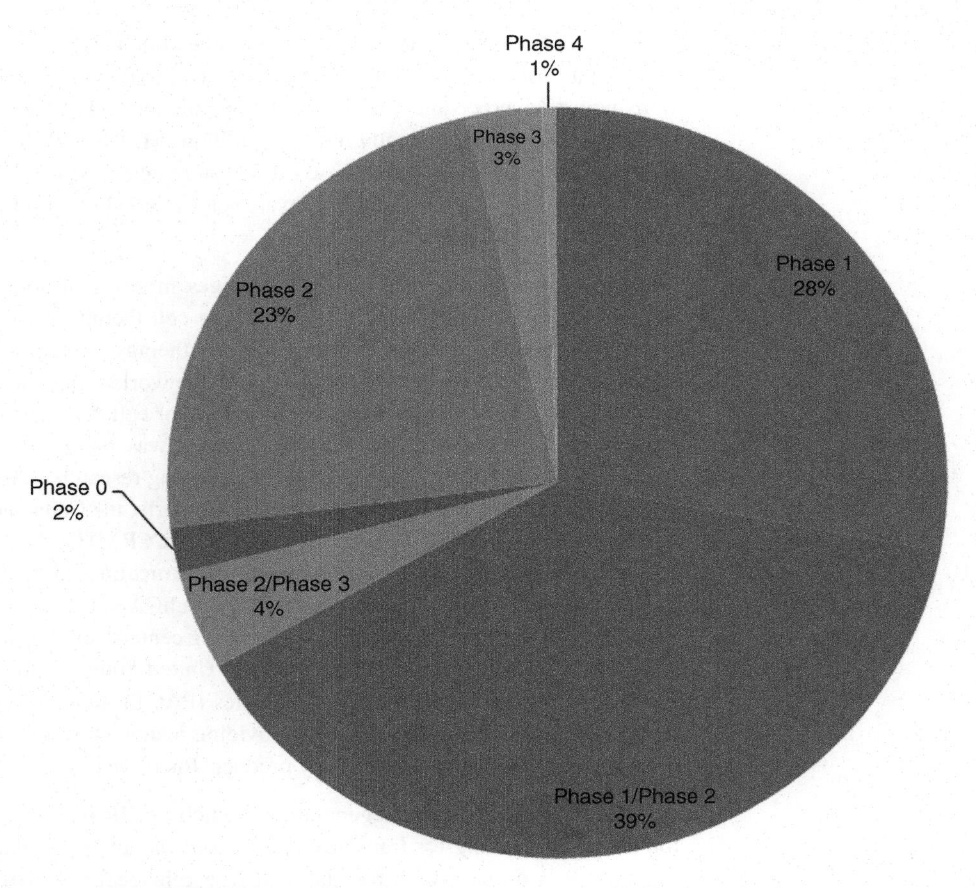

FIGURE 1.1 Distribution of MSCs clinical trials per phase. Information gathered from WHO website (http://apps.who.int/trialsearch/) searched for following strings: "MSC," "mesenchymal stromal cells," and "mesenchymal stem cells" on June 10, 2015.

a decade and costs over 2 billion (B) US dollars from synthesis to market approval [5]. These figures can be extrapolated to MSC-based therapies.

In addition to these costs, there is a significant timeline investment to navigate through the required preclinical and clinical phases to gather substantial evidence of safety and efficacy. For example, the 2012 approval of Prochymal for Osiris Therapeutics' allogeneic MSC therapy for steroid resistant graft versus host disease (GVHD) was preceded by a decade of developmental and clinical testing [6–8]. Similarly, CARTIȘTEM, an allogeneic umbilical cord-derived MSC product for knee cartilage defects in osteoarthritis (OA) patients by MEDIPOST, was approved in South Korea more than 7 years after it was first cleared for testing in clinical trials [9].

Funding for both academic- and industry-initiated clinical trials remains a significant obstacle. Costs for conducting clinical trials vary with jurisdiction, indication, procedure, extent of cell manipulation and cell dose, patient numbers, etc., but are typically over 1–2 million (M) US dollars for Phase I/II trials and over 5–10 M for Phase II trials and higher [10,11]. Given that 73% of recruiting MSC trials in 2015 were academic-led [3], grants are still a major source of funding [4].

Academic investigators face unique challenges in terms of accessing required infrastructure and expertise to navigate cell therapy products. The recognized need for support in academic cell therapy centers has led to several government-funded initiatives around the world, aimed at bridging funding gaps and providing subsidized services for cell therapy studies. In the United Kingdom, the Cell Therapy Catapult was established in 2012 and has been funding cell therapy projects, assisting researchers with grant applications, and providing fee for services. Similarly, the Canadian Centre for Commercialization of Regenerative Medicine (CCRM) has been providing scientists with access to a centralized infrastructure and business development expertise. The recently established CellCAN, another Canadian not-for-profit, is a network of cell manufacturing centers across Canada that provides cell manufacturing services. In the United States, until 2015 the Production Assistance for Cellular Therapies (PACT), funded by the National Institutes of Health (NIH), was providing manufacturing and regulatory support for cell and gene therapy products for 10 years.

The California Institute of Regenerative Medicine (CIRM) remains a major source of funding for translating cell-based therapies. To date it has funded 1.9 B dollars, with translational research receiving 630 M dollars. The next iteration of CIRM 2.0 has earmarked 800 M in funding for late-stage preclinical projects, clinical trials, and accelerating activities in California [12].

Importantly, CIRM funding has been leveraged with other sources of money including disease foundations (Juvenile Diabetes Research Foundation has contributed to CIRM funding toward ViaCyte's embryonic stem cell-based device product for the treatment of Type 1 diabetes), and pharmaceutical companies (Janssen, a division of Johnson & Johnson, has cofunded Capricor's trial on myocardial infarction along with CIRM). This leveraging is critical for increasing the academic funding investment, and for allowing translational work to progress in a limited funding environment.

The NIH also remains an important source of peer-reviewed funding in the United States. However, the typical research project (R01) and program project (P01) grants are insufficient for clinical trial implementation. The

American Society of Cell and Gene Therapy has put out a position statement recommending a new streamlined structure for peer-review with appropriate clinical and product development review panel expertise [13]. Indeed, there is general acknowledgement that most funding bodies are insufficiently positioned for enabling translational and clinical research of cell-based therapies.

The European Union (EU) funds research through framework programs that specifically support translational research and clinical trials. The recently concluded seventh framework program (FP7) from 2007 to 2013 focused on cancer, cardiovascular diseases, diabetes and obesity, rare diseases, and severe chronic disease [14]. The eighth framework program, Horizon 2020 (to run from 2014 to 2020), will have a larger budget 7.4 billion (B) Euros with funding available for clinical research on regenerative medicine in the first 2 years [15].

Although there has been excitement about several companies raising venture capital (VC) including Semma Therapeutics ($USD 44 M [16]), Juno Therapeutics ($USD 310 M [17] with a market cap of $USD 4.6 B [18]), Bluebird Bio, and Kite Pharma ($USD 5.6 B and $USD 2.5 B market caps respectively [19,20]), the landscape of private funding for early-stage cell therapy products has generally been sparse. However, patient advocacy foundations, philanthropists, and pharmaceutical companies are increasingly filling the gap [21]. As a result, companies are turning to more innovative ways to acquire funding. For example, Cardiff-based Cell Therapy Limited (CTL) raised almost £700,000 in early 2015 using crowd funding in exchange for less than 1% of the company. CTL is developing MSC-like cells and has early phase clinical data on the treatment of myocardial infarctions.

1.1.2 Regulatory Classification of MSCs

MSCs are considered more than minimally manipulated and are typically not used in a homologous manner, that is, they do not perform the same function(s) in the recipient and the donor, and thus qualify as medicinal products under most jurisdictions including under the European Medicine Agency (EMA) [22], the Food and Drug Administration (FDA) [23], and Health Canada [24]. They thus require authorization for clinical investigations by regulatory bodies including National Competent Authorities (NCA) in Europe, the FDA in the United States, and Health Canada in Canada. In the United States, MSCs are regulated under section 351 of the Public Health Services (PHS) Act (42 U.S.C. 262) [25], while in Canada they are regulated under the Food and Drug Regulations [26]. In Europe, MSCs fall under Directive 2001/83/EC [27] and Regulation EC (No) 1394/2007 [28].

Table 1.2 Listing of Combination Product (MSCs and Scaffolds) Clinical Trials

Scaffold Type	ClinialTrials.gov Identification	Title of Study
Granulated biphasic calcium phosphate	NCT01842477	Evaluation of Efficacy and Safety of Autologous MSCs Combined to Biomaterials to Enhance Bone Healing (OrthoCT1)
Periodontium gel	NCT00221130	Clinical Trials of Regeneration for Periodontal Tissue
Ceramic granules	NCT01742260	Cranial Reconstruction Using Mesenchymal Stromal Cells and Resorbable Biomaterials
Mixture of collagen and fibrin	NCT02449005	Transplantation of Autologous Bone Marrow Mesenchymal Stem Cells for the Regeneration of Infrabony Periodontal Defects (PerioRegen)
Cross-linked matrix of autologous plasma	NCT01389661	Treatment Of Maxillary Bone Cysts With Autologous Bone Mesenchymal Stem Cells (MSV-H) (BIOMAX)
Collagen	NCT00850187	Autologous Transplantation of Mesenchymal Stem Cells (MSCs) and Scaffold in Full-thickness Articular Cartilage
Collagen	NCT02313415	Treatment of Infertility by Collagen Scaffold Loaded With Umbilical Cord Blood-derived Mesenchymal Stem Cells
Collagen	NCT02352077	Nerve Regeneration-guided Collagen Scaffold and Mesenchymal Stem Cells Transplantation in Spinal Cord Injury Patients
Hyaluronan	NCT01290367	Safety and Preliminary Efficacy Study of Mesenchymal Precursor Cells (MPCs) in Subjects With Lumbar Back Pain
Fibrin	NCT02037204	IMPACT: Safety and Feasibility of a Single-stage Procedure for Focal Cartilage Lesions of the Knee
Demineralized bone matrix	NCT00250302	Autologous Implantation of Mesenchymal Stem Cells for the Treatment of Distal Tibial Fractures
Hyaluronan	NCT01981330	Pilot Study of Stem Cell Treatment of Patients With Vocal Fold Scarring
Master graft resorbable ceramic granules	NCT00549913	Study of 3 Doses of NeoFuse Combined With Master Graft Granules in Subjects Requiring Posterolateral Lumbar Fusion (PLF)
Hyaluronan	NCT01088191	Safety and Efficacy Study of MSB-CAR001 in Subjects 6 Weeks Post an Anterior Cruciate Ligament Reconstruction

Clinicaltrials.gov searched for the terms, "mesenchymal," "scaffold," and "carrier." Terminated and withdrawn studies excluded. Studies with minimally manipulated bone marrow derived mononuclear cells excluded; searched on June 2015.

Scaffolds and carriers are often used to facilitate MSC transplantation, locally retain cells, and/or provide mechanical support for tissue regeneration, typically in orthopedic indications. Examples of scaffold and carrier materials include collagen, hyaluronic acid, decellularized tissue, ceramic granules, and a variety of absorbable and nonabsorbable polymers (Table 1.2). Most jurisdictions including the FDA regulate combination products based on their primary mode of action, either as a device or as a drug.

Market authorization for sale of MSC products varies slightly from jurisdiction to jurisdiction, but typically requires submission of substantial product clinical and preclinical evidence for the regulatory agencies to perform a thorough risk/benefit analysis. The FDA requires a Biologics License Application (BLA) to introduce a biologic product for sale into interstate commerce under 21 CFR (Code of Federal Regulations) 601.2 [29]. A centralized application can be submitted to EMA, which makes recommendation based on review by the agency's scientific committee, Committee for Advanced Therapies (CAT) to the European Commission (EC). EC has the authority to grant market authorization in all member states. Health Canada requires submission of a New Drug Submission (NDS) under Division 8, Part C of the Food and Drug Regulations [30].

Depending on the indication of use, MSC products may qualify for accelerated market authorization. Prochymal or Remestemcel-L, a non-human leucocyte antigen (HLA) matched, cultured, expanded, cryopreserved allogeneic MSC product, received conditional market authorization from Health Canada in 2012 for the treatment of acute pediatric GVHD that is refractory to corticosteroid or other immunosuppressive agents. Prochymal is also available through the Extended Access Program (EAP) in the United States, and in other jurisdictions including New Zealand.

In some cases, there is continued evaluation of marketed products. For example, an autologous bone marrow-derived MSC product, HeartiCellGram-AMI, received marketing approval from South Korea in 2011 for treatment of acute myocardial infarction (AMI), 1 month post-AMI. A postmarket study on 58 patients showed modest changes in left ventricular ejection fraction (LVEF) by single photon emission computed tomography (SPECT) from baseline to 6 months posttreatment; however, there were no significant differences in % LVEF values at the 6-months posttreatment between the groups [31]. It is unclear how the regulators will respond to this postmarket data and how commercialization and use of this product by the clinical community will be affected.

1.1.3 Regulatory Process for Clinical Investigations

To initiate a clinical investigation, the sponsor needs to demonstrate scientific rationale for the clinical use of MSCs via sufficient preclinical studies typically involving some combination of in vitro and in vivo data. Regulators also need to review chemistry, manufacturing, and control (CMC) of the MSC product. A fully detailed clinical protocol is the third important component of a regulatory review process.

There are multiple guidelines on submission, contents, and review process of an investigational new drug (IND) filing with the FDA through the Center for Biologics Evaluation and Research (CBER)'s Office of Cellular, Tissue, and Gene Therapies (OCTGT) [30,32,33]. An IND is a formal document with sections outlining the general investigational plan, investigator's brochure (IB), clinical protocol, informed consent, CMC of the product, preclinical data, and safety testing including pharmacology and toxicology, previous human experience, a statement of investigator, and additional information.

Clinical investigations require clinical trial authorization (CTA) from NCAs in Europe, and concurrent ethics review, typically a 60-day review process. Information on submission of CTA is available in section 2 of the European Commission (EC) Guidance CT-1 [34], and through Eudrodex (v10) clinical trials guidelines [35]. Prior to CTA submission, sponsors need to register their trial with the EudraCT Community Clinical Trial System, and obtain a unique EudraCT number. An investigational medicinal product dossier (IMPD) in the common technical document (CTD) format is prepared along with the IB for evaluation by the NCA of each member state. NCAs can also provide regulatory, scientific, or procedural advice for both clinical trials and market authorization; the EMA can also provide consensus scientific advice from member states, especially on market authorization.

In Canada, investigational use of MSCs under the Food and Drug Regulations requires specific Health Canada authorization through a clinical trial application (CTA) process that must be submitted before initiating each of Phases I–III clinical trials. Health Canada has recently issued a draft guidance on preparation of CTAs for use of cell therapy products in humans [36], which along with other existing guidelines helps with CTA submissions [37].

1.2 PRECLINICAL STUDIES

1.2.1 Regulatory Requirements for Preclinical Studies

The goal of conducting preclinical studies is to produce sufficient evidence on safety and efficacy of the new cell therapy to support its use in clinical trials. This includes data on feasibility of the proposed treatment, dose levels and dosing regimen, and characterization of the route of administration. Preclinical studies follow a risk–benefit approach, with proof-of-concept studies focusing on the benefits and safety studies focusing on potential risks. The International Council on Harmonization (ICH) has guidance documents on preclinical safety studies for conventional pharmaceuticals and biologics [38], which are not fully applicable for designing animal studies for cell and gene therapies. In the United States, FDA has a more relevant guidance document on preclinical assessment of cell and gene therapy products [39]. This guidance document calls for the use of the clinical grade investigational products in definitive preclinical studies, where possible, but allows for testing of an analogous animal product, provided the analogy between the animal and human cellular product is characterized and sufficiently justified.

The guideline also provides information on selection of appropriate animal species including the use of models of disease/injury to provide more informative risk–benefit analyses, especially for cell and gene therapy products which may have multiple mechanisms of action. The guidelines recommends the use of a tiered approach recognizing that all animal models have some limitations and that the complex properties of cell and gene therapy products may need to be tested in multiple models to obtain the most relevant information. The guideline also recognizes that many of these proof-of-concept studies in specialized models of disease/injury may not be good laboratory practice (GLP)-compliant, and recommends that deviations that impact the quality of data be justified; an independent quality assurance unit/person is recommended to provide oversight on such studies.

The FDA also provides guidance on preclinical study considerations in defined product areas including knee cartilage repair or replacement [40], cardiac repair [41], allogeneic pancreatic islet cells [42], and xenogeneic products in humans [43].

The EMA provides guidelines on nonclinical studies including pharmacodynamics, pharmacology, toxicology, biodistribution, tumorogenicity, reproductive toxicity and local tolerance of cell-based medicinal products (CBMPs) in its 2007 guidelines on CBMPs [44]. EMA has also put out minimal guidelines on the requirements that nonclinical studies must address for clinical investigation of gene therapy products [45].

1.2.2 Proof-of-Concept Studies/Animal Models

Using disease animal models instead of healthy animals is encouraged by regulators to demonstrate multiple mechanisms of action of MSCs which involve complex interactions with the injured/diseased target tissue. Choosing a suitable disease model is thus crucial and should be based on careful consideration of several factors including comparability of physiology and anatomy of the species of choice with humans, immune tolerance of the selected model to human cell therapy products, and the feasibility of using the same delivery system that is to be used in clinical settings [39].

Although small rodents are generally preferred in MSC preclinical studies due to low cost, abundance of specialized reagents, and easier animal care and handling, certain disease models and delivery methods may require using larger animals. Use of human MSCs in immune-compromised rodents is preferred over use of rodent MSCs, as the two cell types differ significantly in their mechanisms of action [46]. While small animals are better for determining mechanisms of action at cellular/molecular levels, using larger animals allows for a more clinically relevant functional assessment of the endpoints, and facilitates clinical diagnostics and treatment procedures on the animal. However, large animal models may require analogous MSCs, which may have different mechanisms of action or the use of human MSCs along with immunosuppression.

Most companies working with MSC products use multiple animal models. For example, Athersys used analogous cells to their MultiStem product in smaller disease models to conduct proof-of-concept and dosing studies, and then moved to larger animals to model human delivery systems. Typically, the larger animal studies were GLP-compliant [47] pivotal studies, done after extensive consultation with regulatory authorities.

An example of using of larger animal models to demonstrate MSC efficacy can be seen in osteoarthritis (OA), which has necessitated load-bearing animals to more accurately capture the physiology and biomechanics of weight-bearing joints. Dogs, sheep, goats, and horses have been used to demonstrate a regenerative effects of MSCs on cartilage lesions, and antiinflammatory effects on bone remodeling [48]. FDA's guidance document for products intended to repair or replace cartilage includes recommendations for animal model selection, which can be applied to OA [40].

Similarly, larger animal models have been used for cardiac diseases to enable appropriate delivery of cells, as well as for imaging purposes [49–56]. FDA guidelines recommend the use of larger animals for studying arrhythmogenesis, and establishing potentially safe starting doses and delivery methods [57]. MSC products for brain aneurysms have also been evaluated in small and large animal models [58,59], and recommendations from stem cell therapy as an emerging paradigm for stroke (STEPS) suggest using larger animal models when specific neuroanatomical sites such as white matter are being studied [60].

1.2.3 Safety Studies/GLP Requirements

A prospectively designed study protocol with well-defined methods and endpoints, similar to that of the intended clinical study, is usually required by regulators for demonstration of preclinical safety [39]. Preclinical safety studies should include: (1) sufficient number of animals per gender, (2) appropriate control groups, (3) multiple dose levels, (4) dosing schedule as close as possible to the proposed regimen for clinical studies, and (5) routes of administration similar to proposed clinical routes. Safety endpoints should at least include mortality, body or organ weight, gross and histopathology of target and nontarget organs and tissues, behavioral changes such as food and water consumption, blood work, and urinalysis [39]. Safety endpoints should be evaluated at multiple time points to capture "acute, chronic, or delayed-onset responses" and possible resolution of them [39].

Immunogenicity is a concern when human cells are used in animals. Although the immune privileged nature of MSCs allows allogeneic transplantation of cells without serious complications, transplantation of human MSCs into animals may require immunosuppression. If the proposed treatment involves using autologous cells in humans, administration of analogous cell type in the animals may be appropriate, provided that the level of analogy between animal and humans cells is sufficiently characterized [39].

Information about the biodistribution of MSCs after local or systemic administration should also be provided. Intravenously (IV) injected MSCs initially become entrapped in the lungs [61–64] and are redistributed to other filtering organs such as liver, spleen, and kidneys within 24–48 h of injection [65]. The potential risk for pulmonary embolism, especially at higher doses [66], calls for careful control of dose and injection rates, and interpretation of animal dosing data.

Tumorigenicity is generally not a concern with human MSCs, as they have limited replicative potential and are considered nontransformative [67,68]. This is confirmed by a metaanalysis study reviewing data from 36 MSC clinical investigations (1,012 patients) [69], which showed no correlations between MSC and any acute infusional toxicity, organ system complications, infection, death, or malignancy. The only adverse event associated with MSC treatment was transient fever [69].

1.3 CHEMISTRY, MANUFACTURING, AND CONTROL

1.3.1 Regulatory Requirements

Investigational use of MSCs under an IND or CTA in North America requires documentation on the manufacturing process, and compliance with Good Manufacturing Practice (GMP). In the United States, there is a requirement for compliance with GMP for manufacturing, processing, packaging, or holding investigational drugs [70]. However, Phase I investigational drugs including somatic cell therapy products and gene therapy products may be exempt from full cGMP regulations [71], although investigational products are still expected to comply with the statutory GMP requirements. Regulators have increased expectations of GMP compliance as the product matures through clinical trials; full cGMP compliance is a mandatory requirement for product licensure through both the FD&C Act and the PHS Act, and is coded in CFR Parts 210 and 211 [72,73].

Investigational use of MSCs under a CTA in the EU requires compliance with GMP, and licensing of the manufacturing facility by a NCA in addition to meeting donor screening requirements under the European Tissues and Cells Directive (EUTCD) [74]. The 2003/94/EC Directives also lay out guideline principles for GMP compliance for medicinal products [75].

Health Canada has similarly increasing stringency requirements for cGMP compliance according to applicable regulations (Division 2 of the Food and Drug Regulations) and guidelines [76] through clinical development. However, interpretation of how these GMP guidelines are applied, especially during early phases of clinical trials, is flexible without compromising product safety. Product licensure requires full facility inspection and procurement of an Establishment License (EL) [24] by the manufacturer or distributer.

In general, GMP compliance in these jurisdictions includes demonstration of the following: selection of appropriately screened starting and raw materials, ancillary reagents and excipients, qualification and validation of the manufacturing process; quality control of materials; in process control of critical manufacturing steps; product release criteria and specifications for acceptance of products (identity, purity, potency, dose, and viability); product stability; use of appropriate reference materials; use of qualified containers and closure systems; demonstration of facilities that meet appropriate GMP requirements; qualified personnel and an appropriate quality control unit (Fig. 1.2).

1.3.2 MSC Source Material

For MSC products, the starting cellular material is arguably one of the most important determinants of product quality, efficacy, and safety. Therefore, specifications must be clearly defined for the MSC source. This is particularly important as there is heterogeneity of culture expanded MSCs, driven largely by donor variability. Autologous MSCs or allogeneic MSCs which are generated from multiple donors may thus have different functional and therapeutic values. Additionally, MSCs may be derived from different tissue sources using different cell processing techniques or culture conditions amplifying this inherent variability.

All allogeneic donors who meet selection criteria are subject to screening and testing measures aimed at preventing transmission of communicable diseases. Most jurisdictions have extensive guidelines regarding donor screening. Under 21 CFR part 1271 [77], FDA requires using FDA-licensed kits to test for: human immunodeficiency virus types 1 and 2 (HIV-1 and HIV-2), Hepatitis B virus (HBV), Hepatitis C virus (HCV), Treponema pallidum (syphilis), human T-lymphotropic virus types I and II (HTLV I/II) for donors

FIGURE 1.2 Multiple aspects of MSC manufacturing.

of viable leukocyte-rich HCT/Ps, Cytomegalovirus (CMV) for donors of viable leukocyte-rich HCT/Ps, Trypanosoma cruzi (Chagas' disease; T. cruzi), and West Nile virus (WNV). In addition, diseases such as human transmissible spongiform encephalopathy (TSE), vaccinia, and sepsis require medical history screenings as there are no FDA-licensed tests available. The United States currently excludes EU donors due to risks of Creutzfeldt-Jakob disease (CjD).

In both the United States and Canada, transmissible disease screening and testing measures are recommended but not required for autologous donors of MSCs. In Europe, the same level of testing is required for both autologous and allogeneic donors [74]. The EUTCD directs member states on minimal donor testing requirements, but NCAs may require additional testing. There are differences in the testing requirements between the FDA [77] and EU [74], and Health Canada [76].

While bone marrow has been the most common source for MSC preparations, other tissue and cell sources, including adipose tissue, placental/umbilical cord blood, and amniotic fluid have increasingly been used [78]. Collection, banking, testing, and distribution of cord blood is also governed by 21 CFR part 1271 [77] and must be compliant with Good Tissue Practices (GTP). In EU, establishments that process, bank, and distribute tissues must be licensed by an NCA under directive 2004/23/EC [74]. Additional considerations for collection of umbilical cord blood derived cells or tissues include maternal screening and testing, typically within 7 days of collection using validated test methods.

1.3.3 Ancillary Materials

To isolate and expand MSCs, ancillary materials (AMs) that are not part of the final product formulation are used. The US Pharmacopeia has a chapter <1043> on AMs including their testing and qualification [79]. ICH has quality guidelines (Q7) on the manufacturing of active pharmaceutical ingredients (API) [80], and there are EU guidelines [44] on AMs. All of these guidelines provide recommendations on the qualification, characterization, and testing of AMs to appropriate standards suitable for clinical use. If the AMs are high-risk materials, such as those of human or animal source, or have known or potential toxicities associated with them, final MSCs products should be tested for presence of residual AMs. This may be done as a qualification study and/or as part of routine lot release testing for MSCs, as appropriate.

As an example of AM, GMP-grade Percoll and Ficoll is used for isolating MSCs from bone marrow (BM) [81]. Different enzymes are used for MSC isolation from other sources including collagenase, hylarunidase, dispase, and trypsin. Collagenase is frequently used for isolation from adipose tissue, while hylarunidase and dispase are used along with collagenase for isolation from umbilical cord [82] and gingival connective tissue [83] respectively. Trypsin is a serine protease of porcine origin, and requires additional testing for porcine adventitious agents. All enzymes as AMs require demonstration of safety, sterility, and absence of pryogenicity and adventitious agents.

1.3.3.1 Medium and supplements

Medium composition affects proliferation and differentiation of MSCs [81], and typically needs to be optimized depending on the tissue source [84].

A majority of clinical investigations use DMEM-low glucose (53%) for MSC expansion, 15% use DMEM, and only 11% use Minimal Essential Medium (MEM) [81]. Sotiropoulou et al. [85] systemically compared the expansion efficacy of MSCs and showed that among serum-containing medium, MEM containing Glutamax achieved the greatest expansion of BM-derived MSCs.

1.3.3.2 Fetal bovine serum versus xenogeneic-free alternatives

Supplementation, especially with factors such as fibroblast growth factor-2 (FGF2), platelet-derived growth factor (PDGF), epidermal growth factor (EGF), transforming growth factor (TGF), and insulin-like growth factor (IGF), enhances MSC proliferation [86–88]. These factors are abundant in fetal bovine serum (FBS), which has been used frequently in MSC investigational trials [69% (25 of 36) of trials] and more than 80% of IND submissions reported using 2–20% FBS [79,90]. Commercial suppliers of GMP grade FBS need to provide documentation on the country of origin and testing of adventitious agents. The European Directorate for the Quality of Medicines and HealthCare (EDQM) certifies FBS, and EDQM-certified FBS is a requirement for investigational use in the EU and Canada. The FDA also has regulations [89] regarding the use of bovine materials, the certificate of analysis (CoA) needs to additionally be compliant with requirements described in 9 CFR 113.53 [90]. However due to batch-to-batch variability of FBS, latent immunogenicity especially upon repeat infusion [91], risk of prion and adventitious viral transmission, regulators and sponsors alike are keen to move toward serum- and xenogeneic-free medium.

Pooled human serum [92] and platelet lysate [93] are sometimes used as replacements to bovine serum, to avoid the risk of hypersensitivity and prion transmission. Platelet lysate can be easily obtained from apheresis products and buffy coat of healthy volunteers and has been used in more than 10% of clinical trials conducted between 2007 and 2013 [81]. However, risk of adventitious agent transmission, and batch-to-batch variation issues, persist.

Commercial vendors have developed serum- and xenogeneic-free medium, which contain a number of defined factors that can replace FBS in expanding MSCs without compromising the quality of the cells [94]. These commercial medium have proprietary formulations, but some of them have drug master files filed with the FDA to facilitate IND applications. MSCs grown in some serum-free medium have shown superior population doubling times, homogeneity, and colony-forming capacity when compared with traditional serum-containing medium [95]. Clinical trials are now beginning to use serum- and xenogeneic-free medium (personal communications to Viswanathan, S).

1.3.3.3 *Adhesion proteins and microcarriers*

MSCs are typically grown on plastic tissue culture dishes as monolayers with no additional coatings. The use of xenogeneic-free medium has prompted the development of xenogeneic-free substrates including fibronectin. Fibronectin is obtained from plasma of screened and tested human donors. Laminin and collagen type I have also been shown to support MSC attachment [96]. Proprietary surface-coating formulations are being used on precoated microcarriers to support scalable MSC expansion [97], and on traditional tissue culture plastic [98]. A number of commercially available microcarriers including Cytodex-1, Cytodex-3, Cultispher-S, Cultispher-G, and SoloHill Plastic, SoloHill Plastic Plus, SoloHill Collagen, SoloHill Pronectin F, and SoloHill Hilex II have been investigated for MSC expansion (Table 1.3). Appropriate cell-harvesting techniques for detachment of cells from the microcarrier surface need to be further developed and validated as there have been issues recovering MSCs from microcarriers [99].

1.3.3.4 *Antibiotics*

Regulators recommend manipulation of cells in the absence of exposure to beta-lactam antibiotics to prevent adverse events in patients with sensitivities to such antibiotics. However, if MSCs are sourced from placental, adipose, umbilical cord blood, or other sources requiring open manipulations it may be appropriate to use antibiotics to promote the aseptic expansion of MSCs, at least for the initial passages. Importantly, the decision to use beta-lactam antibiotics as AM may need to be appropriately justified and precautions taken to inform patients, and screen them on the basis of hypersensitivity.

1.3.4 Excipients Used in Final MSC Product Formulation

Excipients are inactive ingredients added to the final product formulation to protect the active ingredient. Regulators require information on source, concentration, qualification, and characterization of all excipients, in a similar manner to AMs. Human serum albumin (HSA) purified from plasma of appropriately screened and tested donors is widely used as an excipient. It is not associated with safety or immunogenicity concerns, and as a stable protein makes for easy storage and handling. Sodium hyaluronate, a glycosaminoglycan found in various connective, epithelial, and neural tissues, is

Table 1.3 Propagation of MSCs on Microcarriers

Microcarrier Type	MSC Expansion With Microcarriers	Microcarriers Matrix/ Composition (Brand Name)	Manufacturer	Reference
SoloHill Plastic	~16-Fold	Cross-linked polystyrene (Rexolite), no coating	Pall Corporation	[99]
SoloHill Plastic	~18-Fold	Collagen	Pall Corporation	[99]
SoloHill Plastic Plus	~12-Fold	Cross-linked polystyrene (Rexolite), cationic charge, no coating	Pall Corporation	[99]
SoloHill Pronectin F	~16-Fold	Recombinant RGD-containing protein	Pall Corporation	[99]
SoloHill Hilex II	~7-Fold	Modified polystyrene, no coating	Pall Corporation	[99]
Cultispher S	6–15-Fold	Gelatin	Thermo Scientific	[100–103]
Cytodex 1	4-Fold	Dextran	GE Healthcare	[104]
Cytodex 3	20-Fold	Type I collagen/Dextran	GE Healthcare	[105]
Cultispher S and polystyrene beads	14–18-Fold	Gelatin	Thermo Scientific	[106]
Gelatin beads	Not available	Not available	Not available	[107]
SoloHill C102	16.7-Fold	Type I porcine collagen	Pall Corporation	[108]
SoloHill F102	8.2-Fold	Type I porcine collagen, cationic charge, no coating	Pall Corporation	[108]
SoloHill G102	8.8-Fold	Glass, no coating	Pall Corporation	[108]
Microcarriers fabricated from decellularized adipose tissue (DAT)	6-Fold	Not available	Not available	[109,110]

another approved excipient, that has been used in CARTISTEM, a MEDIPOST MSC product.

Dimethyl sulfoxide (DMSO) has been used as an excipient, for cryopreservation, in off-the-shelf MSC products (Prochymal) and in investigational use (approximately 35% of trials [69] used cryopreserved MSCs typically in the range of 5–10%) [81]. Adverse events such as seizures have been associated with intracranial administration of DMSO [111–113] in patients with preexisting conditions, and caused concerns. Removal of DMSO by washing and resuspension, dilution to lower concentrations, or the use of defined cryopreservation medium with cyropreservants such as glycerol and polyethylene glycol (PEG) or sugars such as sucrose, trehalose, and dextrose are all being considered.

1.3.5 MSC Manufacturing Process

Regardless of the source of MSCs, cells must be obtained from appropriately consented, screened and tested donors, isolated and purified, expanded to obtain appropriate cell numbers and formulated for storage/shipping or fresh infusion. MSCs may be isolated from BM cells by density gradient centrifugation or from adipose or umbilical cord tissue by enzymatic digestion

[114,115]. The isolated MSCs are typically washed and plated on adherent surfaces including microcarriers in suspension cultures. Cells can be expanded in flasks, cell factories, and other bioreactors including hollow-fiber bioreactors that are coated with fibronectin (eg, Quantum device from Terumo BCT) [116], stirred suspension bioreactors using coated microcarriers (eg, MobiusCellReady bioreactor [117]), and passaged to remove contaminating hematopoietic and endothelial cells. Cells are harvested with the aid of enzymes such as trypsin and ethylenediaminetetraacetic acid (EDTA), and replated at controlled densities that can vary depending on the cell source, cell medium supplementation, and flasks or bioreactors used. MSCs are typically used up to 30 population doublings, or prior to P7 as quality of MSCs seems to diminish with passage [118]. MSCs may be formulated in excipients consisting of buffers, serum proteins, and cryoprotectants such as DMSO, glycerol, PEG, or sugars. If a master cell bank (MCB) is used, additional testing and documentation on adventitious agent testing is needed [30,119,120]. Typically, the use of multiple donors results in the creation of multiple MCBs, which need to be compared to show consistency between lots. A stability program with validated cell recovery data at periodic intervals, documentations on storage conditions for banked cells, and effect of cryopreservation process and medium should be in place for cryopreserved MSCs.

1.3.5.1 MSC in-process and final product release testing

The MSC-manufacturing process, intermediate products and final product are controlled by testing for various safety (sterility and mycoplasma), identity, purity (endotoxin free), and potency attributes throughout the process and as part of lot release. Guidelines for safety and endotoxin testing are well established and standardized between various jurisdictions [121–123]. Pyrogen and mycoplasma testing are also relatively standardized or methods can be validated per existing guidelines [121,124–126]. There is much less consensus on other assays including those that characterize MSCs identity and potency.

In the absence of definitive functional markers for MSCs, the MSC committee of International Society of Cellular Therapy (ISCT) has minimally defined MSCs [127] as CD105+, CD73+, CD90+, and CD45−, CD34−, CD14− or CD11b−, CD79α- or CD19- and HLA-DR−. However, these markers are common to fibroblasts and stromal cells, and only serve to discriminate MSCs from hematopoietic cells. MSCs from different tissue sources may additionally exhibit different sets of cell surface marker. For example, adipose-derived MSCs are CD36-positive and CD106-negative while nonculture-expanded, stromal vascular fraction (SVF) cells are CD45, CD235a, and CD31 negative, and CD34 positive [128].

Potency assays need to quantitatively measure biological activity based on an attribute of the MSC product which is linked as closely as possible to a relevant biological property. Potency assays should represent a mechanism of action, be quantitative, include appropriate reference standards/materials, and characterize relevant product attributes. There are multiple guidelines available on developing appropriate potency assays [44,129–133]. Again, there is little consensus on defining MSC functionality given their multiple modalities of activity. ISCT had minimally defined MSC functionality based on in vitro trilineage differentiation [127], but there is little evidence of MSC cell replacement in most indications. The MSC committee at ISCT has additionally put out recommendations on relevant assay parameters measuring MSC modulation of immune cells in classical mixed lymphocyte reactions (MLR) or using purified immune effectors [134]. The MSC committee at ISCT has an opinion paper on developing multiple functional rather than potency assays for MSCs, recognizing the complexity and challenges of defining simple in vitro read-outs that are predictive of product performance with respect to desired clinical outcomes [135]. Developing multiple functional assays for MSCs will additionally help with product release, comparison between batches, comparability between products after process changes, and confirming aspects of stability including lot expiration.

Commercial MSC products such as Prochymal have developed quantitative functional assays, such as measurement of tumor necrosis factor receptor 1 (TNFR1) levels, which are associated with in vivo functionality. Other MSC functional assays could include secretion of TNF-α stimulated gene/protein 6 (TSG-6), secretion of proangiogenic factors (eg, VEGF, IL-8, CXCL5) [136,137], enabling endothelial tube formation (in angiogenic assays), and more.

1.4 CLINICAL PROTOCOL

1.4.1 Regulatory Requirements

Investigational use of MSCs requires adherence to principles of Good Clinical Practice (GCP) in the United States [138–141], EU [142,143], and Canada [144]. ICH has also put out a consolidated GCP guidance for industry [145], which highlights the roles and responsibilities of investigators, sponsors, monitors, and ethics board and spells out the requirements for the safe and responsible conduct of clinical trials.

1.4.2 Clinical Protocol Consideration

MSC protocols should conform to GCP [146] as outlined in Section 2.3. The study protocol should generally address these topics: (1) study rationale,

(2) nonclinical and clinical study summary, (3) safety, (4) route of administration and dosage, (5) study objectives, (6) trial design, (7) study population, (8) treatment plan, (9) study assessments, (10) statistics, (11) stopping rules/data and safety monitoring board, (12) serious adverse event reporting, (13) data handling and record keeping [146]. Clinical protocols should include instructions for investigators on potential adverse events, for example, measures in case of pulmonary embolism with systemic administration of MSCs, measures for dealing with transient fever [69], or mild DMSO-related toxicity events (pretreatment with α-histamines).

Investigational use of MSCs also requires clearly defined end-point measures, and definitions of success criteria. For early phase trials however, outcomes may largely focus on safety of MSCs, and not necessarily a direct therapeutic effect; patient awareness and consent is therefore important. While intravenous infusions of MSCs are typically cleared through lung and spleen [147], there is some evidence that cells might persist for longer periods; gene-marked autologous MSCs were detected in marrows of patients up to 8 months postinfusion [148]. There is thus a need to track patients long-term, and maintain clinical records in a secure fashion. Record keeping should also include traceability of recipients, donors, ancillary reagents, and excipients to manufacturing and processing records.

Each route of administration has potential safety concerns and must be appropriately addressed in the clinical protocol. For example, distal microvessel occlusion may occur when MSCs are administered intra-arterially [149] or potential neurotoxicity may occur upon intra-cranial injections of cryopreserved MSCs due to the DMSO administration [111–113,150,151]. Appropriate instructions to investigators in case of expected adverse events should be also included in the protocol.

1.5 IMPLEMENTATION OF MSC CLINICAL INVESTIGATION

1.5.1 Implementation of Regulatory Process

Assembling the regulatory package with the necessary preclinical safety and efficacy information, CMC data and clinical protocol is the first step in obtaining regulatory approval and initiating the clinical trial. The IND review process takes at least 30 days for Phase 1, 2, and 3 clinical trials. There are opportunities before submission of the IND to meet and consult with the FDA, in pre-IND or pre-pre-IND, end-of-phase I, or end-of-phase 2 meetings (type B). Other meeting types include dispute resolution (type A) or other meetings (type C). The FDA has issued guidance on the format and requirements governing these exchanges [152]. Institutional review board (IRB) approval is required in the United States for research involving human subjects, and is governed by Department of Health and Human Subjects (DHHS) regulations [153] and implemented by the Office for Human Research Protection (OHRP). IRB approval of research involving FDA-regulated products must additionally comply with other regulations, although there is significant overlap between the two [154].

The CTA review process in Europe can vary with member state and usually takes 60 days. There is concurrent regulatory and ethics review to expedite the process. The CTA review process in Canada takes 30 days, and is a default review. If there are no objections, a no objection letter (NOL) is issued, which authorizes initiation of clinical investigation, pending local ethics board review; else a Non-Satisfactory Notice (NSN) is issued. Local research ethics board review (REB) review is mandatory [26], and must adhere to guidelines in the Tri-council policy statement [155]. Length of REB review and approval varies with institution.

1.5.2 Implementation—Contracts

Implementation of a clinical trial typically involves multiple parties that fund, manage, and conduct the clinical trial. Each party needs to have a clear reference of their roles and responsibilities, and this is usually handled by having contracts and agreements in place prior to the initiation of a clinical trial. There may be different types of agreements including sponsorship agreements, funding agreements, clinical trial agreements, collaboration agreements, commercialization agreements, and material transfer agreements. These agreements require clarification on confidentiality of information, ownership of data, publication rights, liability, indemnification, etc. Service agreements may also be in place with contract research organization (CRO) that may be hired to implement the trial, monitor data handling and reporting. The manufacturing of the cell product may be conducted at an academic manufacturing facility or at a contract manufacturing facility, and the scope and responsibility for product manufacturing will need to be specified. Ultimately, compliance with regulatory requirements still falls on the sponsor, and thus appropriate selection of contract organizations to outsource the work requires due diligence.

1.5.3 Clinical Implementation

Logistics such as the coordination of the timing of final product preparation in relation to timing of administration to patient need to be considered, whether fresh cells with limited shelf-life or thawed cells with or without cryoprotectants, are used. The time required for performing final product release testing and obtaining QA/QC release of the MSC product, and the duration and rate of the infusion of the MSC product also need to be considered. Patients are usually brought in prior to the MSC infusion and may or may not require overnight stay. Research nurses need to be on-hand, coordinate with hospital staff to conduct premedications, baseline tests, vital signs, and synchronize timing of MSC infusion with the other clinical, manufacturing, and QA/QC team.

Clinical trials need to be monitored to oversee data collection and ensure compliance with GCP [144]. Case report forms and source documentation need to be reviewed for accuracy. CROs can be contracted to conduct onsite monitoring visits; this can also be handled internally at academic center but needs to be appropriately planned and resourced. Records regarding the adverse events, informed consent, enrollment, shipment, receipt, return and destruction of the drug product and other records pertaining to investigator's brochure, delegation log, etc. need to be kept for varying periods of time, depending on the jurisdiction. In Canada, records need to be maintained for a period of 25 years [156]. In the United States, under IND regulations [153,157] investigators are required to retain clinical research records, research misconduct records, and completion of clinical research study records for a period of two years following the date of market authorization. If the product is not approved, records must be kept until two years after the investigation is discontinued. In the EU, Directive 2005/28/EC Article 17 and 18 [141] sets out the retention duration of essential documents and medical files. Typically, patient identification code records are retained for 15 years after completion or discontinuation of a trial. Other source data are kept for the maximum period of time permitted by the hospital or institution.

1.6 STANDARDIZATION EFFORTS AROUND MSCs

Although MSC safety profile has been largely established, at least in the near term, questions about their efficacy persist in most indications. This is compounded by disparity between outcome results in various trials. For example, while a series of phase II European MSC trials for GVHD proved to be successful [158], Prochymal phase III trial in the United States (NCT00366145, sponsored by Osiris Therapeutics Inc.) failed to produce positive results [159]. An analysis of the Prochymal trial has identified some design issues that could have been responsible for the failure of the trial [159] including differences in isolation, processing and manufacturing methods, different passage numbers [160], and differences in the use of fresh versus thawed cells [161,162].

These mixed clinical outcomes and disparity in processing and analytical methods have prompted calls for standardization in the field. This includes standardizing definitions and terminology used for MSCs, and standardization of assays used to assay MSC functionality. A useful starting place would be to have a reference material for MSCs which could be used in preclinical evaluations as a ruler to benchmark findings within and between labs, as we recently proposed [163].

The ISCT MSC committee has also been interested in harmonization with the minimal criteria proposed in 2006 [164]. They recognize the need to update this definition and are looking to provide an opinion paper on functional definitions and characterizations of MSCs [135].

Standards organizations such as National Institute of Standards and Technology (NIST) and Unites States Pharmacopeia (USP) are also increasingly looking to develop tools to enable standardizing of cell therapies; NIST has been hosting workshops to develop strategies on developing measurement assurances for assays that measure potency or critical quality attributes of cellular products. USP has been working with ISCT to identify and prioritize topics related to key quality issues for cell and tissue products, to eventually develop standards to be published in the USP compendium. American Society for Testing and Materials (ASTM) has a subcommittee that has been developing standard tests, practices, and guides for characterizing cell and tissue-engineered constructs, ASTM F2149-01(2007) [165], ASTM F2210-02(2010) [166], ASTM F2944-12 [167]. These standardization efforts are necessary to bring harmonization and to allow the MSC field to continue its advancement.

1.7 SUMMARY

In this chapter, we have provided an overview of translational research process for bringing MSCs into the clinic. Briefly, we provided insights into some of the funding challenges that both academic investigators and industry face, the regulatory requirements for MSCs in the United States, EU and Canada, the required preclinical POC and safety studies, the CMC information and characterization needed to validate the MSC manufacturing process, and the requirements for developing an adequate clinical protocol. We also reviewed steps in implementing a clinical trial including securing regulatory and ethics approval, executing on contracts and coordinating with a clinical team for timely infusion of MSCs. It is clear that the MSC field is maturing as reflected in the commercialization of several MSC products, the use of novel manufacturing technologies, evolving global regulations, and increasing calls and collaborations between research societies and standards organization on standardization of nomenclature, reagents and assays to characterize MSCs. Challenges still remain, but the field is well situated on a translational trajectory that promises to launch novel MSC cellular products for various disease indications.

ABBREVIATIONS

AB	Alberta
AM	Ancillary material
AMI	Acute myocardial infarction
ANZCTR	Australian New Zealand Clinical Trials Registry
API	Active pharmaceutical ingredients
ASTM	American Society for Testing and Materials
B	Billion
BLA	Biologics License Application
CAT	Committee for Advanced Therapies
CBER	Center for Biologics Evaluation and Research
CBMPs	Cell-based medicinal products
CCRM	Canadian Centre for Commercialization of Regenerative Medicine
CFR	Council on Foreign Relations
CFU-F	Colony-forming unit-fibroblast
CHiCTR	Chinese Clinical Trial Register
CIRM	California Institute of Regenerative Medicine
CMC	Chemistry, manufacturing and control
CMV	Cytomegalovirus
CRIS	Clinical Research Information System
CRO	Contract Research Organization
CTA	Clinical Trial Authorization
CTD	Common Technical Document
CTL	Cardiff-based Cell Therapy Limited
CTO	Cells, tissues and organs
CTRI	Clinical Trials Registry India
CXCL	Chemokine (C-X-C motif) Ligand
DHHS	Department of Health and Human Subjects
DMEM	Dulbecco's Modified Eagle Medium
DMSO	Dimethyl sulfoxide
EAP	Extended access program
EC	European Commission
EDQM	European Directorate for the Quality of Medicines & HealthCare
EDTA	Ethylenediaminetetraacetic acid
EFIPA	European Federation of Pharmaceutical Industry Association
EGF	Epidermal growth factor
EL	Establishment License
EMA	European Medicine Agency
EU	European Union
EUTCD	European Tissues and Cells Directive
EUTCD	European Union Tissues and Cells Directive
FBS	Fetal bovine serum

FD&C	Federal Food, Drug, and Cosmetic
FDA	Food and Drug Administration
FGF2	Fibroblast growth factor-2
FP7	Seventh Framework Program
GCP	Good clinical practice
GLP	Good Laboratory Practice
GMP	Good Manufacturing Practices
GTP	Good tissue practices
HBV	Hepatitis B Virus
HCT/Ps	Human cells, tissues, and cellular and tissue-based products
HCV	Hepatitis C virus
HIV-1	Human immunodeficiency virus types 1
HIV-2	Human immunodeficiency virus types 2
HLA	Human leucocyte antigen
HSA	Human serum albumin
HTLV	Human T-Lymphotropic Virus
IB	Investigator's Brochure
ICH	International Council on Harmonization
IFN-γ	Interferon-Gamma
IGF	Insulin-like growth factor
IL	Interleukin
IMDM	Iscove's Modified Dulbecco's Medium
IMI	Innovative Medicines Initiative
IMPD	Investigational Medicinal Product Dossier
IND	Investigational New Drug
IRB	Institutional Review Board
IRCT	Iranian Registry of Clinical Trials
ISCT	International Society of Cellular Therapy
ISRCTN	International Standard Randomized Controlled Trial Number
IV	Intravenous
JPRN	Japan Primary Registries Network
LVEF	Left ventricular ejection fraction
M	Million
MCB	Master cell bank
MEM	Minimal Essential Medium
MLR	Mixed Lymphocyte Reaction
MSC	Mesenchymal stromal cells
NCA	National Competent Authorities
NCA	National Confectioners Association
NDS	New Drug Submission
NIH	National Institutes of Health
NIST	National Institute of Standards and Technology
NOL	No Objection Letter
NSN	Non-Satisfactory Notice
OA	Osteoarthritis
OCTGT	Office of Cellular, Tissue and Gene Therapies
OHRP	Office for Human Research Protection
ON	Ontario

P01	Program Project
PACT	Production assistance for cellular therapies
PDGF	Platelet-derived growth factor
PEG	Polyethylene glycol
PHS	Public Health Services
POC	Proof-of-concept
QA	Quality assurance
QC	Quality control
R01	Research project
REB	Research Ethics Board
SPECT	Single photon emission computed tomography
STEPS	Stem cell therapy as an emerging paradigm for stroke
SVF	Stromal vascular fraction
TGF-β	Transforming growth factor
TNFR1	Tumor necrosis factor receptor 1
TSE	Transmissible spongiform encephalopathy
TSG-6	TNF-α stimulated gene/protein 6
USD	United States Dollar
USP	Unites States Pharmacopeia
VEGF	Vascular endothelial growth factor
WNV	West Nile virus

REFERENCES

[1] Augello A, Kurth TB, De Bari C. Mesenchymal stem cells: a perspective from in vitro cultures to in vivo migration and niches. Europ Cells Mater 2010; 20:121–33.

[2] Le Blanc K, Mougiakakos D. Multipotent mesenchymal stromal cells and the innate immune system. Nat Rev Immunol 2012;12(5):383–96.

[3] ClinicalTrials.gov. [Internet].

[4] California Institute for Regenerative Medicine (CIRM). Where CIRM funding goes. Available at: https://www.cirm.ca.gov/about/where-cirm-funding-goes.

[5] DiMasi J, Grabowski HG, Hansen RW. Briefing – cost of developing a new drug. Boston, MA: R&D Cost Study Briefing; 2014. p. 30.

[6] Abbott S, Mackay G, Durdy M, Solomon S, Zylberberg C. Twenty years of the International Society for Cellular Therapies: the past, present and future of cellular therapy clinical development. Cytotherapy 2014;16(Suppl. 4):S112–9.

[7] History – OsirisTherapeutics. Available at: http://www.osiris.com/about/#history.

[8] McIntosh KR, JDM, Klyushnenkova EN, inventor Osiris Therapeutics, Inc., assignee. Mesenchymal stem cells for prevention and treatment of immune responses in transplantation. US patent US6,328,960 B1. 1999.

[9] History – MEDIPOST. Available at: http://www.medi-post.com/history/.

[10] Buckler L. Cell therapy manufacturing market. Available at: http://celltherapyblog.blogspot.ca/2008/08/cell-therapy-manufacturing-market.html, 2008.

[11] Buckler L. Clinical trials costs. Available at: http://celltherapyblog.blogspot.com/2011/07/clinical-trial-costs.html, 2011.

[12] Randal M. CIRM 2.0. 2015.

[13] American Society of Gene and Cell Therapy. Policy/position statement: NIH funding of gene therapy trials. Available at: http://www.asgct.org/position_statements/nih_funding.php, 2006.

[14] CORDIS. The 7th framework programme funded European Research and Technological Development from 2007 until 2013. Available at: http://cordis.europa.eu/fp7/home_en.html.

[15] Gancberg D, Draghia-Akli R. Gene and cell therapy funding opportunities in horizon 2020: an overview for 2014-2015. Hum Gene Ther 2014;25(3):175–7.

[16] Semma Therapeutics. Semma therapeutics announces $44 million in funding led by mpm capital to develop cell therapy for treating type 1 diabetes; signs agreement with global pharmaceutical company. Available at: http://www.semma-tx.com/media1/semma-therapeutics-announces-44-million-in-funding, 2015.

[17] Gormlet B. Mutual funds flood into health care with juno Therapeutics' $134M round. Available at: http://blogs.wsj.com/venturecapital/2014/08/06/mutual-funds-flood-into-health-care-with-juno-therapeutics-134m-round/, 2014.

[18] NASDAQ. Juno Therapeutics, Inc. Stock Quote & Summary Data. Available at: http://www.nasdaq.com/symbol/juno, 2015.

[19] NASDAQ. Bluebird Bio, Inc. Stock quote & summary data. Available at: http://www.nasdaq.com/symbol/blue, 2015.

[20] NASDAQ. Kite Pharma, Inc. Stock quote and summary data. Available at: http://www.nasdaq.com/symbol/kite, 2015.

[21] Ford D, Nelsen B. The view beyond venture capital. Nature Biotechnology 2014; 32(1):15–23.

[22] Parliament E. Directive 2001/83/EC of the European Parliament and of the Council of 6 November 2001 on the Community code relating to medicinal products for human use. 2001.

[23] 351 Public Health Service Act. Title 42 United States Code 262.

[24] Justice Mo. Food and Drug Regulations.

[25] FDA. Title 42—the public health and welfare chapter 6a—public health service; subchapter ii—general powers and duties; part f—licensing of biological products and clinical laboratories; subpart 1—biological products. 1998.

[26] Mo J. Food and drug regulations. In: Mo J, editor; amended 2014.

[27] Directive 2001/83/ec of the European parliament and of the council of 6 November 2001 on the community code relating to medicinal products for human use. Official J 2001:46–94.

[28] EU. Regulation (ec) no 1394/2007 of the European parliament and of the council of 13 November 2007 on advanced therapy medicinal products and amending directive 2001/83/ec and regulation (ec) no 726/2004. In: union Tepatcote, editor: Official J Europea 2007;L324/121–L324/137.

[29] FDA. Food and drugs chapter i—food and drug administration department of health and human services subchapter f—biologics. Code of Federal Regulations—Title 21 (21CFR601.2); 2014.

[30] FDA. Guidance for FDA reviewers and sponsors: Content and review of chemistry, manufacturing, and control information for human somatic cell therapy investigational new drug applications (INDs). 2008.

[31] Lee J-W, Lee SH, Youn YJ, Ahn MS, Kim JY, Yoo BS, et al. A randomized, open-label, multicenter trial for the safety and efficacy of adult mesenchymal stem cells after acute myocardial infarction. J Korean Med Sci 2014;29(1):23–31.

[32] FDA. Guidance for industry: guidance for human somatic cell therapy and gene therapy. 1998.
[33] FDA. Guidance for industry considerations for the design of early—phase clinical trials of cellular and gene therapy products. 2015.
[34] EU. Information from European Union Institutions, bodies, offices and agencies. In: Commission, E., editor. Official J Eur Union; 2010.
[35] EU. EudraLex – Volume 10 Clinical trials guidelines. In: Legislation, E., editor; 2013.
[36] Health Mo. Draft guidance document: guidance for sponsors: preparation of clinical trial applications for use of cell therapy products in humans. In: Branch HPaF, editor; 2014.
[37] Health Mo. Guidance for clinical trial sponsors: clinical trial applications. In: Branch HPaF, editor; 2013.
[38] ICH. ICH harmonised tripartite guideline preclinical safety evaluation of biotechnology-derived pharmaceuticals s6(r1) 2011.
[39] FDA. Guidance for industry preclinical assessment of investigational cellular and gene therapy products. In : CBER/OCTGT, editor. Rockville, MD; 2013.
[40] FDA. Guidance for industry: preparation of IDEs and INDs for products intended to repair or replace knee cartilage. In: C.B.E.R., editor; 2011. p. 1–17.
[41] FDA. Guidance for industry: cellular therapy for cardiac disease. In: CBER/CDRH, editor; 2010. p. 1–24.
[42] FDA. Guidance for industry: considerations for allogeneic pancreatic islet cell products. In: C.B.E.R., editor; 2009.
[43] FDA. Guidance for industry: source animal, product, preclinical, and clinical issues concerning the use of xenotransplantation products in humans. In: C.B.E.R., editor; 2003. p. 1–60.
[44] EMEA. Guideline on human cell-based medicinal products (Doc. Ref. EMEA/CHMP/410869/2006). In: C.H.M.P., editor; 2007. p. 1–24.
[45] EMEA. Guideline on the non-clinical studies required before first clinical use of gene therapy medicinal products (Doc. Ref. EMEA/CHMP/GTWP/125459/2006). In: C.H.M.P., editor; 2007. p. 1–9.
[46] Ren G, Su J, Zhang L, Zhao X, Ling W, L'huillie A, et al. Species variation in the mechanisms of mesenchymal stem cell-mediated immunosuppression. Stem Cells 2009;27(8):1954–62.
[47] Feigal EG, Tsokas K, Viswanathan S, Zhang J, Priest C, Pearce J, et al. Proceedings: international regulatory considerations on development pathways for cell therapies. Stem Cells Transl Med 2014;3(8):879–87.
[48] Wolfstadt JI, Cole BJ, Ogilvie-Harris DJ, Viswanathan S, Chahal J. Current concepts: the role of mesenchymal stem cells in the management of knee osteoarthritis. Sports Health 2015;7(1):38–44.
[49] Samper E, Diez-Juan A, Montero JA, Sepulveda P. Cardiac cell therapy: boosting mesenchymal stem cells effects. Stem Cell Rev 2013;9(3):266–80.
[50] Halkos ME, Zhao ZQ, Kerendi F, Wang NP, Jiang R, Schmarkey LS, et al. Intravenous infusion of mesenchymal stem cells enhances regional perfusion and improves ventricular function in a porcine model of myocardial infarction. Basic Res Cardiol 2008;103(6):525–36.
[51] Medicetty S, Wiktor D, Lehman N, Raber A, Popovic ZB, Deans R, et al. Percutaneous adventitial delivery of allogeneic bone marrow-derived stem cells via

infarct-related artery improves long-term ventricular function in acute myocardial infarction. Cell Transplant 2012;21(6):1109–20.
[52] Poh KK, Sperry E, Young RG, Freyman T, Barringhaus KG, Thompson CA. Repeated direct endomyocardial transplantation of allogeneic mesenchymal stem cells: safety of a high dose, "off-the-shelf", cellular cardiomyoplasty strategy. Int J Cardiol 2007;117(3):360–4.
[53] Price MJ, Chou CC, Frantzen M, Miyamoto TZ, Kar S, Lee S, et al. Intravenous mesenchymal stem cell therapy early after reperfused acute myocardial infarction improves left ventricular function and alters electrophysiologic properties. Int J Cardiol 2006;111(2):231–9.
[54] Wang D, Jin Y, Ding C, Zhang F, Chen M, Yang B, et al. Intracoronary delivery of mesenchymal stem cells reduces proarrhythmogenic risks in swine with myocardial infarction. Ir J Med Sci 2011;180(2):379–85.
[55] Zhou Y, Wang S, Yu Z, Hoyt RF Jr, Qu X, Horvath KA. Marrow stromal cells differentiate into vasculature after allogeneic transplantation into ischemic myocardium. Ann Thorac Surg 2011;91(4):1206–12.
[56] Zhou Y, Wang S, Yu Z, Hoyt RF Jr, Sachdev V, Vincent P, et al. Direct injection of autologous mesenchymal stromal cells improves myocardial function. Biochem Biophys Res Commun 2009;390(3):902–7.
[57] FDA. Summary minutes meeting # 37, March 18–19, 2004. In: Center for biologics evaluation, research BRMAC, editor: biological response modifiers advisory committee; 2004.
[58] Rouchaud A, Journe C, Louedec L, Ollivier V, Derkaoui M, Michel JB, et al. Autologous mesenchymal stem cell endografting in experimental cerebrovascular aneurysms. Neuroradiology 2013;55(6):741–9.
[59] Kuwabara A, Liu J, Lee J-W, Hashimoto T. Human mesenchymal stem cells reduce the rupture rate of intracranial aneurysm. ANESTHESIOLOGY(2014 – Annual Meeting of the American Society of Anesthesiologists (ASA); 2014; New Orleans, LA: American Society of Anesthesiologists; 2014.
[60] Savitz SI, Chopp M, Deans R, Charmichael T, Phinney D, Wechsler L, et al. Stem cell therapy as an emerging paradigm for stroke (STEPS) II. Stroke 2011;42(3):825–9.
[61] Schrepfer S, Deuse T, Reichenspurner H, Fischbein MP, Robbins RC, Pelletier MP. Stem cell transplantation: the lung barrier. Transplant Proc 2007;39(2):573–6.
[62] Barbash IM, Chouraqui P, Baron J, Feinberg MS, Etzion S, Tessone A, et al. Systemic delivery of bone marrow-derived mesenchymal stem cells to the infarcted myocardium: feasibility, cell migration, and body distribution. Circulation 2003;108(7):863–8.
[63] Gao J, Dennis JE, Muzic RF, Lundberg M, Caplan AI. The dynamic in vivo distribution of bone marrow-derived mesenchymal stem cells after infusion. Cells Tissues Organs 2001;169(1):12–20.
[64] Fischer UM, Harting MT, Jimenez F, Monzon-Posadas WO, Xue H, Savitz SI, et al. Pulmonary passage is a major obstacle for intravenous stem cell delivery: the pulmonary first-pass effect. Stem Cells Dev 2009;18(5):683–92.
[65] Kraitchman DL, Tatsumi M, Gilson WD, Ishimori T, Kedziorek D, Walczak P, et al. Dynamic imaging of allogeneic mesenchymal stem cells trafficking to myocardial infarction. Circulation 2005;112(10):1451–61.

[66] Furlani D, Ugurlucan M, Ong L, Bieback K, Pittermann E, Westien I, et al. Is the intravascular administration of mesenchymal stem cells safe? Mesenchymal stem cells and intravital microscopy. Microvasc Res 2009;77(3):370–6.
[67] Knoepfler PS. Deconstructing stem cell tumorigenicity: a roadmap to safe regenerative medicine. Stem Cells 2009;27(5):1050–6.
[68] Tarte K, Gaillard J, Lataillade JJ, Fouillard L, Becker M, Mossafa H, et al. Clinical-grade production of human mesenchymal stromal cells: occurrence of aneuploidy without transformation. Blood 2010;115(8):1549–53.
[69] Lalu MM, McIntyre L, Pugliese C, Fergusson D, Winston BW, Marshall JC, et al. Safety of cell therapy with mesenchymal stromal cells (SafeCell): a systematic review and meta-analysis of clinical trials. PLoS One 2012;7(10):e47559.
[70] FDA. Title 21—food and drugs; chapter 9—federal food, drug, and cosmetic act; subchapter v—drugs and devices; part a—drugs and devices; sec. 351—adulterated drugs and devices. United States Code, 2006 Edition, Supplement 4, Title 21—food and drugs; 2011. p. 141–3.
[71] FDA. Guidance for industry: CGMP for phase 1 investigational drugs. In: CDER/CBER/ORA, editor; 2008. p. 1–17.
[72] FDA. Title 21-food and drugs chapter I-food and drug administration department of health and human services; subchapter c-drugs: General; part 210 current good manufacturing practice in manufacturing, processing, packing, or holding of drugs; general. 21CFR210: Code of Federal Regulations; 2014.
[73] FDA. Title 21-food and drugs; chapter i--food and drug administration department of health and human services; subchapter c-drugs: General; part 211 current good manufacturing practice for finished pharmaceuticals (21CFR211). Code of Federal Regulations; 2014.
[74] EU. Directive 2004/23/ec of the European parliament and of the council of 31 march 2004 on setting standards of quality and safety for the donation, procurement, testing, processing, preservation, storage and distribution of human tissues and cells. In: union Tepatcote, editor: Official J European Union; 2004:L102/48– L/58.
[75] EU. Commission directive 2003/94/ec of 8 October 2003 laying down the principles and guidelines of good manufacturing practice in respect of medicinal products for human use and investigational medicinal products for human use. In: communities Tcote, editor: Official J Eur Union; 2003: L262/22–L/26.
[76] Canada H. Guidance document. Annex 13 to the current edition of the good manufacturing practices guidelines drugs used in clinical, trials, GUI-0036. In: Inspectorate HPaFB, editor; 2009.
[77] FDA. Part 1271 human cells, tissues, and cellular and tissue-based products. 21CFR1271: Code of Federal Regulations; 2014.
[78] Mendicino M, et al. MSC-based product characterization for clinical trials: an FDA perspective 2014;14. **2**(141–145).
[79] Pharmacopeia US. <1043> Ancillary materials for cell, gene and tissue engineered products.
[80] ICH. Good manufacturing practice guide for active pharmaceutical ingredients q7. In: guideline Iht, editor; 2000.
[81] Ikebe C, Suzuki K. Mesenchymal stem cells for regenerative therapy: optimization of cell preparation protocols. BioMed Res Int 2014;2014:951512.
[82] Azandeh S, Orazizadeh M, Hashemitabar M, Khodadadi A, Shayesteh AA, Nejad DB, et al. Mixed enzymatic-explant protocol for isolation of mesenchymal stem

cells from Wharton's jelly and encapsulation in 3D culture system. J Biomed Sci Eng 2012;5:580–6.
[83] Jin SH, Lee JE, Yun JH, Kim I, Ko Y, Park JB. Isolation and characterization of human mesenchymal stem cells from gingival connective tissue. J Periodontal Res 2015;50(4):461–7.
[84] Inamdar A, Inamdar AC. Culture conditions for growth of clinical grade human tissue derived mesenchymal stem cells: comparative study between commercial serum-free media and human product supplemented media. J Regenerative Med Tissue Eng 2013;2(10).
[85] Sotiropoulou P, Perez SA, Salagianni M, Baxevanis CN, Papamichail M. Characterization of the optimal culture conditions for clinical scale production of human mesenchymal stem cells. Stem Cells Develop 2006;24(2):462–71.
[86] Bianchi G, Banfi A, Mastrogiacomo M, Notaro R, Luzzatto L, Cancedda R, et al. Ex vivo enrichment of mesenchymal cell progenitors by fibroblast growth factor 2. Exp Cell Res 2003;287(1):98–105.
[87] Ng F, Boucher S, Koh S, Sastry KS, Chase L, Lakshmipathy U, et al. PDGF, TGF-2, and FGF signaling is important for differentiation and growth of mesenchymal stem cells (MSCs): transcriptional profiling can identify markers and signaling pathways important in differentiation of MSCs into adipogenic, chondrogenic, and osteogenic lineages. Blood 2008;112(2):295–307.
[88] Tamama K, Kawasaki H, Wells A. Epidermal growth factor (EGF) treatment on multipotential stromal cells (MSCs). Possible enhancement of therapeutic potential of MSC. J Biomed Biotechnol 2010;2010:795385.
[89] FDA. Use of materials derived from cattle in medical products intended for use in humans and drugs intended for use in ruminants. 72 FR 1581; 2007.
[90] FDA. Requirements for ingredients of animal origin used for production of biologics. 9 CFR 113.53. 1995.
[91] Sakamoto N, Tsuji K, Muul LM, Lawler AM, Petricoin EF, Candotti F, et al. Bovine apolipoprotein B-100 is a dominant immunogen in therapeutic cell populations cultured in fetal calf serum in mice and humans. Blood 2007;110(2):501–8.
[92] Pérez-Simon JA, Lopez-Villar O, Andreu EJ, Rifon J, Muntion S, Diez Campelo M, et al. Mesenchymal stem cells expanded in vitro with human serum for the treatment of acute and chronic graft-versus-host disease: results of a phase I/II clinical trial. Haematologica 2011;96(7):1072–6.
[93] Lucchini G, Introna M, Dander E, Rovelli A, Balduzzi A, Bonanomi S, et al. Platelet-lysate-expanded mesenchymal stromal cells as a salvage therapy for severe resistant graft-versus-host disease in a pediatric population. Biol Blood Marrow Transplant 2010;16:1293–301.
[94] Agata H, Wantanabe N, Ishii Y, Kubo N, Oshima S, Yamazaki M, et al. Feasibility and efficacy of bone tissue engineering using human bone marrow stromal cells cultivated in serum-free conditions. Biochem Biophys Res Commun 2009;382(2):353–8.
[95] Jung S, Panchalingam KM, Rosenberg L, Behie LA. Ex vivo expansion of human mesenchymal stem cells in defined serum-free media. Article ID 123030. Stem Cells Int 2012;21.
[96] Mittag F, Falkenberg EM, Janczyk A, Gotze M, Felka T, Aicher WK, et al. Laminin-5 and type I collagen promote adhesion and osteogenic differentiation of animal serum-free expanded human mesenchymal stromal cells. Orthop Rev (Pavia) 2012;4:160–4.

[97] Rafiq Q, Brosnan KM, Coopman K, Nienow AW, Hewitt CJ. Culture of human mesenchymal stem cells on microcarriers in a 5 l stirred-tank bioreactor. Biotechnol Lett 2013;35:1233–45.
[98] CELLstart™CTS™: CELLstart™ humanized substrate for stem cell culture. In: Technologies, L., editor. Available at: www.lifetechnologies.com/ca/en/home/life-science/stem-cell-research/stem-cell-culture/stem-cell-research-misc/cellstart.html.
[99] Szczypka M, Splan D, Woolls H, Brandwein H. Single-use bioreactors and microcarriers scalable technology for cell-based therapies. BioProcess Int 2014;12(3).
[100] Chen M, Wang X, Ye Z, Zhang Y, Tan W-S. A modular approach to the engineering of a centimeter-sized bone tissue construct with human amniotic mesenchymal stem cells-laden microcarrier. Biomaterials 2011b. 32;7532–42.
[101] Eibes G, et al. Maximizing the ex vivo expansion of human mesenchymal stem cells using a microcarrier-based stirred culture system. J Biotechnol 2010;146:194–7.
[102] Sun L, et al. Cell proliferation of human bone marrow mesenchymal stem cells on biodegradable microcarriers enhances in vitro differentiation potential. Cell Prolif 2010;43:445–56.
[103] Timmins N, et al. Closed system isolation and scalable expansion of human placental mesenchymal stem cells. Biotechnol Bioeng 2012;109:1817–26. (160).
[104] Schop D, et al. Expansion of human mesenchymal stromal cells on microcarriers: growth and metabolism. J Tissue Eng Regen Med 2010;4:131–40.
[105] Hewitt C, et al. Expansion of human mesenchymal stem cells on microcarriers. Biotechnol Lett 2011;33:2325–35.
[106] Santos F, et al. Toward a clinical-grade expansion of mesenchymal stem cells from human sources: a microcarrier-based culture system under xeno-free conditions. Tissue Eng Part C Methods 2011;17:1201–10.
[107] Zhou Y, et al. Expansion and delivery of adipose-derived mesenchymal stem cells on three microcarriers for soft tissue regeneration. Tissue Eng Part A 2011;17:2981–97.
[108] Jung S, et al. Expansion of human bone marrow-derived mesenchymal stem cells in microcarrier culture. Poster Presentation, MSC 2007 regenerative medicine and adult stem cell therapy, Cleveland, Ohio, USA (August 27–29, 2007); 2007. Available from: http://www.mscconference.net/abstracts/jung.expansion.pdf.
[109] Turner AE, et al. Design and characterization of tissue-specific extracellular matrix-derived microcarriers. Tissue Eng Part C Methods 2012;18:186–97.
[110] Turner AEB, et al. The performance of decellularized adipose tissue microcarriers as an inductive substrate for human adipose-derived stem cells. Biomaterials 2012;33:4490–9.
[111] Junior AM, Arrais CA, Saboya R, Velasques RD, Junqueira PL, Dulley FL. Neurotoxicity associated with dimethylsulfoxide-preserved hematopoietic progenitor cell infusion. Bone Marrow Transplant 2008;41(1):95–6.
[112] Mueller LP, Theurich S, Christopeit M, Grothe W, Muetherig A, Weber T, et al. Neurotoxicity upon infusion of dimethylsulfoxide-cryopreserved peripheral blood stem cells in patients with and without pre-existing cerebral disease. Eur J Haematol 2007;78(6):527–31.
[113] Bauwens D, Hantson P, Laterre PF, Michaux L, Latinne D, De Tourtchaninoff M, et al. Recurrent seizure and sustained encephalopathy associated with dimethylsulfoxide-preserved stem cell infusion. Leuk Lymphoma 2005;46(11):1671–4.

[114] Zuk PA, Zhu M, Ashjian P, De Ugarte DA, Huang JI, Mizuno H, et al. Human adipose tissue is a source of multipotent stem cells. Mol Biol Cell 2002;13(12):4279–95.

[115] Sarugaser R, Lickorish D, Baksh D, Hosseini MM, Davies JE. Human umbilical cord perivascular (HUCPV) cells: a source of mesenchymal progenitors. Stem Cells 2005;23:220–9.

[116] Lechanteur C, Baila S, Janssen ME, Giet O, Briquet A, Baudoux E, et al. Large-scale clinical expansion of mesenchymal stem cells in the GMP-compliant, closed automated quantum cell expansion system: comparison with expansion in traditional T-flasks. J Stem Cell Res Ther 2014;4(8).

[117] Kaiser SC, Eibl R, Eibl D. Engineering characteristics of a single-use stirred bioreactor at bench-scale: the Mobius CellReady 3 L bioreactor as a case study. Eng Life Sci 2011;11:359–68.

[118] Choi M, Kim HY, Park JY, Lee TY, Baik CS, Chai YG, et al. Selection of optimal passage of bone marrow-derived mesenchymal stem cells for stem cell therapy in patients with amyotrophic lateral sclerosis. Neuro Sci Lett 2010;472:94–8.

[119] FDA. Points to Consider in the Characterization of Cell Lines Used to Produce Biologicals. 1993.

[120] ICH. Derivation and Characterisation of Cell Substrates Used for Production of Biotechnological/Biological Products (Q5D). Quality Guidelines; 1997.

[121] FDA. Guidance for FDA reviewers and sponsors: content and review of chemistry, manufacturing, and control information for human gene therapy investigational new drug applications (INDs), 2008.

[122] Pharmacopeia US. <71> Sterility Tests. 2009.

[123] FDA. General biological products standards (610.12 sterility). 21CFR61012; 2014.

[124] Pharmacopeia US. <151> Pyrogen test.

[125] Pharmacopeia US. <85> Bacterial Endotoxin test.

[126] Pharmacopeia US. <63> Mycoplasma Tests.

[127] Dominici M, Le Blanc K, Mueller I, Slaper-Cortenbach I, Marini F, Krause D, et al. Minimal criteria for defining multipotent mesenchymal stromal cells. The International Society for Cellular Therapy position statement. Cytotherapy 2006;8(4):315–7.

[128] Bourin P, Bunnell BA, Casteilla L, Dominici M, Katz AJ, March KL, et al. Stromal cells from the adipose tissue-derived stromal vascular fraction and culture expanded adipose tissue-derived stromal/stem cells: a joint statement of the International Federation for Adipose Therapeutics and Science (IFATS) and the International Society for Cellular Therapy (ISCT). Cytotherapy 2013;15:641–8.

[129] FDA. Guidance for industry—potency tests for cellular and gene therapy products. 2011.

[130] ICH. Validation of Analytical procedures: text and methodology Q2(R1). Quality guidelines; 1994.

[131] ICH. Specifications: test procedures and acceptance criteria for biotechnological/biological products Q6b 1999.

[132] ICH. Comparability of biotechnological/biological products subject to changes in their manufacturing process Q5E 2004.

[133] EMEA. Guideline on potency testing of cell based immunotherapy medicinal products for the treatment of cancer (Doc. Ref. EMEA/CHMP/BWP/271475/2006). In: C.H.M.P., editor; 2008.

[134] Krampera M, Galipeau J, Shi Y, Tarte K, Sensebe L. MSC Committee of the International Society for Cellular Therapy (ISCT). Immunological characterization

of multipotent mesenchymal stromal cells—The International Society for Cellular Therapy (ISCT) working proposal. Cytotherapy 2013;15:105–6.
[135] Galipeau J, et al. International Society for Cellular Therapy perspective on immune functional assays for mesenchymal stromal cells as potency release criterion for advanced phase clinical trials. Cytotherapy 2016;18(2):151–9.
[136] Roddy G, Oh JY, Lee RH, Bartosh TJ, Ylostalo J, Coble K, et al. Action at a distance: systemically administered adult stem/progenitor cells (MSCs) reduce inflammatory damage to the cornea without engraftment and primarily by secretion of TNF-α stimulated gene/protein 6. Stem Cells 2011;29(10):1572–9.
[137] Lehman N, Cutrone R, Raber A, Perry R, Van't Hof W, Deans R, et al. Development of a surrogate angiogenic potency assay for clinical-grade stem cell production. Cytotherapy 2012;14(8):994–1004.
[138] FDA. Title 21—food and drugs chapter i—food and drug administration department of health and human services subchapter a—general part 11 electronic records; electronic signatures. 21CFR11: Code of Federal Regulations; 2014.
[139] FDA. Title 21-food and drugs; chapter i-food and drug administration department of health and human services; subchapter a-general part 50 protection of human subjects. 21CFR50: Code of Federal Regulations; 2014.
[140] FDA. Title 21—food and drugs chapter i—food and drug administration department of health and human services subchapter a—general part 54 financial disclosure by clinical investigators. 21CFR54: Code of Federal Regulations; 2014.
[141] FDA. Title 21—food and drugs chapter i—food and drug administration department of health and human services subchapter a—general part 56 institutional review boards. 21CFR56: Code of Federal Regulations; 2014.
[142] EU. Commission directive 2005/28/ec of 8 April 2005 laying down principles and detailed guidelines for good clinical practice as regards investigational medicinal products for human use, as well as the requirements for authorisation of the manufacturing or importation of such products. In: communities Tcote, editor: Official J Eur Union 2005:L91/13–L9/9.
[143] EU. Directive 95/46/EC of the European Parliament and of the Council of 24 October 1995 on the protection of individuals with regard to the processing of personal data and on the free movement of such data. In: union Tepatcote, editor: Official J Eur Parliament 1995:L281/ 31–50.
[144] Justice Mo. Food and Drug Regulations (C.R.C., c. 870) Part: C Drugs. 2015.
[145] FDA. Guidance for industry E6 good clinical practice: consolidated guidance. In: CBER/CDRH, editor; 1996.
[146] FDA. Guidance for industry: E6 good clinical practice: consolidated guidance. In: Services USDoHaH, editor; 1996.
[147] Wagner B, Henschler R. Fate of intravenously injected mesenchymal stem cells and significance for clinical application. Adv Biochem Eng Biotechnol 2013;130:19–37.
[148] Keating A, et al. Safe engraftment of gene-marked non-hematopoietic marrow mesenchymal cells: a platform for cell therapy. Blood 2001;98(11):832a.
[149] Yavagal DR, Lin B, Raval AP, Garza PS, Dong C, Zhao W, et al. Efficacy and dose-dependent safety of intra-arterial delivery of mesenchymal stem cells in a rodent stroke model. PLoS One 2014;9(5):e93735.
[150] Hanslick JL, Lau K, Noguchi KK, Olney JW, Zorumski CF, Mennerick S, et al. Dimethyl sulfoxide (DMSO) produces widespread apoptosis in the developing central nervous system. Neurobiol Dis 2009;34(1):1–10.

[151] Windrum P, Morris TCM. Severe neurotoxicity because of dimethyl sulphoxide following peripheral blood stem cell transplantation. Bone Marrow Transplant 2003;31(315).

[152] FDA. Guidance for Industry Formal Meetings Between the FDA and the Sponsors or Applicants. 2009.

[153] FDA. Title 45 public welfare department of health and human services part 46 protection of human subjects. In: C.B.E.R., editor. 45CFR part 46. Code of Federal Regulations 2009.

[154] FDA. Comparison of FDA and HHS human subject protection regulations. 2000.

[155] Canada H. The Tri-council policy statement: guidelines on research involving human subjects (TCPS-2). In: Research CIoH, editor; 2010.

[156] Canada H. Guidance for records related to clinical trials guide 0068 interpretation of section c.05.012 of the food and drug regulations—division 5 "drugs for clinical trials involving human subjects". 2006.

[157] FDA. Title 21—food and drugs chapter i—food and drug administration department of health and human services subchapter d—drugs for human use part 312 investigational new drug application. 21CFR 312; 2014.

[158] Le Blanc K, Frassoni F, Ball L, Locatelli F, Roelofs H, Lewis I, et al. Mesenchymal stem cells for treatment of steroid-resistant, severe, acute graft-versus-host disease: a phase II study. Lancet 2008;371(9624):1579–86.

[159] Galipeau J. The mesenchymal stromal cells dilemma—does a negative phase III trial of random donor mesenchymal stromal cells in steroid-resistant graft-versus-host disease represent a death knell or a bump in the road? Cytotherapy 2013;15(1):2–8.

[160] Karen B, Kinzebach S< Karagianni M. Translating research into clinical scale manufacturing of mesenchymal stromal cells. Stem Cells Int 2010;11:11.

[161] Galipeau J. Concerns arising from MSC retrieval from cryostorage and effect on immune suppressive function and pharmaceutical usage in clinical trials. ISBT Sci Series 2013;(8):100–1.

[162] Moll G, Alm JJ, Davies LC, von Bahr L, Heldring N, Stenbeck-Funke L, et al. Do cryopreserved mesenchymal stromal cells display impaired immunomodulatory and therapeutic properties? Stem Cells 2014;32(9):2430–42.

[163] Viswanathan S, Keating A, Deans R, Hematti P, Prockop D, et al. Soliciting strategies for developing cell-based reference materials to advance mesenchymal stromal cell research and clinical translation. Stem Cells Develop 2014;23(11):1157–67.

[164] Horwitz EM, Le Blanc K, Dominici M, Mueller I, Slaper-Cortenbach I, Marini FC, et al. Clarification of the nomenclature for MSC: The International Society for Cellular Therapy position statement. Cytotherapy 2005;7(5):393–5.

[165] ASTM. Standard test method for automated analyses of cells-the electrical sensing zone method of enumerating and sizing single cell suspensions. ASTM F2149-01(2007) 2007.

[166] ASTM. Standard guide for processing cells, tissues, and organs for use in tissue engineered medical products. ASTM F2210-02(2010) 2010.

[167] ASTM. Standard test method for automated colony forming unit (CFU) assays—image acquisition and analysis method for enumerating and characterizing cells and colonies in culture. ASTM F2944-12; 2012.

Chapter 2

Preclinical Animal Testing Requirements and Considerations

J.M. Centanni

Regulatory Affairs Scientist, Institute for Clinical and Translational Research, University of Wisconsin School of Medicine and Public Health, University of Wisconsin, Madison, WI, United States

CHAPTER OUTLINE

Mesenchymal Stromal Cells. http://dx.doi.org/10.1016/B978-0-12-802826-1.00002-7

2.1 INTRODUCTION

Preclinical testing of mesenchymal stromal cells (MSC) intended for human clinical application should include proof-of-concept, and early safety and efficacy studies. Complete descriptions of prospectively planned preclinical studies and adequate experimental documentation should not be overlooked during early phase of MSC product development. If it is determined that the MSC product needs to be tested in an animal model of disease, then the following important considerations need to be evaluated. For example, selection of the most appropriate animal model and use of the intended MSC delivery method and administration schedule is essential to adequately demonstrate safety and potential early efficacy in support of future human clinical development. When reasonably possible, all preclinical studies intended to support an Investigational New Drug (IND) application should be conducted using clinical-grade MSC product or at a minimum, product that has been demonstrated to be of comparable quality with respect to identity, strength, purity, and potency. Consideration of the information provided in this chapter will facilitate the development of scientifically sound, cost-effective, and efficient preclinical studies designed and implemented to support the evaluation in human clinical studies.

2.2 PRECLINICAL STUDY CONDUCT

The best approach in obtaining preclinical safety data to support the use of MSCs in clinical research is to conduct carefully controlled and well-documented preclinical studies. This approach is the most cost-effective and efficient strategy to collect the necessary preclinical data in support of demonstrating safety and early efficacy required to initiate human clinical studies. Preclinical studies that provide essential information in the Chemistry, Manufacturing, and Controls (CMC) section of an IND application to US Food and Drug Administration (FDA) are presented in Fig. 2.1. It is essential for any study that is not conducted in compliance with Good Laboratory Practice (GLP) to be conducted as diligently as possible to avoid performing an inadequate study with poor documentation or lack of protocol adherence. This is especially true in the academic research environment where documentation expectations and adequate controls may not be as rigorous as those that are typically emphasized in industry.

2.2.1 Documentation

All preclinical studies intended to support an IND submission or justify early biological product safety and efficacy should be performed using a prospectively defined study protocol. A bill of testing or certificate of

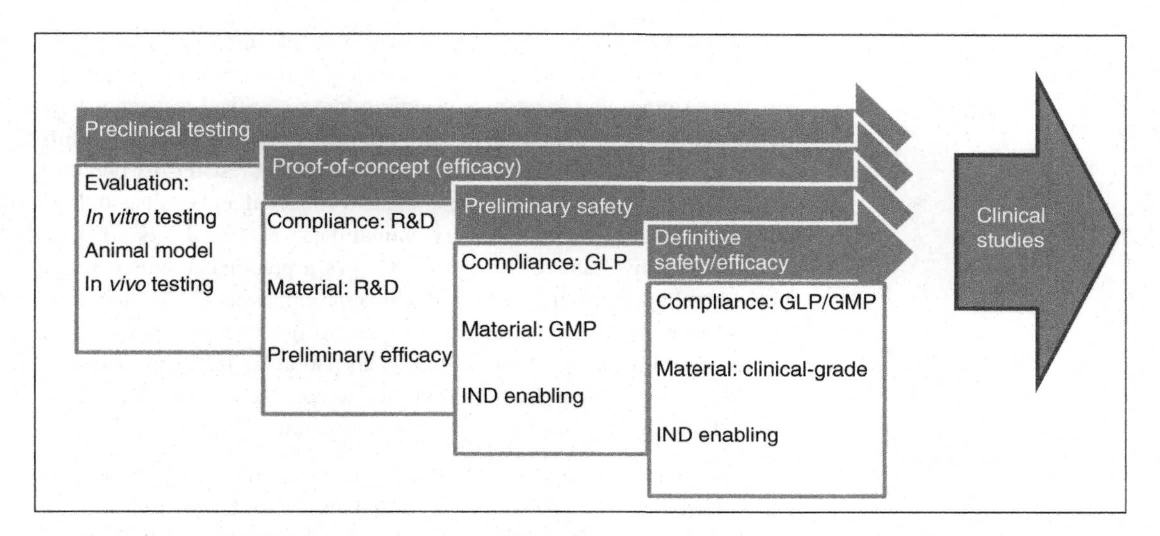

FIGURE 2.1 Product evaluation flow diagram.

analysis provided by the manufacturer of the cellular product and other critical reagents may be needed to support study credibility. Additional documentation to support a well-conducted preclinical study typically includes: (1) certificate of analysis of raw materials and supplies used to manufacture the cellular product, (2) certificate of testing and country of origin for specified animal-derived products, (3) animal bill of health, daily capture of animal weights, appearance, and appetence for all animal studies conducted, (4) expiration dates of materials and reagents, (5) use of written test methods, (6) training files or representative documentation of qualifications of the persons participating in the study conduct, (7) calibration records of critical instrumentation, (8) list of any deviations from the intended experimental plan, (9) raw data capture, and (10) the generation of final study reports.

Specific to animal facilities, documentation should include approved animal transfer, receipt, quarantine, pathogen testing, and housing of healthy animals, which are essential prior to animals being placed on study.

Preclinical safety animal studies should be designed and conducted in such a way to capture general health assessments as well as functional assessments to include evaluation of animal weight, appetences, routine clinical observations such as body condition and activity, bleeding, and morbidity or mortality. Functional assessments may include key body system functional tests such as pulmonary, heart, systemic, and other important functional endpoints to evaluate safety of cellular product delivery and sustainability to be defined in the study protocol. For example, an infused MSC product should include monitoring heart rate, breath rate, and blood oxygen levels (SpO_2) during the duration of cellular product infusion. Subsequent monitoring may include the addition of serum chemistry and blood coagulation parameters. The level of assessment and documentation to support these studies can seem overwhelming but are critical to demonstrate preclinical safety of a cellular therapy product. Final safety study reports should include analyses and interpretation of all the test results obtained such as complete data sets for all parameters collected and evaluated, both individual (tabulated) and summary animal data [1]. Fig. 2.2 is provided as a representative table of contents (TOC) of a preclinical animal safety study Final Report which demonstrates an appropriate detail of information including study design, data capture, and justification to support a well-documented preclinical animal safety study. This particular TOC was generated as a representative Final Study Report used to demonstrate the safety of an MSC cellular therapy product using a rat lung injury model.

The FDA regulations require preclinical studies, to sufficiently support an IND application, to be performed in compliance with GLP regulations. It is the investigator's obligation to describe how the preclinical studies were conducted and whether they were in compliance with GLPs. Usually if a preclinical study is prospectively defined, well documented, and controlled, it will be considered acceptable for use in supporting an early evaluation of safety and efficacy. This decision is usually taken on a case-by-case basis by the FDA and is dependent on a risk–benefit analysis along with evaluation of any study deficiencies that may impact the validity of data [1].

2.2.2 Staff Qualifications

All study staff should be adequately trained with applicable experience and associated supporting documentation to demonstrate qualifications in training files. No disqualified study staff should be allowed to participate in any aspect of study conduct. Just as the veterinary staff who oversees the animal facility operations during the in-life portion of a preclinical animal study, the engagement of a certified veterinary pathologist is essential for credible postlife evaluation of animal health. This blinded evaluation by the veterinary pathologist usually includes thorough documentation of gross pathological observations, notable observations, as well as microscopic or histological evaluation of specified organs and/or tissues. Additionally, the use and documentation of certified and adequately trained animal care personnel and research team members are essential in demonstrating the ability to perform dependable preclinical safety studies. Again, for continued credibility of study conductance, data capture, reporting, and documenting of study results it is important to ensure that no disqualified individuals are involved in the conduct of the study. The involvement of quality assurance (QA) personnel in the prospective study design, documentation capture, and

Preclinical Animal Safety Study Report – Table of Contents (sample)

FIGURE 2.2 Preclinical animal safety study report—table of contents (sample).

FIGURE 2.2 *(Continued)*.

review process is also beneficial to maintain integrity of a well-conducted preclinical safety study. There are multiple aspects of any animal study with many processes to be performed, monitored, and captured so having QA oversight is highly recommended, if not essential, for conducting well-documented and controlled preclinical safety studies.

2.2.3 Facilities

All laboratories used to conduct or support preclinical studies should allow for sufficient work space that is clean with access restricted to qualified laboratory personnel. Systems should be in place to avoid contamination, cross-contamination, or mix-ups within the laboratory area.

The animal facility needs to have all appropriate accreditations (eg, Public Health Service, Animal and Plant Health Inspection Service, Association for Assessment and Accreditation of Laboratory Animal Care) and should be conducted under an approved and current Institutional Animal Care and Use Committee animal protocol. Appropriate control of animal gender, diet (food and water), individual animal rooms with monitoring, and controls (eg, temperature, humidity, air changes, and light–dark cycle) is essential in justifying and reporting the validity of the animal study results. Animals are to be obtained from approved vendors with appropriate level of documentation and bill of animal health before being placed on study. The animal receipt process should consist of quarantine and release procedures to ensure adequate animal acclimation along with continuous monitoring of animal colonies for common infectious diseases. In addition to rodent quarantine and serology testing, surveillance programs typically include the use of sentinel animals to monitor and control the spread of pathogens in the vivarium.

2.2.4 Equipment, Supplies, and Reagents

All equipment used to support preclinical safety testing needs adequate associated documentation to support that it has been properly used, maintained, and calibrated. The use of appropriate equipment that are fully traceable, properly maintained, documented, and obtained from a reputable vendor is essential. Procedures and practices should be established to avoid contamination, cross-contamination, or mix-ups of products, samples, or results. Supplies and reagents need appropriate documentation to demonstrate that they come from reputable and qualified vendors with a history of providing quality products. A history of sufficient documentation may include a certificate of analysis and appropriate labeling with pertinent product information, which may include a quality assessment (clinical-grade or research-grade), expiration date, lot number, and identity, strength, purity, and potency information. A representative Certificate of Analysis is provided in Fig. 2.3 for a Mesenchymal Stromal Cells (MSC) Product.

Where possible clinical-grade materials should be used, but when not possible then the use of material with demonstrated comparability to clinical material is usually acceptable. Overall study conduct will be evaluated on all these important aspects of demonstrating validity of data and final data analysis.

Adaptation of GLP processes in the investigator's laboratory begins by defining minimal elements that will be in compliance with GLP [2].

INFUSED TEST MATERIAL:
Human bone marrow-derived MSC (*Manufacturer*)
Cryopreserved in PlasmaLyte + 2.5% DMSO + 10% Human Serum Albumin

Lot Numbers:	DAD-P-34; Date of MFG: 03/30/2012, Passage 5
Viable Cell Concentration (Post Thaw)	Low dose: 1.3×10^6 cells/mL, 96% viable, 87% recovery High dose: 2.5×10^6 cells/mL, 95% viable, 83% recovery
Sterility	No contamination detected
Endotoxin	< 0.5 EU/mL
Identity/Purity:	Flow cytometry for MSC markers
Potency:	Immunopotency assay (IPA)
Storage Conditions:	Nunc cryovials, vapor phase liquid nitrogen

INFUSED CONTROL MATERIAL:
Vehicle Control (*Manufacturer*)

Purity:	PlasmaLyte and HSA were 0.2 μm sterile filtered. Sterile DMSO was added to formulation.
Composition:	PlasmaLyte/0.5% DMSO/2.0% HSA
Storage Conditions:	Vapor phase Liquid Nitrogen

INFUSED DILUENT:
PlasmaLyte A (*Manufacturer*)

Lot Numbers:	C852681 (Ex 4/13); C896852 (Ex 9/14)
Purity:	sterile
Storage Conditions:	Room temperature

■ **FIGURE 2.3** Representative mesenchymal stromal stem cells (MSC) product certificate of analysis.

Ultimately, the investigator should strive for a well-controlled and well-documented study that is scientifically sound in the approach, execution, and completion.

2.2.5 Preclinical Study Cellular Product

The quality of material used to support preclinical testing is a very important decision and not always an easy one because of constraints such as cost or cellular product availability, the result of necessary adjustments to manufacturing scale or final formulation. Determination of the most appropriate cellular product (research-grade, comparable clinical-grade, or clinical-grade), the use of human products or animal-derived products are critical decisions that weigh heavily in defining the most appropriate preclinical study design. A general rule of thumb would be to use human cellular (clinical-grade) product in studies intended to support the collection of preclinical safety data. However, the use of animal cells in an animal model of disease may be the most appropriate cellular product choice to demonstrate preclinical efficacy or proof-of-concept. In any case, it is critically important to confirm the rationale and justification of the preclinical testing strategy with the assigned FDA review team before initiating a costly and resource-intensive study that is intended to support an IND application.

2.2.6 Preclinical Study Protocol

A prospectively defined study protocol should be written, signed, and dated at a minimum by a Study Director, Study Monitor, and Quality Assurance representative prior to the initiation of the study. All protocol deviations are to be captured and assembled for incorporation into the final study report. In addition, the capture and management of all raw data is important for future analysis and report generation. A prospectively defined procedure for the archival/disposal of final product, in-process materials, tissues and organs, documentation, and other valuable study materials and data is of high importance.

As previously mentioned, the use of an established documentation system will prove useful for collection of certificates of analysis, batch production records, test methods, material expiry information, lot numbers, and references such as scientific papers, methods, and associated dates/signatures. Ultimately, this level of documentation is of utmost importance in contributing to the generation of a complete final study report. A representative TOC defining the major elements of a complete Final Study Report for a preclinical safety study is provided for reference in Fig. 2.2.

2.3 PROOF-OF-CONCEPT STUDIES

As basic research projects progress along the product development pathway, one of the key elements is to evaluate the product for future development and clinical utility by performing proof-of-concept (POC) studies. The primary focus of a POC study is on demonstrating preliminary efficacy [3]. Collection of dose response data is helpful in evaluating assay validity with respect to identifying the range or limit of detection and specificity of the assay. Once the clinical potential of a product is realized the transition to assess preclinical safety is initiated as the logical next step and normal progression of product development. The better understanding and characterization of a cellular product during this early stage of development often leads to a more efficient and cost-effective product development effort in the short term. Potency/functional assay development and identification of the most appropriate assay or combination of assays for continued cellular product development is a key component of product characterization.

2.3.1 Preliminary Efficacy Studies

These early efficacy studies are often initiated as research and development (R&D) studies since they are usually performed before formal product development begins. It is worth noting that, if these early efficacy studies are prospectively designed and documented appropriately, the data can be very powerful in justifying later development decisions and strategies.

Both in vitro and in vivo biological activity assessments will need to be developed and used for further cellular product characterization. The identification or development of biological activity assays may include a measure of cellular product quality attributes such as differentiation potential of an MSC product or the measure of cell migration or growth that may be essential in evaluating cellular product quality characteristics. Ideally, one would want to measure an attribute that would be representative of an intended clinical benefit of the cellular product. Additionally, the identification of surrogate markers such as specific cytokine expression levels or preliminary potency assessments such as an immune potency evaluation in a mixed cell culture assay may be useful in contributing to the establishment of a product characterization profile. Due to the inherent variability of biological activity and utilization of cell-based assay formats, it is important to demonstrate assay validity by including appropriate controls and serial dilutions of the cellular product being tested to generate results consistent with a dose response.

The generation and use of in vitro assays provide a valuable resource during cellular product development which is especially useful early on in establishing an extended product characterization profile. As previously described, these in vitro assays may support POC study data by generating additional functional data to verify product biological activity. Throughout the course of product development, the use of these preliminary assays will enable these methods to become well demonstrated and instill confidence as reliable indicators of cellular product quality. These data can also enable baseline specifications, which can be used throughout product development and ultimately support and justify future process changes.

Preclinical efficacy animal studies need to be designed using the most appropriate cell product in an effort to avoid potential cellular product influence on the immune system or immune system effects on the activity of the cellular therapy product. In other words, consideration of studying xenogenic products (cells or constituents) that contain antigens likely to elicit an immune rejection in the host animal will need to be addressed in designing preclinical studies. An obvious approach to avoid this potential immune rejection issue is to use a syngeneic cellular product; although this may be a good strategy to evaluate efficacy endpoints it does not address safety of the human cellular product. Therefore, most animal efficacy studies are designed using syngenic cells and animals; whereas safety studies predominantly evaluate the use of human cells in healthy or diseased animal model systems. Examples of early efficacy endpoints may include an overall improvement of health or functional assessments such as pulmonary, cardiac, and other physiological conditions that can be evaluated in a compromised or animal disease model. Also, in vitro assessments of biological activity may provide powerful data to be used to compliment POC study data and provide a link between product activity and early efficacy outcomes.

2.3.2 **Animal Model of Disease**

Usually in an academic environment, there is an abundance of expertise and resources available to develop appropriate animal models that are representative of a particular clinical disease or disorder. In industry, however, the development or even the transfer of a newly developed animal disease model is usually more challenging and not usually a service routinely offered by a Contract Research Organization (CRO). The technology transfer required to develop a new disease model requires a notable level of specialized training, instrumentation and usually results in a logical extension of the timeline devoted to preclinical testing. In the event that a CRO is willing to adopt or offer a customized animal model of disease, it will likely not be without specialized training and an intensive technology transfer effort. Since animal disease models are developed for specific clinical indications, they do not typically provide a strong business platform opportunity for

CROs and usually end up being a costly and time-consuming endeavor. This is in contrast to routine or well-established animal models and testing platforms already developed and typically offered by a CRO to evaluate the safety and efficacy of standard pharmaceutical products. Thus, if cellular therapy product testing requires the development of an appropriate animal model of disease the general advice is to maintain control of an animal disease model by developing, retaining, and using it "in-house." This strategic approach provides the most flexibility in the utilization and potential future adjustments to this model during later phases of cellular product development. Future modifications to a cellular production process will likely be encountered as a result of scale-up efforts, lowering cost of goods, or process simplifications. In many cases, evaluation of process modifications on the cellular product will require the use of an already established animal model system to confirm cellular product comparability. A comparability study using an animal disease model may be the most desirable evaluation of a significant process change to demonstrate the change has no impact on quality attributes of the cellular therapy product.

2.3.3 Selection of an Appropriate Animal Model

Scientific merit should prevail in the justification of the selection and use of an animal model. Selection of an appropriate animal model should initially be based on how closely the model may reasonably mimic the human disease or disorder of interest. Another key decision in selecting an acceptable animal disease model should include the ease or reproducibility to replicate the animal model without resorting to heroic efforts. Strong consideration in the selection of an appropriate animal model system should avoid elaborate, time-consuming, and expensive measures to develop what may seem like the "ideal" model system. An animal model needs to provide a viable, highly reproducible, and predictable test system that can demonstrate the ability to sustain the disease state, be conducive to evaluating cellular product delivery, and allow for reasonable methods to monitor the disease progression. It is also important to be able to evaluate critical organ systems and assess the immune status of the animals as they may elicit an immune response associated with the cellular product or animal-derived excipients associated with the cellular product.

Any anticipated animal attrition due to the creation of the model needs to be considered in the experimental design and selection of an animal model system being used. To enhance the therapeutic outcome and minimize the number of animals to be studied, it is beneficial to eliminate high degrees of variability as it relates to animal model creation. This evaluation may suggest using more appropriate models that can avoid systemic error and result in the rejection of an initially identified inconsistent model system. Of particular importance is the consideration given to the appropriate stage of disease progression, and if the clinical intent is prophylactic intervention then the timing of cellular product administration is also essential in the appropriate animal model selection. Other elements to consider when selecting an appropriate animal model may include the need or desire to include repeat dosing, evaluate a comparator treatment such as the current standard of care for that particular disease indication, a placebo, and/or other controls as deemed appropriate.

It may be reasonable to anticipate slightly different efficacy outcomes in an animal model compared to human clinical outcomes, especially since an animal model is not likely to be fully correlative with the underlying human clinical pathophysiology.

2.3.4 Preliminary Evaluation of Safety

The most cost-effective and efficient approach in determining the safety profile of a cellular therapy product is to evaluate safety as part of this preliminary evaluation of product development. This preliminary safety evaluation may include a compilation of safety data including in vitro assays, cell-based assays, and most importantly from POC studies. POC studies, although performed mainly to identify preliminary efficacy, are strategically designed to include some obvious elements of overall animal health and safety. Usually following the establishment of an appropriate animal disease model, the model is used to more broadly evaluate safety which is either directly associated with the disease model or to evaluate unanticipated results of animal exposure to the cellular product. This general safety evaluation of cellular product in the POC study may include adverse reactions to delivery methods, premature death, respiratory distress, cardiac arrest, moribund conditions, general appearance, appetence, dehydration, noticeable weight change, and importantly tumor formation. A concerted effort using additional research-based studies in the early stages of product development could prove extremely valuable in supporting POC study results. It is more efficient and cost effective to run these studies to support preclinical safety in parallel but performed in a prospective and well-documented manner. In addition, alternative assays and associated results that are developed and collected during the evaluation and selection of ideal assay systems may be used to support POC data and results. Typically, assay development efforts are performed to identify the most biologically responsive model system for use in demonstrating optimal effects of the cellular product. These results should lead to establishing the most appropriate assays that can detect a robust biological effect of the cellular product in vivo. Additional test methods may be needed to support product comparability or stability in future product development efforts. Acceptable methods are suggested in relevant guidance documents such as FDA and International Conference on Harmonization (ICH) guidance documents [4,5]. Both during and after the development of these alternative test methods, used to evaluate biological activity, it is important to consider evaluation of cellular product from multiple manufacturing campaigns. Usually during the early stages of product development very few manufacturing runs have been performed so it becomes increasingly more important to gain manufacturing experience as product development progresses toward production of clinical-grade material. It is highly encouraged to conduct several manufacturing runs during the early stages of product development to resolve any unanticipated cellular product deficiencies such as product yields, purity, stability, or meeting testing specifications. Lot release testing on these early manufacturing campaigns can avoid prematurely setting tight product specifications that may be hard to achieve. Of importance, cellular product specifications may vary as a result of future changes to the manufacturing process or from necessary adjustments to a quality control assay method. In any case, this additional assay development effort will provide the opportunity to evaluate lot-to-lot MSC cellular product variability. Just as important, if an assay method is determined to be suitable in discerning lot-to-lot differences in biological activity then it may be an ideal candidate for use as a final product lot release assay. These are all factors to consider when determining the overall safety profile and potential biological activity of a cellular therapy product.

2.3.5 In Vitro Assessments

Additional in vitro testing of MSC products during the early POC stage may include maximum cellular product growth and expansion to demonstrate integrity of normal growth and stable genetic characteristics. For MSC products, similar to evaluation of cells in cell banking efforts, a commonly accepted evaluation includes end of production (EOP) cell testing [6]. EOP cell testing may represent a "worst case scenario" evaluation of cellular product safety with respect to evaluation of genetic stability and normal growth characteristics at the maximum (or usually extended beyond anticipated maximum) intended cell expansion or growth.

Evaluation of a biological response such as increased or decreased regulation of a specific protein or factor (eg, cytokine) is very powerful in demonstrating a functional property or characteristic of a cellular product [1]. These types of assays are relatively simple and reproducible, and are amenable to serve as a reliable in vitro bioassay for use in evaluation of future cellular therapy production in establishing lot release criteria. Also, these supportive or confirmatory data will prove essential to solidifying the utility of functional assays in later product development efforts. Other types of in vitro assays may include cell migration assays, cell differentiation assays, and various assay formats that evaluate gene expression either at the RNA or protein level.

Use of in vitro assays in the later stages of development will become increasingly important for evaluation of MSC product quality. In many cases, surrogate endpoint assays are used to determine final cellular product quality or the selection of a specific assay to serve as a lot release assay. The most appropriate lot release assay must be evaluated to demonstrate specificity, sensitivity, and robustness. Development of a potency assay is in its infancy at this stage, but becomes increasingly important as product characterization and development matures. Several assays that have the potential to evaluate the biological activity of the cellular product are recommended before initiation of Phase 1 human clinical studies. Although utilization of a potency assay is not typically required for cellular therapy products until Phase 3 human clinical studies, the only way to evaluate the ability of a potency assay to predict human clinical outcome is to have several putative in vitro assay results to compare to preliminary human data [7]. This is precisely why more than one biological assay should be developed and used throughout early clinical evaluation, thereby allowing for the selection of a single scientifically sound reproducible predictor of in vivo efficacy [1].

2.3.6 **Cell-Based Assays**

Cell-based assay formats are not typically amenable for use as a lot release assay due to lengthy culture process, inherent variability, and expense. Nonetheless, cell-based assays are invaluable for the evaluation of preclinical safety and may serve to some degree as a product potency assessment (ie, biological function). Use of a cell-based culture assay to measure a functional response in the presence of a cellular product may be vital in assessing the quality and function of a cellular product. Assessment of cellular product activation or inability to become activated may be important to study in preclinical evaluation of a cellular product [1].

These types of assays may be employed to evaluate biological activity of a cellular product either as a surrogate measure to identify any correlation with a less complex lot release assay or a potency assay [7]. Use of cell-based assays may identify overall product quality attributes or provide a measure of biological activity of a cellular product. Collectively, the use of an in vitro with results of a cell-based assay may lend support to POC data and evaluation of cellular product activity in an appropriate animal model of disease.

2.4 PRECLINICAL SAFETY STUDIES

To best position the development of a cellular product for first-in-human clinical studies, a series of generally accepted preclinical safety studies are typically performed that are specific to cellular therapy products. The cell type and nature of the cell selection and complexity of the manufacturing process and cell product will influence the recommended panel of preclinical tests to be performed. Some test methods are standard and performed most efficiently by contract testing laboratories, while other test methods may need to be developed to meet the unique requirements of a particular cellular therapy product. In this case, cell product specific methods need to be qualified as part of a formal quality control procedure before routine use in testing and release of the cellular product for human clinical use. Examples of these types of assays or test methods include those to measure cell viability, sterility, purity, endotoxin, and potency as is presented in the representative Certificate of Analysis for MSC in Fig. 2.3.

2.4.1 Preclinical Safety Assays

As part of the cellular product characterization, a series of assays will be employed which may include a variety of assay formats such as flow cytometry, specialized cell product staining, turbidometric plate assays, and perhaps specific cell fixation methods or the use of polymerase chain reaction methods. These assay formats will likely be adopted to perform cellular characterization such as profiling of cell surface expression markers, cell viability, cell storage conditions, and adventitious agent testing. Other studies may include the evaluation of nontumorigenic potential in soft agar, karyotype analysis, mutagenicity testing, cell differentiation and stability, cellular product identity, product impurity profile, and preliminary potency assay(s). Additional specialized assays may be developed to evaluate residual undifferentiated cells, assess effectiveness of a cell replication inactivation process (ie, "feeder cells"), or evaluation of cell proliferative capacity [1,8–10].

Overall in the development of these many testing methods it is essential that the appropriate positive and negative controls are included to assist in assessing test method validity. In addition, the inclusion of serial dilutions is highly encouraged whenever possible in an effort to obtain a "dose–response" curve. Collection of dose response data is helpful in evaluating assay validity with respect to identifying the range or limit of detection and specificity of the assay.

2.4.2 Preclinical Assessments in Animals

Choice of an appropriate animal model will likely be one of the biggest decisions that will impact future safety testing, cellular product delivery, dosing, and administration methods throughout product development. This scientific decision needs to be carefully chosen based on good data and justification of model selection over other models. During the early stage of product development an indication-appropriate model system is needed to evaluate short-term safety which is preferably performed to mimic the anticipated clinical delivery method. In addition, longer term safety studies are performed to evaluate potential cell product effects on overall animal health, target tissues, and organ systems. Some unique aspects of preclinical testing of cellular products in animal models include the evaluation of (1) route of administration, (2) cell persistence or fate postadministration, (3) host immune response, (4) tumor formation, and (5) potential local or target tissue inflammatory response [1]. Depending on the cellular product characteristics and any potential variability in cell characteristics (ie, induced pluripotent stem cells), it may be necessary to evaluate tumorigenic potential in animals. Studies to demonstrate freedom from tumorigenic potential, if required, are usually performed using immune compromised animals in an effort to achieve maximum detection levels. The benefits of utilizing a whole animal system for this type of evaluation include the ability to assess cumulative effects associated with studies requiring multiple dosing schemes to reflect the intended human clinical dosing regimen.

2.4.3 Preliminary Safety in Animals

Animal study design for cellular therapy products typically includes justification for in-life study duration, in-life study assessments, postlife study assessments, selection of appropriate study endpoints, qualifications of the individuals selected to evaluate study endpoints, blinding for data capture and interpretation, evaluation of cellular product biodistribution, and cell fate over time. The selection of animal species and gender should be justified and the determination for the number of animals required for a safety study should be mainly based on statistically relevant criteria. Selection of appropriate animal species and disease model depends on the putative mode of action and target specificity of the investigational product [1]. The MSC safety profile may be distinctly different when studied in normal healthy animals as compared to that of a diseased or compromised animal model. Therefore, animal studies should be designed to reasonably mimic the human clinical condition. Overall consideration needs to be given to the choice of whether or not to use a large or a small animal, appropriate animal species, and animal gender or determination that no gender differences are anticipated. The components of the cellular product will greatly impact the animal selection process. For example, if the cellular product contains immunogenic components or would create a xenogeneic reaction, then the selection of an immune compromised animal would likely be preferred over an immune competent one for obvious reasons of immune rejection and potential cellular product clearance by an intact animal model immune system.

To adequately address preclinical safety of a human cellular product, it is essential to test the human cells in either a healthy animal or animal disease model system. Although evaluating safety in a complex living biological system has limitations, this is generally accepted as the way to directly assess the in vivo safety of a human cellular product before it is evaluated clinically. Of importance is the most appropriate selection of a healthy animal or animal model of disease along with the cellular product being studied which has the potential to greatly influence the validity of this safety assessment. Case in point, the use of an immune compromised animal may be necessary to evaluate the systemic effects of a human cellular product within the context of an animal model system. In cases where a cellular product like human MSC recognized has having immunomodulatory activity and deemed immune privileged, then the human cellular product may be adequately evaluated in an immune competent animal. Conversely, if a human cellular product is rapidly cleared by the host (animal) immune system then evaluation of long-term safety endpoints may be futile in this combination of human cells in an animal model system with an intact immune system.

It is highly recommended that preclinical safety studies are performed in accordance with regulatory expectations associated with current Good Laboratory Practice (cGLP) [2]. Some US FDA regulatory discretion may be allowed and full compliance with cGLP (21 CFR Part 58) may not be required for early stages of product development. Nonetheless, every effort should be made to comply with as many elements of cGLP as possible to include (1) the use of appropriately qualified personnel, (2) quality assurance oversight, (3) use of appropriate facilities to conduct animal research and to provide appropriate care for animals, (4) standard operating procedures in place for the management of the animal facility, (5) individual animal data collection and evaluation, (6) test product to be appropriately identified and handled, (7) the study to be conducted under an approved protocol, (8) records and raw data to be retained as described in the protocol, and (9) a final and complete study report to be prepared. Demonstration of safety will be required using either healthy animals or a well-defined animal disease model. General Biological Product Standards (21 CFR 610) are generally applicable to the development of MSC product for human clinical evaluation. These include evaluation of potency, sterility, purity, identity, mycoplasma, adventitious agent testing, test requirements for donors, manufacturing, and labeling. Although the General Safety Test is required for biological products to demonstrate preliminary safety prior to the initiation of a first-in human clinical study, this testing is not always scientifically feasible or informative and as a result not typically warranted for MSC products [11]. In some cases, evaluation of tumorigenicity in animals may be justified and can be performed. These studies may be performed by simple subcutaneous injections of the cellular therapy product in immune compromised rodents followed by monitoring the local microenvironment for subcutaneous growth potential. Selection of cell dose and human dose equivalents and to include an estimation of the margin of safety is important in the design of these studies. In these types of studies, controls are essential to demonstrate appropriate administration technique and confirm utility of this method in the animal species selected to perform these studies.

2.4.4 **Use of Hybrid Pharmacology/Toxicity Study Design**

Hybrid animal safety studies are becoming an accepted approach where a single animal study is designed to include both pharmacology and toxicology endpoints, rather than performing independent pharmacology and toxicology studies. In part, the goal of this hybrid approach is to use fewer animals while collecting more comprehensive endpoint data to evaluate both safety and efficacy in a single study. These types of combined studies are best developed to demonstrate a dose response along with appropriate positive and negative controls. If appropriate, the design of hybrid studies should include the evaluation of cellular product biodistribution.

In general terms, the evaluation of safety in animals and observation of toxicities may be the result of a number of possibilities including components of final cellular product formulation, route of administration, or the accumulation of cellular product/cellular product components in nontarget tissues or organs. Cell tracking or biodistribution is typically an FDA requirement for MSC products to demonstrate preclinical safety [1]. Cell tracking studies are best initiated during early stages of product development to assess cellular therapy product longevity and the ability of the cellular product to engraft and survive after administration. These studies may incorporate the evaluation of exogenous cell migration, cell proliferation, cell differentiation, intended or unintended cell phenotypes, and the potential for germline transmission if detected in reproductive organs. Lastly, these studies may be designed to evaluate in vivo responses such as immune activation or suppression, elicitation of an immune response to the cellular product or perhaps indirectly related to the cellular product constituents. There are several inherent challenges associated with the development and design of cell tracking studies that should be taken into consideration when designing such studies. These notable challenges include real-time evaluation of live cells, cell product labeling, and available imaging modalities.

2.5 PRECLINICAL SAFETY PROTOCOL DESIGN

Ideally, the design of a preclinical safety protocol should mimic, as closely as possible, the intended design of the human clinical trial protocol. This is especially important with respect to cellular product delivery method, cell dosing, number of administrations, study endpoints, and other relevant safety study evaluations.

2.5.1 Protocol Design

Several important considerations are taken into account when designing an appropriate representative preclinical study protocol.

The preclinical studies need to be prospectively defined with all the essential elements as is typical for a human clinical protocol and to include: dose escalation and schedule, clinically relevant route of administration, appropriate positive and negative controls, clinically appropriate time points and study endpoints, justification for proposed study duration, calculated human dose equivalent, comparable rate of cell product administration, and attempt to evaluate cell product biodistribution and tracking in the animal model of disease. If designed appropriately, the preclinical animal studies should have been performed to reasonably mimic that intended for the human clinical study to include the previously stated study parameters outlined in the clinical protocol. Similarly, all study assessments, endpoints, and outcome analysis should be planned to capture and support the parameters intended for the human clinical study.

2.5.2 Cellular Therapy Product Dose Selection

Selection of the cellular product dose should be justified and supported by research studies, peer-reviewed publications, or personal experience. Ideally, the design of a preclinical safety study will include a close approximation of the intended human clinical study design with respect to dose (human dose equivalent), dosing schedule, route of administration, and rate of administration. Additional dosing and dose strategy information may be found in a US FDA guidance document on estimating the maximum safe starting dose in initial clinical trials for therapeutics [12]. It is advised to consider a preclinical dosing strategy that includes bracketing the targeted human dose equivalent to provide an additional margin of evaluating safety. Preclinical safety data that exceeds the intended human clinical dose equivalent may prove invaluable. It may assist with justifying increasing the dose or exposure to exceed an initially intended dose with minimal additional risk based on already existing preclinical safety data.

Several methods have been proposed in determining human dose equivalents in a variety of animal models [12]. Some methods are more appropriate for specific clinical indications than others; for example, in the development of a dermal applied wound care product it would be most appropriate to used total body surface area as a way to normalize or equate to human dose equivalents [12]. In other similar clinical indications targeting specific organs, it may be most appropriate to use organ size (volume or area) as the best comparator to the analogous human tissue or organ. Another example may include the use of bone length, more specifically tibia or femur length may be used to determine relationship to respective human bone length in appropriate orthopedic indications. Human dose equivalent approximations have been used historically in the pharmaceutical industry for decades with the use of body weight and weight conversions resulting in dose per kilogram of animal weight compared to human dose per patient weight; however, this type of allometric scaling is of limited utility in determining equivalent dosing of cellular products. Approximate or direct scaling based on organ size or volume (eg, vocal fold) may be appropriate in those specialized organ systems. Of importance is the prospective nature and justification of approximating the dose equivalent in preclinical studies and the scientific rationale for the comparison to human clinical anatomy and physiology.

2.5.3 Study Endpoints

Prospective study endpoints should be clinically meaningful with respect to duration of cell product exposure and proposed study endpoints. In-life evaluation would ideally include interim and final functional endpoints and other safety endpoints such as overall animal health with documented evaluation of activity, diet, weights, and appearance.

A variety of observations may be important to capture in dose escalation or dose exploration studies. A safety study may be designed and conducted to identify a maximum tolerated dose (MTD), to identify the low or minimum anticipated biological effect level which can provide dose equivalent information necessary to support the proposed clinical cellular therapy product dose. Note that an MTD value may be representative of cellular product volume limitations rather than a direct result of cellular product toxicity. Some safety studies are designed to evaluate MTD to define the maximum cell dose in an animal model, which can demonstrate no observable adverse effect level. Additional information on the preclinical design, dosing, and utility of these endpoints is provided in a variety of resources which include FDA guidance documents and articles in the scientific literature [3,13].

The postlife evaluation of all preclinical safety studies intended to support an IND application should include the measurements of body and tissue weights, gross and histological observation, and evaluations of prospectively specified organs and tissues. Study endpoints should also include the collection, analysis, storage, and archival of appropriate fluids, tissues, and organs. Of note, the collection of organ appearance, weight, and blood chemistry evaluations are expected to be performed by a certified veterinary pathologist, when possible.

2.5.4 Materials

The use of cellular product manufactured in compliance with current Good Manufacturing Practices is strongly encouraged [14]. Note that sometimes the use of alternative but comparable material to clinical-grade product may be acceptable to use in preclinical safety studies. Raw materials will also need to be evaluated and deemed acceptable for use in preclinical safety studies [6]. It is best practice to utilize the same manufacturing or production process that has demonstrated to be comparable to that intended for generation of human clinical product. Early consideration of specific materials may help to avoid or alleviate the need to perform comparability studies. In any case, comparability studies may be necessary to demonstrate or justify differences in materials or processes used to generate cellular product for preclinical safety studies that are intended to support human clinical studies.

2.6 CONCLUSIONS

The information provided in this chapter is intended to provide an appreciation for the elements important to the design and conduct of preclinical studies that will ultimately be used to support an IND application for a cellular MSC product. Preclinical studies should include proof-of-concept studies, early efficacy studies, and sufficient safety data to justify the conduct of future human clinical studies. It is essential to plan, document, and conduct preclinical studies using sound scientific practices to ensure the capture of appropriate data to support future clinical development. Important elements that are often overlooked in the design and implementation of preclinical studies include: the quality of cellular product, selection of an appropriate animal model of disease, route and duration of cellular product administration, product administration schedule, clinically relevant study endpoints, biodistribution and longevity of the cellular product in vivo. These essential elements that contribute to preclinical study design and implementation simply cannot be overlooked as they are required to determine clinical relevance of preclinical study results.

REFERENCES

[1] Food and Drug Administration. Guidance for industry: preclinical assessment of investigational cellular and gene therapy products. Rockville, MD. Available at: http://www.fda.gov/biologicsbloodvaccines/guidancecomplianceregulatoryinformation/guidances/cellularandgenetherapy/ucm376136.htm; November 2013.

[2] Food and Drug Administration. Good laboratory practice for nonclinical laboratory studies. Title 21 Code of Federal Regulations, Part 58. Available at: http://www.accessdata.fda.gov/scripts/cdrh/cfdocs/cfcfr/CFRsearch.cfm?CFRPart=58; April 1, 2013.

[3] Lee MH, Arcidiacono JA, Bilek AM, et al. Considerations for tissue-engineered and regenerative medicine product development prior to clinical trials in the United States. Tissue Eng Part B Rev 2010;16:41–54.

[4] Food and Drug Administration. Guidance for industry: comparability protocols—protein drug products and biological products—chemistry, manufacturing, and controls information. Rockville, MD. Available at: http://www.fda.gov/downloads/Drugs/GuidanceComplianceRegulatoryInformation/Guidances/ucm070262.pdf; September 2003.

[5] Food and Drug Administration. Guidance for FDA reviewers and sponsors: content and review of chemistry, manufacturing, and control (CMC) information for human gene therapy investigational new drug applications (INDs). Rockville, MD. Available at: http://www.fda.gov/BiologicsBloodVaccines/GuidanceComplianceRegulatoryInformation/Guidances/CellularandGeneTherapy/ucm072587.htm; April 2008.

[6] Food and Drug Administration. Guidance for industry: characterization and qualification of cell substrates and other biological materials used in the production of viral vaccines for infectious disease indications. Rockville, MD. Available at: http://www.fda.gov/downloads/biologicsbloodvaccines/guidancecomplianceregulatoryinformation/guidances/vaccines/ucm202439.pdf; February 2010.

[7] Food and Drug Administration. Guidance for industry: potency tests for cellular and gene therapy products. Rockville, MD. Available at: http://www.fda.gov/downloads/BiologicsBloodVaccines/GuidanceComplianceRegulatoryInformation/Guidances/CellularandGeneTherapy/UCM243392.pdf; January 2011.

[8] Food and Drug Administration. Points to consider in the characterization of cell lines used to produce biologicals. Available at: http://www.fda.gov/downloads/biologicsbloodvaccines/safetyavailability/ucm162863.pdf; July 1993.

[9] Food and Drug Administration. Current Good Tissue Practice. Title 21 Code of Federal Regulations, Part 1271, Subpart D. Available at: http://www.accessdata.fda.gov/scripts/cdrh/cfdocs/cfcfr/CFRSearch.cfm?CFRPart=1271&showFR=1&subpartNode=21:8.0.1.5.55.4; April 1, 2013.

[10] Food and Drug Administration. Guidance for industry: source animal, product, preclinical, and clinical issues concerning the use of xenotransplantation products in humans. Rockville, MD. Available at: http://www.fda.gov/biologicsbloodvaccines/guidancecomplianceregulatoryinformation/guidances/xenotransplantation/ucm074354.htm; April 2003.

[11] Food and Drug Administration. General Biological Products Standards. Title 21 Code of Federal Regulations, Part 610. Available at: http://www.accessdata.fda.gov/scripts/cdrh/cfdocs/cfcfr/CFRSearch.cfm?CFRPart=610; April 1, 2013.

[12] Food and Drug Administration. Guidance for industry: estimating the maximum safe starting dose in initial clinical trials for therapeutics in adult healthy volunteers. Rockville, MD. Available at: http://www.fda.gov/downloads/Drugs/Guidance/UCM078932.pdf; July 2005.

[13] Food and Drug Administration. Draft guidance for industry: considerations for the design of early-phase clinical trials of cellular and gene therapy products. Rockville, MD. Available at: http://www.fda.gov/downloads/biologicsbloodvaccines/guidancecomplianceregulatoryinformation/guidances/cellularandgenetherapy/ucm359073.pdf; July 2013.

[14] Food and Drug Administration. Current good manufacturing practice for finished pharmaceuticals. Title 21 Code of Federal Regulations, Part 211. Available at: http://www.accessdata.fda.gov/scripts/cdrh/cfdocs/cfcfr/CFRSearch.cfm?CFRPart=211&showFR=1; April 1, 2013.

Chapter 3

Delivery and Tracking Considerations for Cell-Based Therapies

E.G. Schmuck* and A.N. Raval*,**

*Division of Cardiovascular Medicine, Department of Medicine, University of Wisconsin School of Medicine and Public Health, Madison, WI, United States; **Department of Biomedical Engineering, University of Wisconsin, Madison, WI, United States

CHAPTER OUTLINE

3.1 INTRODUCTION

Significant effort is underway worldwide to unlock the therapeutic potential of mesenchymal stem cells for tissue repair. These investigations are fueled by a desire for better and more durable treatments for a variety of diseases. Direct and indirect cell–cell and cell–protein interactions are credited for the favorable effects observed in preclinical and clinical trials to date. Despite notable progress in this field, significant optimization is required for

Mesenchymal Stromal Cells. http://dx.doi.org/10.1016/B978-0-12-802826-1.00003-9

successful translation into clinical practice. Two interdependent concepts deserve mention: (1) cell delivery and (2) cell tracking. In this chapter, we review cell delivery and tracking methods and draw upon examples from the experience in heart disease.

3.2 CELL DELIVERY

"Cell delivery" is the act of transplanting a particular cell type or group of cells from its source to a recipient. The method of cell delivery is critical for determining cell dose, fate, and function. Successful clinical adoption of a therapeutic cell type will hinge on a delivery approach that (1) offers optimal cell retention at the target site, (2) is noninvasive or minimally invasive, (3) can be administered at low cost, and (4) can easily be integrated into clinical practice (Fig. 3.1). Paracrine signaling is believed to be the main mechanism of how mesenchymal stromal cells (MSCs) exert a positive effect for local tissue repair [1,2]. Confining the delivery of MSCs for uptake in the injured tissue may improve cell survival, improve cell engraftment, minimize extraneous diffusion, avoid "off-target" toxicity, and maximize therapeutic efficacy.

The safety of the cell delivery approach is paramount and therefore, noninvasive or minimally invasive approaches are preferred over more invasive approaches. The rising cost of healthcare in the United States has resulted in increased scrutiny over how novel therapies, and specialized instruments

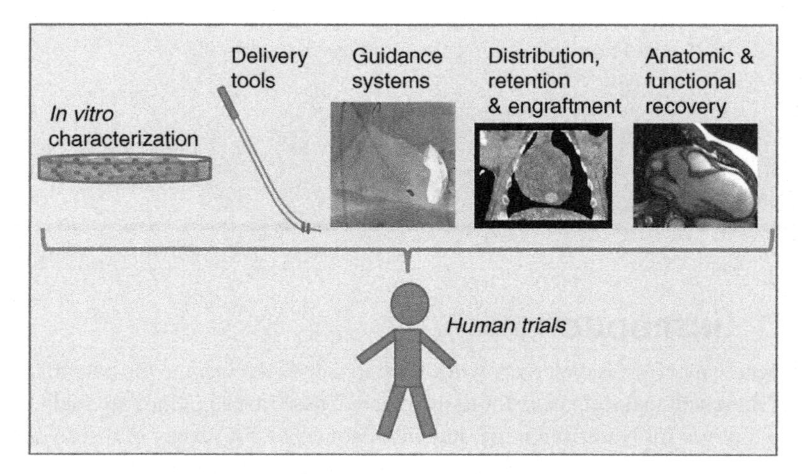

■ **FIGURE 3.1** The pathway to human trials and subsequent clinical adoption of cell therapy requires in vitro discovery and characterization, delivery tools, and imaging systems to guide cell therapy, and methods to measure key anatomic and functional effects.

required for their delivery, are reimbursed. Clinical practicality of the delivery method must also be considered. Does the delivery approach require serial dosing? Is the delivery equipment and expertise found in only a few highly specialized centers? These concepts and regulatory considerations will be discussed with examples drawn from MSC therapy for cardiac disease.

Cell delivery methods can be broadly categorized as:

1. Intravascular infusion
2. Direct injection
3. Scaffold application
4. Trans-mucosal delivery

3.2.1 Intravascular Infusion

3.2.1.1 *Intravenous infusion*

Intravenous or "IV" cell infusion enjoys four centuries of experience, primarily with blood product transfusion [3]. IV bone marrow mononuclear cell infusion has been employed in clinical trials for acute myocardial infarction [4], stroke [5–7], inflammatory bowel disease [8], graft versus host disease [9], and other conditions. IV infusion is simple to perform and can be done at any clinic or infusion center. Typically, a cannula is inserted in a peripheral vein in the hand or arm and the cell product is infused using an infusion pump to control the infusion rate. The infusion rate is dependent upon inner diameter of the cannula, cell size, viscosity, and volume of the cell product.

Disadvantages of IV cell administration include the potential for site inflammation, infection, and thrombosis, especially with wider diameter lines such as central lines. This delivery approach would not be appropriate when there is flow obstruction such as venous thrombosis. If the patient has an intermittent or continuous right to left shunt from an atrial septal defect or arteriovenous malformation for example, then clumped cells have the potential to cross over to the systemic circulation, potentially resulting in ischemic stroke or other embolic events.

Short-term cell retention is low with intravenous infusion. The majority of cells are sequestered within lung, liver, and spleen within a few hours of administration [10–13]. Acute tissue injury such as with myocardial infarction or stroke may attract systemically delivered cells such as MSCs through homing pathways [14–16]. However, only a sparse number of cells are detectable >72 h in the heart based on cell tracking animal studies [13,17].

Safety, retention, and efficacy are key considerations when deciding about the best route of cell delivery. Hare et al. demonstrated the safety of intravenously delivery allogeneic MSCs in 60 subjects with recent myocardial infarction (MI) [4]. There was no evidence of infusional toxicity in this trial. There were fewer episodes of ventricular tachycardia and improved pulmonary function in the treated group. Improved ejection fraction was also seen in a subgroup with anterior MI. These trends resulted in a US FDA Phase II Osiris 403 trial for the treatment of acute MI, which remains unpublished to date. The trial completion date for the primary outcome was Dec. 2011 and a 5-year follow-up plan is estimated to complete in Feb. 2016. Intravenous infusion of cells has generally fallen out of favor for cardiac repair due to low retention, which likely prevents a meaningful therapeutic effect.

IV infusion of allogeneic MSC product has been used successfully for the treatment of pediatric acute graft versus host disease [18]. On the other hand, Prasad et al. showed that IV infused autologous bone marrow mononuclear cells were not effective for the treatment of ischemic stroke [7]. There are numerous ongoing trials sponsored by companies and academic institutions testing variations of intravenous allogeneic and autologous MSCs for a variety of conditions: inflammatory bowel disease, chronic obstructive lung disease, diabetes mellitus, and spine disorders (refer to www.clinicaltrials.gov). However, it is possible that for some of these indications, a different delivery method could be more effective.

Direct portal vein cell infusion and retrograde coronary sinus infusion has been tested in humans with liver disease [19] and heart disease respectively [20,21]. The retrograde coronary sinus approach provides for safe access to most of the myocardium for homogeneous delivery; however, variable coronary sinus anatomy, coronary sinus rupture, and vascular trauma are significant challenges.

3.2.1.2 ***Intraarterial infusion***

Human trials have employed intraarterial infusion of unfractionated bone marrow mononuclear cells which contain MSCs, and circulating proregenerator cells for acute and chronic ischemic heart disease [22–44]. There has been far less human experience with this cell delivery method for peripheral artery disease [45–47], ischemic stroke [48–55], and other neurodegenerative diseases [56]. Intracoronary artery infusion is performed by placing a coronary guide catheter at the ostium of the coronary artery and then directing an over-the-wire balloon catheter into the target coronary artery. The guide wire is removed and cells are infused through the wire port of the catheter. This approach requires the coronary artery to be open.

The experience to date in a recent myocardial infarction population suggests that intracoronary cell infusion is generally safe. Embolization, dissection, perforation, thrombosis, arrhythmias, and vascular access complications are all potential disadvantages of this approach. Further, myocardial cell retention with this approach is very low and consistently less than 5% after 24 h [57,58]. Cell retention may be improved if the balloon is inflated to arterial occlusion during the cell infusion to avoid cell wash-out and encourage cell migration into the myocardial interstitium [22,59]. Applying extracorporeal shock waves prior to intracoronary infusion of bone marrow mononuclear cells appears to modestly improve treatment effect [30], potentially through boosted cell retention from increased expression of certain chemokines [60].

Despite initial enthusiasm, a well-performed, recent metaanalysis of individual patient data from bone marrow cell randomized trials in acute myocardial infarction showed no benefit in terms of clinical events or changes in left ventricular function [61]. Subgroup analyses suggest that trial subjects with very low baseline ejection fraction may benefit from cell therapy the most [62]. An international trial called BAMI (bone marrow derived mononuclear cells on all-cause mortality in acute myocardial infarction) is underway to determine survival benefit from intracoronary infusion of unfractionated bone marrow mononuclear cells (see www.clinicaltrials.gov).

3.2.2 Direct Tissue Injection

3.2.2.1 Intramuscular injection

Intramuscular (IM) injection is commonly performed for vaccine immunizations for long-lasting effect and the same principal has been applied to progenitor cell therapy. Intramuscular injection of proregenerative cell product has been performed primarily for cardiac repair and ischemic limb repair. In the heart, IM injection can occur via direct epicardial approach and through a sternotomy or thoracotomy window with a short needle and syringe. The epicardial injection approach is considered the most reliable method of delivery. Advantages of this approach include a relatively greater level of cell retention [63]. Limitations of this approach include its invasiveness, inability to reach certain walls of the left ventricle such as the septum and posterior wall. Due to the invasiveness of this delivery approach, trials testing epicardial therapy are typically piggy backed onto procedures where a thoracotomy is required, such as coronary artery bypass grafting or left ventricular assist device implantation [64,65].

Catheter-based transendocardial cell delivery is a minimally invasive alternative to epicardial intramuscular delivery (Fig. 3.2). There are a number of industry-sponsored trials in the United States that are exploring this delivery approach for variations of allogeneic and autologous MSCs. After obtaining arterial access, a long, needle-tipped catheter is advanced to the heart and across the aortic valve to enter the left ventricular chamber. Injections are

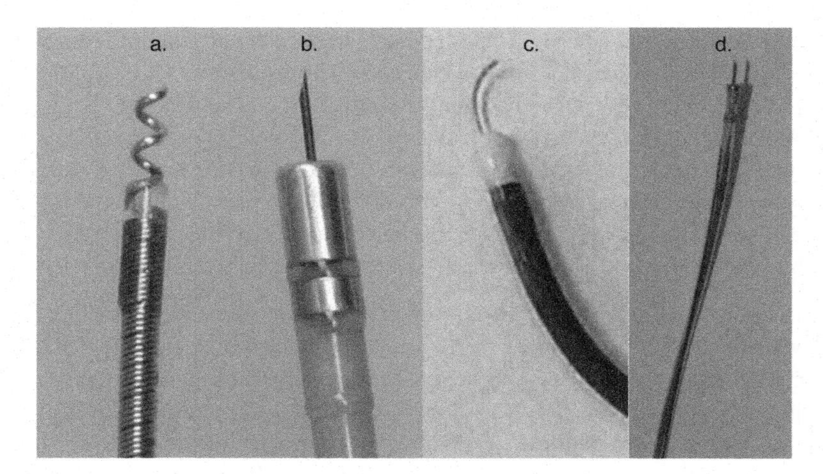

■ **FIGURE 3.2** Transendocardial catheter needle configurations.
a. Corkscrew-shaped Helical Infusional Catheter (Biocardia Inc, San Carlos, CA)
b. Electromagnetic tracked Myostar Catheter (Biosense Webster, Diamond Bar, CA)
c. Curve-shaped C-Cath (Celyad, Mont-Saint-Guibert, Belgium)
d. Stiletto injection needle (Boston Scientific, Marlborough, MA)
Of these, the Helical Infusional Catheter and the Myostar are most commonly employed in human cell therapy trials in the United States.

then performed by steering the catheter and driving the needle tip into the myocardium. There have been at least five different transendocardial injection catheter designs and three interventional imaging platforms to guide the injections. The corkscrew-shaped needle Helix (Biocardia, San Carlos, CA) [66,67], the curve-tipped C-cath (Cardio3Biosciences), the MyoCath (Bioheart Inc, Sunrise, FL), and the straight needle Stilletto (Boston Scientific, Cambridge, MA) have all utilized conventional two-dimensional projection X-ray fluoroscopy in clinical trials (Fig. 3.3). The straight needle Myostar (Biologic Delivery Systems, Biosense Webster) [68–70] uses an electron-anatomic mapping approach with hardware mounted to the cardiac catheterization laboratory table that creates local magnetic fields for tracking purposes [71] (Fig. 3.3). The Helix and Stilletto have been tested in multiple large animal studies. Both types of catheters are guided by three-dimensional MRI coregistered onto X-ray fluoroscopy [72–74] (Fig. 3.4). The Helix has been used for three-dimensional ultrasound-guided injections [75], and the Stilletto has been modified for real-time MRI-guided injections [76]. Cardiac perforations, ventricular arrhythmias, and ischemic stroke are potential adverse events linked to transendocardial catheter injection. Factors such as type and volume of injectate, dwell time of the catheter in the left ventricle, and catheter stiffness/needle protrusion length are probably linked

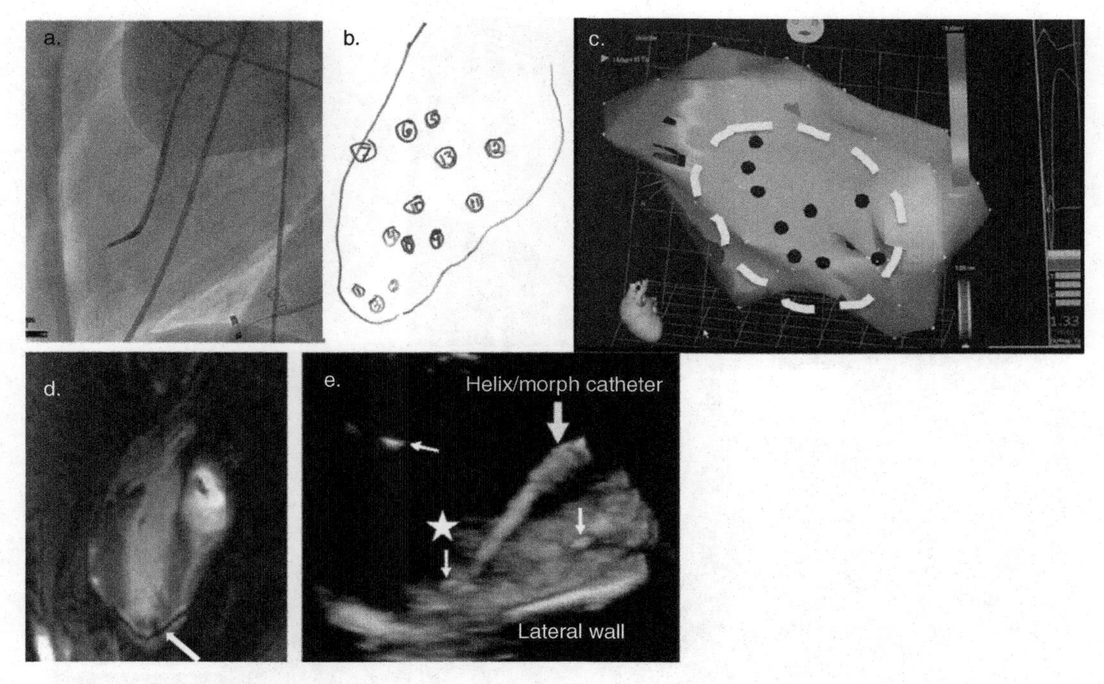

FIGURE 3.3 Transendocardial catheter single modality interventional imaging systems.

a. X-ray fluoroscopic projection showing needle tipped catheter and a "shadow" of the cardiac silhouette.
b. A physical transparency can be overlaid onto the image display in the catheterization laboratory and a contrast ventriculogram can be performed. The endocardial border can be manually contoured in the diastolic phase using a marker. Individual injection locations can be marked during the procedure.
c. Electroanatomic map and injection locations (*black dots*) within a predetermined target area [*broken line* (*broken yellow line* in web version)] using the Myostar catheter and NOGA mapping system (Biosense-Webster, Diamond Bar, CA).
d. Real-time MRI-guided transendocardial injections using an actively coupled MRI tracked injection catheter (modified Stilletto, Boston Scientific, Marlborough, MA) [76].
e. Real-time 3D ultrasound-guided transendocardial injections using the Helix infusional catheter telescoped within the Morph steerable guide (yellow arrow). Several echo contrast injections are visible and persistently stain the endocardium (white arrows) [75].

to the aforementioned serious adverse events. Similar to intraarterial infusion, vascular access complications can also occur.

An alternative transcatheter solution for intraadventitial and intramuscular delivery is the Bullfrog catheter (Mercator MedSystems, Emeryville, CA), and Transvascular injection needle (Vascular Solutions, Minneapolis, MN). The Bullfrog needle is concealed on the end of catheter with a small balloon. The needle becomes conspicuous upon inflation of the balloon, puncturing the coronary artery and permitting cell delivery through the needle into the adventitial space. The Transvascular catheter enters the coronary sinus and

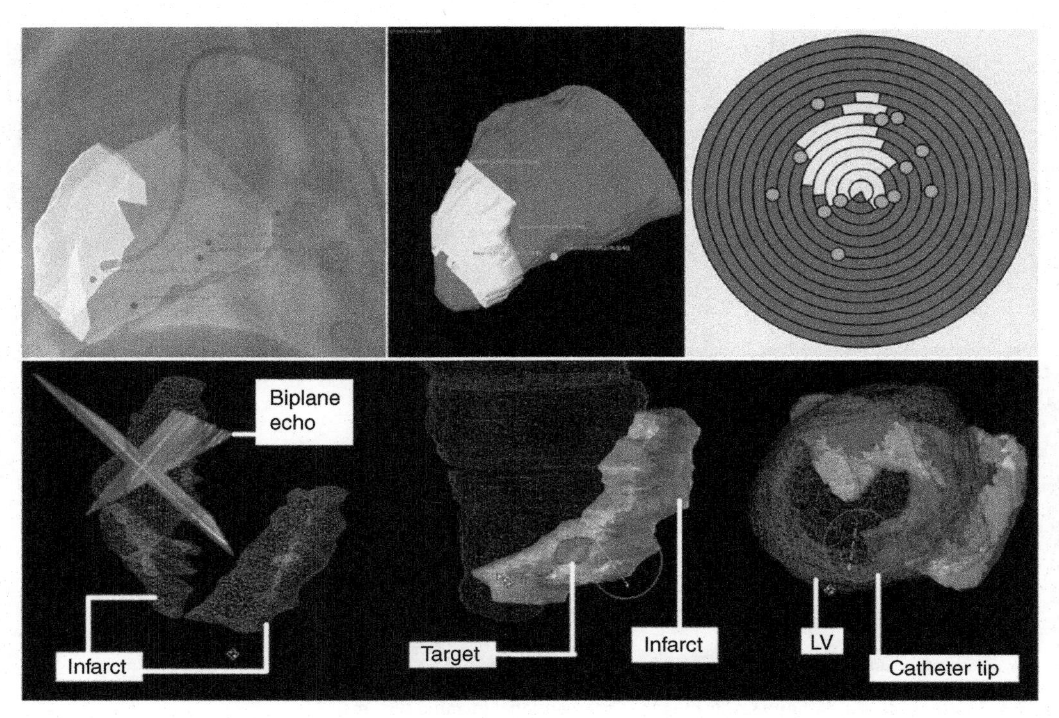

■ **FIGURE 3.4 Multimodality imaging for transendocardial catheter injections.** Top row shows MRI coregistered with X-ray fluoroscopy in vivo. Endocardial and infarct surface model is reconstructed from a preacquired MRI, then registered and overlaid onto biplane X-ray fluoroscopy. An interactive bulls-eye plot shows the distribution of injection locations relative to the infarct zone [*light gray* (*yellow* in web version)] [70]. Bottom row shows 3D multiplanar ultrasound coregistered with MRI in an electromagnetic field. Surface models of the ventricle are generated from MRI showing the infarct and endocardial surfaces. A custom transendocardial injection catheter modified with a 6 degree of freedom electromagnetically tracked sensing coil [*gray arrow* (*yellow arrow* in web version)] is steered to the injection target [*dark gray* (*red* in web version)]. All elements of real-time three-dimensional catheter navigation are provided in an accurate roadmap [77].

a needle can be advanced through the wall of the coronary sinus and into the myocardium. Limitations of these catheters include potential risk of hemorrhage into the pericardium or pleural space, limited target areas, and vessel dissection/thrombosis.

The most common cell delivery route for peripheral artery disease and diabetic neuropathy is intramuscular injection [78]. Typically the injections occur in the calf, and less frequently in the thigh. Risk of local infection and venous thrombosis are potential rare complications.

3.2.2.2 *Intraparenchymal injection*

Intraparenchymal injection has been performed primarily for neurologic conditions such as spinal cord disease, stroke, and other neurodegenerative diseases. Typically, preoperative magnetic resonance imaging and intraoperative stereotactic mapping is used to guide these injections. Feng et al. demonstrated the safety of direct mesenchymal stromal cell (MSC) injection using this guidance approach into the brains of nonhuman primates [79]. Safety and feasibility has also been demonstrated in early human studies using MSCs delivered intrathecally in patients with amyotrophic lateral sclerosis [80,81] and cerebral trauma [82]. Unfortunately, the paucity of clinical trials and small sample sizes prevent conclusions regarding efficacy with these methods.

3.2.3 Scaffold-Based Delivery

Loading MSCs onto scaffolds and applying the construct onto damaged tissues is a unique approach to improve cell retention and potentiate therapeutic effect by exploiting proregenerative properties within the scaffold [83]. Several distinct types of biologic scaffolds have been developed for cell delivery purposes. The ideal cell delivery scaffold offers a favorable, nontoxic microenvironment that is hospitable to cells, offers a predictable and time-dependent dissociation of cells to the site of injury, has high porosity to enable cellular penetration and protein diffusion, and is naturally adhesive without requiring sutures or glues for fixation.

Biologic scaffolds can be broadly categorized as either natural or synthetic [84]. Natural scaffolds are derived from tissues, whereas synthetic scaffolds are derived using inanimate materials. Decellularized adult extracellular matrix has been described as a tool for progenitor cell delivery for cardiac repair [85–87]. Schmuck et al. developed a method to bioengineer cardiac extracellular matrix scaffolds using harvested cardiac fibroblasts [88]. This manufactured scaffold is principally composed of fibronectin and small amounts of collagen. The scaffold is naturally adhesive, and does not require sutures or glues. There are a number of bioactive molecules embedded within the scaffold with potent angiogenic, immunomodulatory, and proregenerative properties. High retention and efficient release of loaded MSCs into the myocardium were observed after applying these scaffolds in a mouse myocardial infarct model [88].

Natural scaffolds for cell delivery have been derived from bovine and human pericardium [89–93]. CardioCel (Amdedus, Australia) is a bovine pericardium derivative and is approved for cardiac applications in general, but there is no published data on its use for therapeutic cell delivery. Other natural scaffolds include serosal intestinal submucosa [94–96] and fibrin [85,97–100]. The Transplantation of Human Embryonic Stem Cell-derived Progenitors in Severe Heart Failure trial is currently underway that combines a fibrin scaffold with human embryonic stem cell-derived cardiac progenitors for patients with advanced ischemic cardiomyopathy.

Finally, there are a number of synthetically derived collagen-based scaffolds that have been manufactured for a variety of indications [98,99,101–108]; however, there is limited experience using them for therapeutic cell delivery. Arana et al. used a bovine collagen-based type I scaffold to deliver adipose-derived MSCs in rat and pig myocardial infarction models [106]. The adipose-MSC seeded collagen improved cardiac function, decreased fibrosis, and increased vasculogenesis. The Myocardial Assistance by Grafting New Bioartificial Upgraded Myocardium trial tested direct application of collagen I matrix seeded with bone marrow mononuclear cells in patients with ischemic cardiomyopathy with prior myocardial infarction [109,110]. Improvements in ejection fraction, end diastolic volume, and reduction in scar thickness area were all seen in the treated group.

3.2.4 Trans-Mucosal Delivery

The pericardium consists of a number of essential matrix proteins that may be hospitable for transplanted cells with direct access to the myocardium by virtue of an extensive lymphatic network. Intrapericardial cell administration can be performed minimally invasively, permits large volume cell transplantation, and is not dependent on intact coronary artery vasculature. Blazquez et al. described the technique in a porcine MI model and observed MSC migration into the infarcted and noninfarcted myocardium, presumably through intact lymphatic channels [111]. Fu et al. described X-ray fused with MRI guidance to permit pericardial localization using percutaneous techniques. Questions remain pertaining to long-term cell retention and whether there is sufficient myocardial cell migration if the delivery scaffold is not in contact with the pericardium such as the ventricular septum. This approach would not be feasible in patients with prior cardiac surgery where the pericardium has been disrupted.

Intranasal cell delivery has been tested rodent models of stroke and other CNS conditions [112]. Wei et al. demonstrated that intranasal delivery of bone marrow-derived MSCs could be found in the ischemic cortex and perivascular space in mice, as early as 1.5 h after administration [113]. Inhalational and intraperitoneal deliveries are forms of trans-mucosal cell delivery that may be explored in the future for lung, liver, and pancreas disorders.

3.3 CELL TRACKING

Cell-based therapeutics have the unique challenge of targeting cellular reagent to the damaged organ or tissue. In contrast to pharmaceuticals, which typically have a half-life of hours to days within the body, therapeutic cells may persist in the body indefinitely by engraftment and/or self-replication;

however, cells such as MSCs are not known to persist for long periods of time. Due to the multipotent or pluripotent potential of many therapeutic cell types, it is imperative from a safety and efficacy standpoint to understand the spatial and temporal biodistribution following transplantation. To facilitate in vivo tracking, cells must be labeled with a contrast agent to make them visible within the body.

There are three broad techniques used for cell labeling including direct labeling, receptor or surface marker labeling, and reporter gene labeling [114]. Direct labeling is carried out by uptake of the contrast agent into the cytoplasm of the cell. Receptor labeling typically involves labeling-specific cell surface receptors. These techniques may be limited as cell differentiation may result in the loss of the contrast label, which may result in under reporting of the cell population, and receptor expression can be transient and may change as stem cells or progenitor cells differentiate. In contrast, reporter gene labeling entails modifying the cell to express genes encoding signal-generating proteins or enzymes which will emit a signal upon interaction with a substrate or molecular probe. Reporter gene labeling is a powerful preclinical tracking technique as it allows for long-term tracking of stem cells. However, it is unlikely to be used in a clinical setting due to the safety concerns associated with genetic modification of the cells.

Noninvasive clinical imaging modalities hold the greatest potential for long-term cell tracking in vivo and rely on two general types of contrast agents. First, tissue contrast agents are used to create a physical distinction between the transplanted cells and the surrounding tissues [114]. These types of contrast agents are used in magnetic resonance tomography (MRI), X-ray computed tomography (CT), and ultrasound imaging. Although CT and ultrasound imaging have been investigated for cell tracking purposes, low sensitivity and poor soft tissue visualization hamper the utility of these imaging modalities [114,115]. Photon-emitting contrast agents generate a signal by emitting light or electromagnetic radiation and are used in conjunction with positron emission tomography (PET), single-photon emission computed tomography (SPECT or SPET), bioluminescence imaging, and fluorescence imaging [114].

Ideal characteristics of labeling contrast agents for tracking stem cells in the clinical settings described by Frangioni and Hajjar are: (1) biocompatible, safe, and nontoxic, (2) no genetic modification or perturbation to the therapeutic cell, (3) single-cell detection at any anatomic location, (4) quantification of cell number, (5) minimal or no dilution with cell division, (6) minimal or no transfer of contrast agent to nonstem cells, (7) noninvasive imaging in the living subject over months to years, (8) no requirement for

Table 3.1 Imaging Modalities

Imaging Modality	Contrast Agents	Limit of Detection	Advantages	Disadvantages
PET	High-energy positron emitters: ^{18}FDG, ^{124}I, ^{64}CU	1×10^4	High sensitivity, 3D imaging	Short tracking window, ionizing radiation, anatomic scan required
SPECT	High-energy gamma emitters (^{111}In, ^{99m}TC)	1×10^5	3D imaging, intramediate tracking window	Low sensitivity, ionizing radiation, anatomic scan required
Bioluminescence	Luciferase substrates	1×10^3	Long-term tracking, longitudinal imaging	Low resolution, 2D imaging, genetic modification required. Limited to small animal imaging
Fluorescence	Fluorophores and quantum dots	In vivo optical imaging: 1×10^6; quantum dot cryoimaging: single cell	Intermediate to long-term tracking, 3D cryoimaging is highly sensitive	In vivo imaging limited by excitation penetration, low resolution, 3D cryoimaging limited to terminal experiments
MRI	Lathanides (Gd^{3+}, Mn) SPIO, ^{19}F	1×10^4	3D imaging, no radiation, high resolution, high sensitivity	SPIO high false-positive rate, cytotoxicity of lanthanides

injectable contrast agent, and (9) inexpensive [114,116,117]. To date, no cell labeling contrast agent and imaging modality fulfill all these criteria. Each imaging modality in conjunction with the appropriate cell labeling contrast agent has its advantages and limitations. We will discuss the most common and clinically relevant in the following section (Table 3.1).

3.3.1 Photon-Emitting Contrast Agents

3.3.1.1 *Positron emission tomography*

PET imaging utilizes positron-emitting radiotracer labels to detect cells within the body. Radiotracer labels such as ^{18}F-fluorodeoxyglucose (^{18}FDG) are readily taken up into the cells. As the radiotracers decay, positrons are emitted and these interact with electrons to produce a "coincident" event emitting two gamma ray photons [116]. The most common contrast label used for PET cell tracking is ^{18}FDG [118]. ^{18}FDG has a half-life of 110 min, which results in a detection window of approximately 10 h [119]. Newer radiolabel tracers such as ^{99m}Tc-exametazime, hexadecyl-4-^{124}I-iodobenzoate, and ^{64}CU-based tracers such as ^{64}CU-pyruvaldehyde-bis (N^4-methylthiosemicarbazone) (^{64}CU-PTSM) have been developed. These newer tracers have longer half-lives (6.02 h, 4.2 days, and 12.7 h respectively) allowing for longer tracking [119].

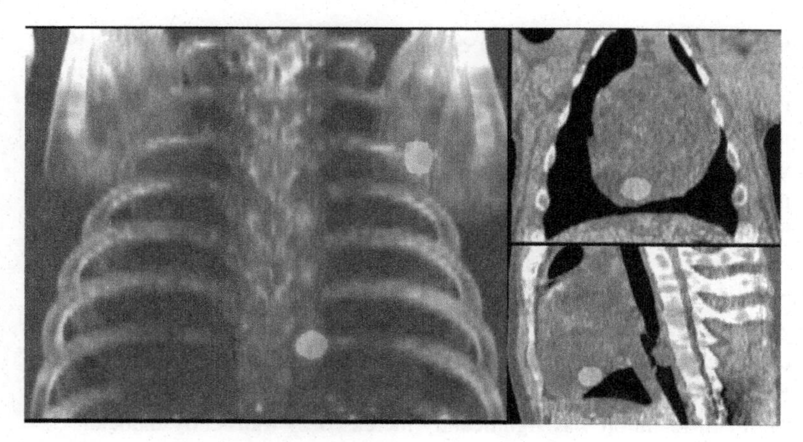

■ **FIGURE 3.5** CT-PET cardiac images of a single-transendocardial injection of ^{18}F-fluorodeoxyglucose labeled mesenchymal stromal cells near the ventricular apex, 1-h postinjection. Left panel is maximal intensity projection, top right panel is coronal slice, and bottom right panel is sagittal slice.

Advantages of PET include three-dimensional imaging with high clinical and preclinical resolution (2–6 mm range and 0.8–1.5 mm range respectively). Sensitivity is high as background is intrinsically low, with the limit of detection of approximately 1×10^4 cells [115,119]. Disadvantages of PET imaging include the use of radioactive tracers and the need for additional anatomical scans such as CT or MRI to colocalize the cell signal with the anatomical location. In addition, current radiotracers have relatively short half-lives which prevent long-term tracking. Finally, radiotracer labeling does not differentiate between living and dead cells in vivo. Thus, this method may overestimate cell retention.

PET imaging has been widely used for tracking therapeutic stem cells in both clinical and preclinical studies and has been adopted to study cardio-regenerative therapies (Fig. 3.5). In a clinical study by Musialek et al., patients with recent myocardial infarction (MI) had autologous CD34+ cells labeled with the radiotracer ^{99m}Tc-exametazime infused during a standard cardiac catheterization through the coronary circulation to the left anterior descending coronary artery. They showed that 5.2% of labeled autologous CD34+ cells were localized to the myocardium after 4 h. Interestingly, colocalizing the PET scans with cardiac MRI, they were able to demonstrate a strong positive correlation between infarct size and myocardial cell uptake, specifically to the periinfarct zone [120]. Similarly, a study by Hoffmann et al. of patients with ST elevated myocardial infarction, autologous CD34+ cells labeled with ^{18}FDG administered by intracoronary infusion resulted in 1.3–2.6% of total radioactivity localized to the myocardium after 1 h, with cells localizing to the border zone of the infarct. Of note, they were not able to detect any cells in the myocardium if they were infused intravenously, indicating that this delivery modality may be less than ideal for cardiac indications [121]. These results were confirmed by Kang et al. who showed an accumulation of autologous hematopoietic stem cells in the myocardium following intracoronary delivery but not intravenous delivery [122]. Blocklet et al. demonstrated that 1 h after intracoronary infusion of autologous CD34+ cells labeled with ^{18}FDG, 5.5% of total radiation was localized to the myocardium, specifically at the border zone of the infarct [123]. Terrovitis et al. demonstrated a method to increase cell retention in a rat MI model. Cardiac-derived stem cells labeled with ^{18}FDG were delivered via intramyocardial injection [124]. Cardiac-derived stem cell retention was significantly increased by transiently inducing cardiac arrest (1 h retention rate of 75.6. ± 18.6%) or bradycardia (35.4 ± 5.3%) during intramyocardial injection. Furthermore, they showed cardiac-derived stem cell retention could also be increased by applying a fibrin glue plug to the injection site likely by reducing cell egress from the needle track [124]. Finally, Zhang et al. demonstrated differences in retention rates between radiotracers [125]. They showed that in rats, intramuscular injection of human cardiac progenitors (hCPC) labeled with ^{18}FDG underestimated cell retention compared to hCPCs labeled with hexadecyl-4-[^{18}F]fluorobenzoate (^{18}F-HFB) (3.1 ± 1.5% compared to 17.2 ± 8.9% respectively). These data suggest that the type of radiolabel is important for quantitative cell tracking [125].

With the development of radiotracers with extended half-lives, PET studies are beginning to track cells for longer periods of time. Kim et al. labeled adipose-derived stem cells (ADSC) with hexadecyl-4-^{124}I-iodobenzoate and injected the cells intramuscularly into healthy and infarcted rat hearts. Cells were detected for up to 3 days in the infarcted myocardium and for up to 9 days in the healthy myocardium. They concluded that the harsh environment found in the infarcted hearts resulted in increased apoptosis and reduced cell retention [126]. Additionally, Kim et al. showed ADSC persistence 18 h after intramyocardial injection in a normal rat heart when using ^{64}CU labeled DOTA-hexadecyl-benzoate [127].

3.3.2 Single-Photon Emission Computed Tomography

SPECT imaging utilizes radiotracer labels that directly emit a single gamma ray. In contrast to PET, coincident events are not required. Instead, SPECT requires mechanical collimation to detect the direction from which the photon was emitted. This reduces detection sensitivity by one to two orders of magnitude below PET imaging. There are two main radiotracers used for stem cell tracking by SPECT imaging, ^{111}In-oxine and ^{99m}Tc-hexamethylpropylene amine oxime (^{99m}Tc-HMPAO) [128]. ^{111}In-oxine is lipophilic in nature and passively diffuses into cells. Once inside the cell the indium-111 stably binds to the intracellular compounds and integrates the isotope inside the cell [118]. ^{111}In-oxine has a half-life of 2.8 days allowing for imaging of labeled cells for up to 14 days [119]. ^{99m}Tc-HMPAO also passively diffuses into cells where it becomes stably trapped inside the cell. It has a half-life of 6 h and has been reported to be imaged for several days after labeling [119].

The main advantage of SPECT imaging for cell tracking is the prolonged half-life of the radiotracers, which permits cell detection up to 14 days [119]. There are also disadvantages to SPECT imaging. Similar to PET, SPECT relies on radiotracers and must be registered to an anatomical scan such as CT or MRI. Although SPECT radiotracers have longer half-lives, they are still limited to a maximum of 14 days postinjection. Therefore, long-term tracking studies are not possible with currently available radiotracers. Furthermore, SPECT offers lower resolution (7–15 mm) with a limit of detection of 1×10^5 cells. Finally, like PET imaging, radiotracers can be passed to resident tissue macrophages upon cell apoptosis which leads to an over estimation of cell retention [119].

SPECT imaging has been used in many clinical and preclinical studies. Using circulating proangiogenic progenitor cells labeled with ^{111}In-oxine, Schächinger et al. demonstrated that in patients who had an MI, 6.9 ± 4.7% of cells were localized to the myocardium at 1-h postintracoronary infusion. This was reduced to 2 ± 1%, 3–4 days after infusion [129]. Furthermore, this group demonstrated that more cells localized to the myocardium acutely (<14 days) following an MI compared to an intermediate (>14 days to 1 year) or a chronic infarct (>1 year) [129]. Following intracoronary infusion of ^{99m}Tc-HMPAO labeled CD133+ cells in patients with ischemic cardiomyopathy, Gouseetis et al. showed that 9.2 ± 3.6% of cells were localized to the myocardium 1 h postinfusion. This was reduced to 6.8% ± 2.4% at 24 h postinfusion. They concluded that CD133+ cells preferentially home to the chronic ischemic myocardium [130]. With only approximately 5% of cells localizing to the myocardium, it is important to determine the tissue distribution of the remaining 90–95% of cells. Multiple studies have shown that the majority of cells localize primarily to the liver, lungs, and spleen following intracoronary or intravenous infusion [130–132].

SPECT imaging has also been used heavily in preclinical cell tracking studies to determine biodistribution following multiple delivery modalities. In a swine MI model, Hou et al. showed that human peripheral blood mononuclear cells labeled with ^{111}In-oxine were preferentially retained in the myocardium following intramyocardial injection (11 ± 3%) compared to intracoronary

(2.6% ± 0.3%) and interstitial retrograde coronary venous delivery (3.2 ± 1%) [64]. Cells were primarily washed out to the lungs with intracoronary delivery (47 ± 1%) and interstitial retrograde coronary venous (43 ± 3%) having greater pulmonary wash-out than intramyocardial injections (26 ± 3%) [64]. In canines, Sabondjian et al. demonstrated that endothelial progenitors labeled with ^{111}In-tropolone and directly injected into the myocardium could be tracked for 11 effective half-lives or 10 days posttransplantation [133]. Similarly, Mitchel et al. looked at epicardial versus endocardial delivery of endothelial progenitors cells labeled with ^{111}In-tropolone in a canine MI model [134]. They found initial retention to be similar between epicardial (57 ± 15%) and endocardial (54 ± 26%) injection. In addition, they demonstrated that ^{111}In-tropolone could be imaged out to 15 days posttransplantation [134].

Additional preclinical cell labeling strategies exist for SPECT imaging. The use of reporter genes to produce a specific receptor, protein, or enzyme to detect transplanted cells can be used in place of direct labeling methods and allow for long-term tracking. Typically, reporter genes are used in conjunction with a reporter probe. Currently, due to genetic manipulation of the cells, this labeling technique is limited to preclinical studies. In a pig MI model, Templin et al. used transgenic human-induced pluripotent stem cells (hiPSC) transfected with sodium iodide symporter (NIS) (NIS^{pos}-hiPSC) [135]. Cells expressing NIS readily take up iodide. Expression of NIS in adults is limited to the thyroid, stomach, choroid plexus, and salivary glands. NIS expressing cells can be visualized after administration of radiotracer ^{123}iodide (^{123}I) due to its uptake into the cell. NIS^{pos}-hiPSCs, prelabeled with ^{123}I, were codelivered with human mesenchymal stromal cells (hMSC) by intramyocardial injection. Strong signal expression in the heart was observed 3–5 h posttransplantation, but was undetectable at 24 h posttransplantation [135]. Between 13 and 15 weeks posttransplantation of NIS^{pos}-hiPSCs and hMSCs, ^{123}I was infused into the left descending coronary artery. One hour postradiotracer infusion, SPECT imaging was carried out and showed strong uptake of ^{123}I in the areas of coinjection (NIS^{pos}-hiPSCs and hMSCs), but not in the control areas (injection of NIS^{pos}-hiPSCs only or hMSCs only) [135]. They concluded that detection of NIS^{pos}-hiPSCs was possible up to 15 weeks posttransplantation and that MSCs support the survival and engraftment of NIS^{pos}-hiPSCs in vitro [135].

3.3.3 Bioluminescence

Bioluminescence is an imaging technique that utilizes luciferase, a family of oxidative enzymes that emit light upon oxidation of a substrate. For cells to be visible, they must be transfected with a luciferase reporter gene such as firefly luciferase (Luc). Other luciferase reporters have been generated from the click beetle, sea pansy, and copepod [136]. Luc will emit light at 620 nm at 37°C for 30–60 min after injection of the substrate D-Luciferin [136]. The light emitted from luciferase can be detected with a cooled charged-coupled device (CCD) camera. A CCD camera can detect the very low levels of light being emitted from the organs of the animal. The image of the detected light can then be superimposed on a digital photograph of the animal to localize the signal. Alternatively, to reduce tissue adsorption of the emitted light, organs can be excised and imaged ex vivo which greatly improves the sensitivity of the technique.

Advantages of bioluminescence imaging are the ability to track stem cell without the need for radiation, accurate cell quantification with a limit of detection approximately 1×10^3 cells [115], and a reporter that is passed to daughter cells without dilution of the signal. The drawback of using bioluminescence is that it has poor spatial resolution and is limited to 2D images. Because of the use of genetic manipulation and the adsorption of reporter signal from host tissues, bioluminescence imaging is limited to small animal studies [136,137].

Bioluminescence has been widely used in many small animal studies. Van der Bogt et al. demonstrated in a mouse MI model that bone marrow-derived MSCs (BM-MSC) and adipose-derived MSCs (AD-MSC) transplanted by intramyocardial injection were present at day 2, but were significantly reduced by day 7 and were undetectable by day 28. They showed that AD-MSCs and BM-MSCs had similar clearance kinetics [138]. Bioluminescence was used by Wang et al. to track human CD34+ cells injected into severe combined immunodeficiency mouse MI model. They were able to demonstrate persistence of CD34+ cells out to 12 months. Additionally, they demonstrated that improvements observed with treatment were due to paracrine effects and not myogenesis [139]. In a mouse hind limb ischemia model, Van der Bogt et al. investigated BM-MSC homing. They observed that BM-MSCs could home to the site of injury, but the cells did not persist for longer than 4 weeks. No significant effects of neovascularization were observed [140]. Finally, in a model of liver damage, Li et al. demonstrated that delivery of MSCs via the superior mesenteric vein resulted in greater cell entrapment in the liver compared to IV infusion in the inferior vena cava or direct intrahepatic injection. They also found that MSCs were persistent in the liver out to 7 days [141].

3.3.4 Fluorescence Imaging

Fluorescence imaging can be subdivided into in vivo or ex vivo imaging. Unlike luciferase, which generates light through a chemical reaction, fluorescent probes require an excitation light source [142]. Fluorescence probes absorb light at a specific wavelength and emit light at another wavelength. In vivo imaging systems consist of a lightproof box, CCD camera, and an excitation light source. Penetration of excitation light is a limiting factor, limiting imaging depth to a few millimeters. Thus, this technique is only amenable to small animal models.

There are numerous fluorescent probes (fluorochromes) available for stem cell tracking with a wide range of excitation/emission profiles. Fluorescent probes are either organic, made up of proteins and small-molecules polymethines or organic/inorganic hybrids consisting of quantum dots [116]. There are over 26 fluorescent reporter proteins with emission peaks ranging from 475 nm (cerulean) and 508 nm [green fluorescent protein (eGFP)] to 650/670 nm (eqFP650/670) [142]. Near-infrared (700–1,000 nm) fluorochromes are more suitable for in vivo imaging as they have a lower tissue adsorption than lower wavelength fluorochromes [118]. Fluorescent reporter proteins can be transfected into cells to allow for long-term imaging [142].

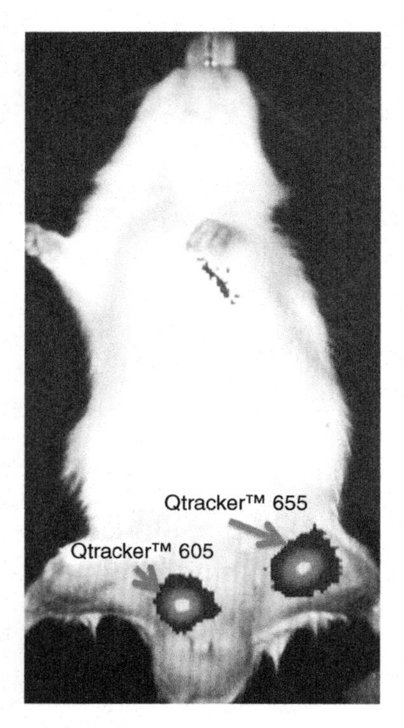

■ FIGURE 3.6 Rat In vivo fluorescent imaging. In vivo fluorescence imaging of intramuscular injected human MSCs labeled with either Qtracker 605 or 655 in a rat and imaged with an IVIS Spectrum (Perkin Elmer). (Unpublished data, Schmuck et al. University of Wisconsin-Madison)

Quantum dots are alternative fluorescent labeling probes that do not require genetic manipulation. Quantum dots are inorganic, fluorescent semiconductor nanoparticles that passively infuse into cells or through peptide-mediated uptake [118,143]. They have a broad excitation range and a narrow emission band and are resistant to photo bleaching. Quantum dots are available in a broad range of excitation/emission profiles including up to 800 nm (near-infrared range) making them highly amenable to multiplexing [142]. Near-infrared quantum dots are more suitable for in vivo imaging as they have a lower tissue adsorption than lower wavelength quantum dots. At low working concentrations, quantum dots are not cytotoxic and do not effect cell viability, proliferation, or differentiation [144–146]. The kinetics of quantum dot clearance is debated. The manufacturer of *QTracker* (Life Technologies) indicates stable expression for a minimum of 7 days, while the literature reports quantum dot expression up to 8 weeks posttransplantation [147]. Quantum dots are not readily transferred to adjacent tissues or phagocytizing macrophages upon apoptosis with a reported transfer rate of 4%. This is in contrast to most other cell-labeling agents that are readily taken up by host tissues and cells [148]. The low transfer rate, broad range of emission peaks, and relatively long-lived quantum dot expression make quantum dots an attractive cell-tracking label (Fig. 3.6).

In vivo fluorescence imaging has been used to track stem cells in several preclinical animal studies. As bioluminescence offers superior deep tissue imaging, fluorescent reporter genes have not been widely used for in vivo fluorescence imaging. Instead, quantum dots have been actively investigated. In a model of hind limb ischemia, Ding et al. showed that ADSCs labeled with quantum dots (655) and transfected with luciferase were present out to 42 days posttransplantation [149]. The quantum dot and bioluminescence signal colocalized together and had similar decay characteristics. They concluded that quantum dots could be used for long-term tracking of stem cells [149]. Embryonic stem cells were tracked in an in vivo mouse model by Lin et al. using quantum dots with emission peaks of 525, 565, 605, 655, 705, and 800 nm. These cells were injected subcutaneously on the back athymic nude mice and imaged for 28 days [145]. They found that for in vivo imaging, 800 nm quantum dots yielded the greatest signal-to-noise ratio and was detectable out to 14 days. Additionally, they showed that quantum dots had no effect on viability, proliferation, or differentiation of embryonic stem cell [145]. Finally, Chen et al. demonstrated that Ag_2S quantum dots (near-infrared window-II, 1.0–1.4 μm) could be used to demonstrate human MSC homing to injured tissues. Collagen sponges impregnated with Stem Cell Derived Factor-1 (SDF-1) were implanted under a cutaneous wound in mice. Human MSCs were then infused intravenously. Within 24 h, human MSCs were visible using in vivo imaging and persisted for at least 48 h [150] (Fig. 3.7).

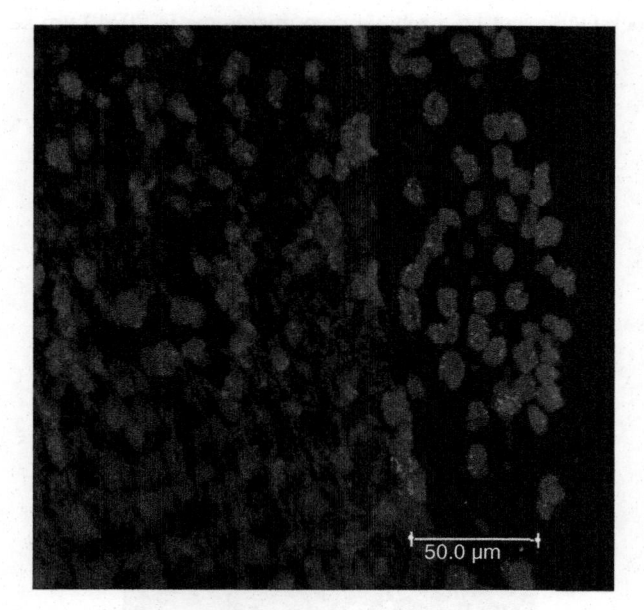

FIGURE 3.7 FISH staining for human MSCs in a mouse MI model. Fluorescent in situ hybridization (FISH) was used to track human MSCs transferred by cardiac fibroblast-derived extracellular matrix in a mouse myocardial infarction model. Human nuclei are distinguished by the presence of human centromeres [*gray* (*pink* in web version) *punctate dots within nuclei*)]. Scale bar = 50 mm. (Unpublished data, Schmuck et al. University of Wisconsin-Madison)

Fluorescent probes can be used ex vivo to track stem cell biodistribution and fate. Traditional immunohistochemistry or immunofluorescence techniques employing antibodies targeted against specific surface antigens have been used to track stem cell location or fate [137,151,152]. This technique is frequently used to confirm the presence of cells following in vivo tracking studies [137]. Fluorescence in vitro hybridization (FISH) is another technique used to track stem cells. FISH can be carried out based on the detection of Y-chromosomes (sex-mismatched) or probing for xenotransplant nuclei (typically human centromers) in xenotransplant models [88,137,153]. FISH is highly specific and has a low false-positive rate, making this technique attractive for semiquantitative applications (Fig. 3.8).

More recently, three-dimensional (3D) cryo-imaging techniques have been developed to provide single-cell sensitivity in whole organ or mouse-sized specimens [154–157]. 3D cryo-imaging combines high-resolution color brightfield imaging with molecular fluorescent imaging of specimens up to mouse size. In contrast to traditional histology techniques which require specimens to be placed onto slides, whole organ or animal specimens are embedded in a cyro-imaging compound and the block face imaged. The block-face is cut by a robotic cryo-microtome in 20–40 μm sections

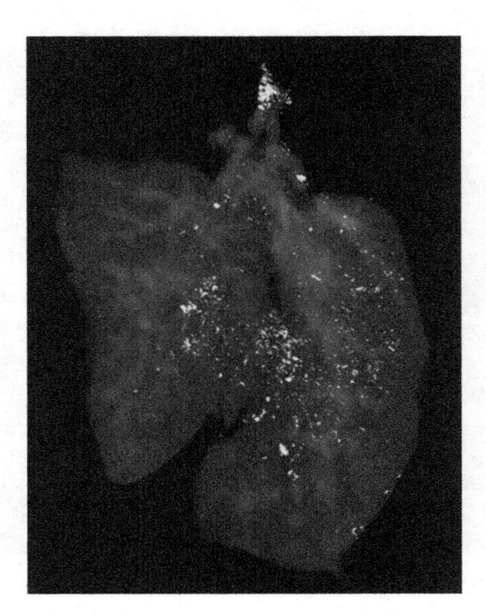

FIGURE 3.8 3D cryo-imaging of rat lungs. Intratracheal delivery of human MSCs to a rat model of bronchiolitis obliterans. Human MSCs were labeled with Qtracker 655 and imaged with 3D cryoimaging. Human MSCs are represented by white dots and overlaid on a 3D reconstructed bright field image of the lungs. (Unpublished data, Schmuck et al. University of Wisconsin-Madison)

allowing for whole specimen volume reconstruction [155]. Cells labeled with a fluorescent tag (protein or quantum dot) are visible in the fluorescent images and coregistered to the bright field images allowing for accurate single-cell quantitation and biodistribution assessment. Auletta et al. used 3D cryo-imaging to track the distribution of human MSCs in a mouse model of graft versus host disease, demonstrating that cells were initially retained in the lungs before migrating to the spleen after 24 h [158]. In a rat MI model, Schmuck et al. demonstrated that cardiac fibroblast-derived extracellular matrix could retain eGFP + MSCs in the heart 48 h after transplantation. Quantitation was not possible due to high concentration of eGFP + MSCs [88]. Multiple cell populations can be tracked with 3D cryo-imaging by multiplexing fluorochromes. This allows for tracking multiple cell populations and/or time points within the same animal. The main drawback of 3D cryo-imaging is the inability to track cells longitudinally in the same animal. The quantitative single-cell sensitivity of this technique makes it appealing for tracking rare cell populations in small animal models.

3.3.5 Tissue Contrast Agents

3.3.5.1 *Magnetic resonance imaging*

MRI has been widely used to track stem cells in both human trials and animal models. Unlike photon emitting contrast agents, MRI contrast agents create a physical distinction between the transplanted cells and the surrounding tissues. Unlike most photon-emitting contrast agents, MRI contrast agents are more suitable for long-term tracking, as signal does not depend on radioactive decay. There are three main categories of MRI contrast agents, paramagnetic, superparamagnetic iron oxide (SPIO), and flourine-19 contrast agents. Paramagnetic agents generate T1 contrast and consist of the lanthanide gadolinium (Gd^{3+}) or manganese [116,118,159]. Paramagnetic contrast agents generate a signal by changing the relaxivity of protons from associated water molecules [116]. This results in a hyper-intense signal in the T1-weighted images. Paramagnetic contrast agents may be limited by high concentration of contrast agent needed to generate a signal and potential cytotoxic effects of the lanthanides [159]. Additionally, internalization of the contrast agent reduces the interaction with extracellular water resulting in low sensitivity of these contrast agents [116].

SPIOs create a negative contrast (hypointensities) in T2/T2*-weighted images. Labeling with SPIOs does not effect cell proliferation, viability, or differentiation and has a limit of detection of approximately 1×10^4 cells [115,116]. In addition, there are FDA-approved clinical formulations which increase the clinical applicability of SPIO tracking by MRI. While SPIOs allow for long-term stem cell tracking, there are drawbacks worth mentioning. First, SPIOs readily transfer to daughter cells, thus diluting signal if cell division occurs. Second, long-term tracking may overestimate cell retention due to deposition of residual iron into the extracellular space or due to iron particle uptake by macrophages [114,160–162]. For example, Huang et al. found in a swine MI model that at 6 months post-MSC transplantation, iron particles were predominately found in the extracellular space or within 68+ macrophage populations [160]. Iron particle signal was not correlated with the presence of transplanted cells by sex mismatch [160]. Ex vivo validation of iron containing cells is necessary to confirm cell retention.

Labeling with Flourine-19-rich perfluorocarbon particles has emerged as an attractive contrast agent for MRI tracking [163,164]. Besides teeth, ^{19}F is not naturally found in the body; thus, there is an absence of background. Anatomical localization of the ^{19}F signal is achieved by a normal proton MRI scan [163]. Furthermore, ^{19}F is nonradioactive, and stable in the intracellular environment as it is not degraded by any known enzymes or at typical lysosomal pH values [163]. ^{19}F perfluorocarbons must be formulated in a biocompatible emulsion to load into and has minimal effects phenotype, morphology, or function of the cells. ^{19}F perfluorocarbon labeling does suffer from several drawbacks. First, low uptake of ^{19}F perfluorocarbon reduces sensitivity to approximately 2000 cells/voxel at 7 T [165]. Second, imaging techniques for ^{19}F perfluorocarbon are approximately 10-fold slower compared with traditional proton imaging. This may limit its use in certain applications such as cardiac MRI imaging [166].

Clinical MRI cell tracking has been carried out in several disease states. Karussis et al. transplanted MSCs labeled with SPIO (Feridex) by lumbar puncture into patients with multiple sclerosis or amyotrophic lateral sclerosis. MSCs were visualized in the occipital horns of the ventricles indicating potential migration of cells [167]. In patients with chronic spinal cord injury, Callera and de Melo injected autologous bone marrow CD34+ cells labeled with magnetic beads into the lumbar space. Hypointensities were found at the lesion site after 20 and 35 days; they concluded that CD34+ cells labeled with magnetic beads could migrate to the site of injury [168]. SPIO labeling has been used to track transplanted cadaver islet cells by Toso et al. in patients with Type I diabetes. They were unable to correlate islet cell retention with hypointensities due to the persistence of spontaneous liver hypointensities [169].

MRI has been widely used in animal studies to track cells. Rat MSCs were labeled with gadolinium diethylenetriamine pentaacetic acid (Gd-DTPA) by Geng et al. and implanted in a rat model of cerebral ischemia. They found that GD-DTPA labeling had no effect on cell proliferation or viability, while providing a detectable signal in vivo [170]. Tachibana et al. tracked rat MSCs labeled with poly (vinyl alcohol)-gadolinium (PVA-Gd) in a rat hind limb ischemia model. MSCs were seeded into a cytocompatible scaffold (Spongel) for 1 day and then transplanted [171]. Clear signal was evident at day 7-post implantation, but was absent by day 14. No evidence of cell migration was observed. In contrast to SPIO's agents, the extinguishing of gadolinium MRI signal indicates that the gadolinium contrast agent is not retained in the tissue or resident macrophage populations upon the death of the transplanted cells [171]. Although not as widely used, gadolinium chelate agents are becoming more popular as issues with cytotoxicity and signal sensitivity become resolved. The rapid clearance of gadolinium following cell death is a major advantage over SPIOs contrast agents and will allow for longer term tracking with reduced false-positive rate of cell detection [171].

SPIOs have been extensively studied as cell-tracking contrast agents. Ris et al. compared two clinical grade SPIOs (ferucarbotran and ferumoxide) to track transplanted human islet cells in a rat model. They showed that ferucarbotran had superior signal out to 8 weeks. Histology revealed that iron was present in and around live engrafted islets; they concluded that ferucarbotran is the superior cell-labeling contrast agent, similar to what was reported by other groups [172]. iPS-derived neural progenitor cells (iPS-NPC) were labeled with SPIOs by Tang et al. and injected into rats with traumatic brain injury and monkeys with spinal cord injury. They showed that iPS-NPCs migrated to the site of injury from the primary injection sites by day 30 posttransplantation. Significant improvements in motor function were also reported in animal (rats and monkeys) treated with iPS-NPCs [173]. Scharf et al. investigated SPIOs labeled MSCs to treat tendon injury in sheep. They found that SPIO signal correlated with the presence of cells at day 7, but by day 14, SPIOs remained embedded in the tissue, despite the absence of transplanted MSCs [161]. Finally, Blocki et al. used SPIOs to label MSCs, which were then encapsulated in microcapsule and injected into a rat MI model. They demonstrated that encapsulated MSCs were retained and integrated into the infarcted rat hearts for several weeks posttransplantation, while cells that were not encapsulated were only detectable by MRI for two days posttransplantation [174].

^{19}F perfluorocarbon has been used to track multiple cell types in preclinical studies. In a rat cerebral ischemia model, Bible et al. demonstrated that human neural stem cells could be labeled and tacked with ^{19}F perfluorocarbon [175]. Gaudet et al. labeled MSCs with ^{19}F perfluorocarbon and transplanted them into either an immune competent or immune compromised mouse transplant model. They found ^{19}F perfluorocarbon signal was initially present in all animals, but by day 17 only two out of seven immune competent animals maintained detectable ^{19}F signal while ^{19}F was detectable in all immune compromised animals [176].

3.4 CONCLUSIONS

How cells are delivered is an important concept that influences the safety and efficacy of cellular therapies. Ideally, the cell delivery method should be minimally invasive, inexpensive, and clinically practical. Tracking cells helps to define cell fate in terms of biodistribution and retention, both of which are linked to the cell delivery method. Currently, no tracking methods permit quantitative single cell in vivo tracking, but advances in clinical and preclinical imaging may make this goal attainable in the future. For example, improved resolution clinical 3D PET and 3D SPECT imaging platforms combined with CT or MRI are under development to help quantify cell retention, biodistribution, and clearance. Long-term, in vivo cell tracking may be enabled in animal models by genetically inserting a sodium iodine symporter, for example. With this system, cells can be tracked longitudinally by administering serial boluses of radioactive tracers. These types of tracking technologies will likely have an important role in future cellular therapeutics.

REFERENCES

[1] Gnecchi M, Zhang Z, Ni A, Dzau VJ. Paracrine mechanisms in adult stem cell signaling and therapy. Circ Res 2008;103(11):1204–19.

[2] Fox JM, Chamberlain G, Ashton BA, Middleton J. Recent advances into the understanding of mesenchymal stem cell trafficking. Br J Haematol 2007;137(6):491–502.

[3] Maluf NS. History of blood transfusion. J History Med Allied Sci 1954;9(1):59–107.

[4] Hare JM, Traverse JH, Henry TD, Dib N, Strumpf RK, Schulman SP, et al. A randomized, double-blind, placebo-controlled, dose-escalation study of intravenous adult human mesenchymal stem cells (prochymal) after acute myocardial infarction. J Am Coll Cardiol 2009;54(24):2277–86.

[5] Bang OY, Lee JS, Lee PH, Lee G. Autologous mesenchymal stem cell transplantation in stroke patients. Ann Neurol 2005;57(6):874–82.

[6] Jeong H, Yim HW, Cho YS, Kim YI, Jeong SN, Kim HB, Oh IH. Efficacy and safety of stem cell therapies for patients with stroke: a systematic review and single arm meta-analysis. Int J Stem Cells 2014;7(2):63–9.

[7] Prasad K, Sharma A, Garg A, Mohanty S, Bhatnagar S, Johri S, et al. Intravenous autologous bone marrow mononuclear stem cell therapy for ischemic stroke: a multicentric, randomized trial. Stroke 2014;45(12):3618–24.

[8] Gazouli M, Roubelakis MG, Theodoropoulos GE. Stem cells as potential targeted therapy for inflammatory bowel disease. Inflammatory Bowel Dis 2014;20(5):952–5.

[9] Filippini P, Rutella S. Recent advances on cellular therapies and immune modulators for graft-versus-host disease. Expert Rev Clin Immunol 2014;10(10):1357–74.

[10] Fischer UM, Hartin MT, Jimenez F, Monzon-Posadas WO, Xue H, Savitz SI, et al. Pulmonary passage is a major obstacle for intravenous stem cell delivery: the pulmonary first-pass effect. Stem Cells Dev 2009;18(5):683–92.

[11] Detante O, Moisan A, Dimastromatteo J, Richard MJ, Riou L, Grillon E, et al. Intravenous administration of 99mTc-HMPAO-labeled human mesenchymal stem cells after stroke: in vivo imaging and biodistribution. Cell Transplant 2009;18(12):1369–79.

[12] Chin BB, Nakamoto Y, Bulte JW, Pittenger MF, Wahl R, Kraitchman DL. 111In oxine labelled mesenchymal stem cell SPECT after intravenous administration in myocardial infarction. Nucl Med Commun 2003;24(11):1149–54.

[13] Barbash IM, Chouraqui P, Baron J, Feinberg MS, Etzion S, Tessone A, et al. Systemic delivery of bone marrow-derived mesenchymal stem cells to the infarcted myocardium: feasibility, cell migration, and body distribution. Circulation 2003;108(7):863–8.

[14] Vandervelde S, van Luyn MJ, Tio RA, Harmsen MC. Signaling factors in stem cell-mediated repair of infarcted myocardium. J Mol Cell Cardiol 2005;39(2):363–76.

[15] Abbott JD, Huant Y, Liu D, Hickey R, Krause DS, Giordano FJ. Stromal cell-derived factor-1alpha plays a critical role in stem cell recruitment to the heart after myocardial infarction but is not sufficient to induce homing in the absence of injury. Circulation 2004;110(21):3300–5.

[16] Ma J, Ge J, Zhang S, Sun A, Shen J, Chen L, et al. Time course of myocardial stromal cell-derived factor 1 expression and beneficial effects of intravenously administered bone marrow stem cells in rats with experimental myocardial infarction. Basic Res Cardiol 2005;100(3):217–23.

[17] Freyman T, Polin G, Osman H, Crary J, Lu M, Cheng L, et al. A quantitative, randomized study evaluating three methods of mesenchymal stem cell delivery following myocardial infarction. Eur Heart J 2006;27(9):1114–22.

[18] Kurtzberg J, Prockop S, Teira P, Bittencourt H, Lewis V, Chan KW, et al. Allogeneic human mesenchymal stem cell therapy (remestemcel-L, Prochymal) as a rescue agent for severe refractory acute graft-versus-host disease in pediatric patients. Biol Blood Marrow Transplant 2014;20(2):229–35.

[19] Gordon MY, Levicar N, Pai M, Bachellier P, Dimarakis I, Al-Allaf F, et al. Characterization and clinical application of human CD34+ stem/progenitor cell populations mobilized into the blood by granulocyte colony-stimulating factor. Stem Cells 2006;24(7):1822–30.

[20] Tuma J, Fernandez-Vina R, Carrasco A, Castillo J, Cruz C, Carrillo A, et al. Safety and feasibility of percutaneous retrograde coronary sinus delivery of autologous bone marrow mononuclear cell transplantation in patients with chronic refractory angina. J Transl Med 2011;9:183.

[21] Raake P, von Degenfeld G, Hinkel R, Vachenauer R, Sandner T, Beller S, et al. Myocardial gene transfer by selective pressure-regulated retroinfusion of coronary veins: comparison with surgical and percutaneous intramyocardial gene delivery. J Am Coll Cardiol 2004;44(5):1124–9.

[22] Strauer BE, Brehm M, Zeus T, Kostering M, Hernandez A, Sorg RV, et al. Repair of infarcted myocardium by autologous intracoronary mononuclear bone marrow cell transplantation in humans. Circulation 2002;106(15):1913–8.

[23] Assmus B, Honold J, Schachinger V, Britten MB, Fischer-Rasokat AU, Lehmann R, et al. Transcoronary transplantation of progenitor cells after myocardial infarction. N Engl J Med 2006;355(12):1222–32.

[24] Wollert KC, Meyer GP, Lotz J, Ringes-Lichtenberg S, Lippolt P, Breidenbach C, et al. Intracoronary autologous bone-marrow cell transfer after myocardial infarction: the BOOST randomised controlled clinical trial. Lancet 2004;364(9429):141–8.

[25] Ge J, Li Y, Qian J, Shi J, Wang Q, Niu Y, et al. Efficacy of emergent transcatheter transplantation of stem cells for treatment of acute myocardial infarction (TCT-STAMI). Heart 2006;92(12):1764–7.

[26] Lunde K, Solheim S, Aakhus S, Arnesen H, Abdelnoor M, Egeland T, et al. Intracoronary injection of mononuclear bone marrow cells in acute myocardial infarction. N Engl J Med 2006;355(12):1199–209.

[27] Huikuri HV, Kervinen K, Niemela M, Ylitalo K, Saily M, Koistinen P, et al. Effects of intracoronary injection of mononuclear bone marrow cells on left ventricular function, arrhythmia risk profile, and restenosis after thrombolytic therapy of acute myocardial infarction. Eur Heart J 2008;29(22):2723–32.

[28] Diederichsen AC, Moller JE, Thayssen P, Videbaek L, Saekmose SG, Barington T, et al. Changes in left ventricular filling patterns after repeated injection of autologous bone marrow cells in heart failure patients. Scand Cardiovasc J 2010;44(3):139–45.

[29] Roncalli J, Mouquet F, Piot C, Trochu JN, Le Corvoisier P, Neuder Y, et al. Intracoronary autologous mononucleated bone marrow cell infusion for acute myocardial infarction: results of the randomized multicenter BONAMI trial. Eur Heart J 2011;32(14):1748–57.

[30] Assmus B, Walter DH, Seeger FH, Leistner DM, Steiner J, Ziegler I, et al. Effect of shock wave-facilitated intracoronary cell therapy on LVEF in patients with chronic heart failure: the CELLWAVE randomized clinical trial. JAMA 2013;309(15):1622–31.

[31] Hirsch A, Nijveldt R, van der Vleuten PA, Tijssen JG, van der Giessen WJ, Tio RA, et al. Intracoronary infusion of mononuclear cells from bone marrow or peripheral blood compared with standard therapy in patients after acute myocardial infarction treated by primary percutaneous coronary intervention: results of the randomized controlled HEBE trial. Eur Heart J 2011;32(14):1736–47.

[32] Traverse JH, Henry TD, Ellis SG, Pepine CJ, Willerson JT, Zhao DX, et al. Effect of intracoronary delivery of autologous bone marrow mononuclear cells 2 to 3 weeks following acute myocardial infarction on left ventricular function: the LateTIME randomized trial. JAMA 2011;306(19):2110–9.

[33] Tendera M, Wojakowski W, Ruzyllo W, Chojnowska L, Kepka C, Tracz W, et al. Intracoronary infusion of bone marrow-derived selected CD34 + CXCR4+ cells and non-selected mononuclear cells in patients with acute STEMI and reduced left ventricular ejection fraction: results of randomized, multicentre Myocardial Regeneration by Intracoronary Infusion of Selected Population of Stem Cells in Acute Myocardial Infarction (REGENT) Trial. Eur Heart J 2009;30(11):1313–21.

[34] Bartunek J, Vanderheyden M, Vandekerckhove B, Mansour S, De Bruyne B, De Bondt P, et al. Intracoronary injection of CD133-positive enriched bone marrow progenitor cells promotes cardiac recovery after recent myocardial infarction: feasibility and safety. Circulation 2005;112(Suppl. 9):I178–83.

[35] Mansour S, Bouchard V, Stevens LM, Gobeil F, Rivard A, et al. One-year safety analysis of the COMPARE-AMI Trial: comparison of intracoronary injection of

CD133 bone marrow stem cells to placebo in patients after acute myocardial infarction and left ventricular dysfunction. Bone Marrow Res 2011;2011:385124.
[36] Chen SL, Fang WW, Ye F, Liu YH, Qian J, Shan SJ, et al. Effect on left ventricular function of intracoronary transplantation of autologous bone marrow mesenchymal stem cell in patients with acute myocardial infarction. Am J Cardiol 2004;94(1):92–5.
[37] Houtgraaf JH, den Dekker WK, van Dalen BM, Springeling T, de Jong R, van Genus RJ, et al. First experience in humans using adipose tissue-derived regenerative cells in the treatment of patients with ST-segment elevation myocardial infarction. J Am Coll Cardiol 2012;59(5):539–40.
[38] Choi JH, Choi J, Lee WS, Rhee I, Lee SC, Gwon HC. Lack of additional benefit of intracoronary transplantation of autologous peripheral blood stem cell in patients with acute myocardial infarction. Circ J 2007;71(4):486–94.
[39] Li ZQ, Zhang M, Jing YZ, Zhang WW, Liu Y, Cui LJ, et al. The clinical study of autologous peripheral blood stem cell transplantation by intracoronary infusion in patients with acute myocardial infarction (AMI). Int J Cardiol 2007;115(1):52–6.
[40] Kang HJ, Kim HS, Zhang SY, Park KW, Cho HJ, Koo BK, et al. Effects of intracoronary infusion of peripheral blood stem-cells mobilised with granulocyte-colony stimulating factor on left ventricular systolic function and restenosis after coronary stenting in myocardial infarction: the MAGIC cell randomised clinical trial. Lancet 2004;363(9411):751–6.
[41] Schachinger V, Assmus B, Britten MB, Honold J, Lehmann R, Teupe C, et al. Transplantation of progenitor cells and regeneration enhancement in acute myocardial infarction: final one-year results of the TOPCARE-AMI Trial. J Am Coll Cardiol 2004;44(8):1690–9.
[42] Assmus B, Fischer-Rasokat U, Honold J, Seeger FH, Fichtlscherer S, Tonn T, et al. Transcoronary transplantation of functionally competent BMCs is associated with a decrease in natriuretic peptide serum levels and improved survival of patients with chronic postinfarction heart failure: results of the TOPCARE-CHD Registry. Circ Res 2007;100(8):1234–41.
[43] Bolli R, Chugh AR, D'Amario D, Loughran JH, Stoddard MF, Ikram S, et al. Cardiac stem cells in patients with ischaemic cardiomyopathy (SCIPIO): initial results of a randomised phase 1 trial. Lancet 2011;378(9806):1847–57.
[44] Makkar RR, Smith RR, Cheng K, Malliaras K, Thomson LE, Berman D, et al. Intracoronary cardiosphere-derived cells for heart regeneration after myocardial infarction (CADUCEUS): a prospective, randomised phase 1 trial. Lancet 2012;379(9819):895–904.
[45] Raval Z, Losordo DW. Cell therapy of peripheral arterial disease: from experimental findings to clinical trials. Circ Res 2013;112(9):1288–302.
[46] Walter DH, Krankenberg H, Balzer JO, Kalka C, Baumgartner I, Schluter M, et al. Intraarterial administration of bone marrow mononuclear cells in patients with critical limb ischemia: a randomized-start, placebo-controlled pilot trial (PROVASA). Circ Cardiovasc Interv 2011;4(1):26–37.
[47] Kirana S, Stratmann B, Prante C, Prohaska W, Koerperich H, Lammers D, et al. Autologous stem cell therapy in the treatment of limb ischaemia induced chronic tissue ulcers of diabetic foot patients. Int J Clin Pract 2012;66(4):384–93.
[48] Correa P, Felix R, Mendonca ML, Freitas G, Azevedo J, Dohmann H, et al. Dual-head coincidence gamma camera FDG-PET before and after autologous bone marrow mononuclear cell implantation in ischaemic stroke. Eur J Nuclear Med Mol Imaging 2005;32(8):999.

[49] Mendonca ML, et al. [Safety of intra-arterial autologous bone marrow mononuclear cell transplantation for acute ischemic stroke]. Arquivos Brasileiros Cardiol 2006;86(1):52–5.
[50] Correa PL, Mesquita CT, Felix RM, Azevedo JC, Barbirato GB, Falcao CH, et al. Assessment of intra-arterial injected autologous bone marrow mononuclear cell distribution by radioactive labeling in acute ischemic stroke. Clin Nucl Med 2007;32(11):839–41.
[51] Barbosa da Fonseca LM, Battistella V, de Freitas GR, Gutfilen B, Dos Santos Goldenberg RC, Maiolino A, et al. Early tissue distribution of bone marrow mononuclear cells after intra-arterial delivery in a patient with chronic stroke. Circulation 2009;120(6):539–41.
[52] Barbosa da Fonseca LM, Gutfilen B, Rosado de Castro PH, Battistella V, Goldenberg RC, Kasai-Brunswick T, et al. Migration and homing of bone-marrow mononuclear cells in chronic ischemic stroke after intra-arterial injection. Exp Neurol 2010;221(1):122–8.
[53] Battistella V, de Freitas GR, da Fonseca LM, Mercante D, Gutfilen B, Goldenberg RC, et al. Safety of autologous bone marrow mononuclear cell transplantation in patients with nonacute ischemic stroke. Regen Med 2011;6(1):45–52.
[54] Rosado-de-Castro PH, Schmidt Fda R, Battistella V, Lopes de Souza SA, Gutfilen B, Goldenberg RC, et al. Biodistribution of bone marrow mononuclear cells after intra-arterial or intravenous transplantation in subacute stroke patients. Regen Med 2013;8(2):145–55.
[55] Friedrich MA, Martins MP, Araujo MD, Klamt C, Vedolin L, Garicochea B, et al. Intra-arterial infusion of autologous bone marrow mononuclear cells in patients with moderate to severe middle cerebral artery acute ischemic stroke. Cell Transplant 2012;21(Suppl. 1):S13–21.
[56] Lee PH, Lee JE, Kim HS, Song SK, Lee HS, Nam HS, et al. A randomized trial of mesenchymal stem cells in multiple system atrophy. Ann Neurol 2012;72(1):32–40.
[57] Vrtovec B, Poglajen G, Lezaic L, Sever M, Socan A, Domanovic D, et al. Comparison of transendocardial and intracoronary CD34+ cell transplantation in patients with nonischemic dilated cardiomyopathy. Circulation 2013;128(11 Suppl. 1): S42–9.
[58] Moreira Rde C, Haddad AF, Silva SA, Souza AL, Tuche FA, Oliveira MA, et al. Intracoronary stem-cell injection after myocardial infarction: microcirculation substudy. Arquivos Brasileiros Cardiol 2011;97(5):420–6.
[59] Dib N, Menasche P, Bartunek JJ, Zeiher AM, Terzic A, Chronos NA, et al. Recommendations for successful training on methods of delivery of biologics for cardiac regeneration: a report of the International Society for Cardiovascular Translational Research. JACC Cardiovasc Interv 2010;3(3):265–75.
[60] Aicher A, Heeschen C, Sasaki K, Urbich C, Zeiher AM, Dimmeler S. Low-energy shock wave for enhancing recruitment of endothelial progenitor cells: a new modality to increase efficacy of cell therapy in chronic hind limb ischemia. Circulation 2006;114(25):2823–30.
[61] Gyongyosi M, Wojakowski W, Lemarchand P, Lunde K, Tendera M, Bartunek J, et al. Meta-analysis of cell-based CaRdiac stUdiEs (ACCRUE) in patients with acute myocardial infarction based on individual patient data. Circ Res 2015;116(8):1346–60.

[62] Arslan Z, Balta S, Demirkol S, Ozturk C, Unlu M, Aparci M, et al. Parameters influencing LVEF improvement with intracoronary bone marrow stem cell delivery in acute myocardial infarction. Int J Cardiol 2014;177(2):644–5.

[63] Hou D, Youssef EA, Brinton TJ, Zhang P, Rogers P, Price ET, et al. Radiolabeled cell distribution after intramyocardial, intracoronary, and interstitial retrograde coronary venous delivery: implications for current clinical trials. Circulation 2005;112(Suppl. 9):I150–6.

[64] Menasche P, Hagege AA, Scorsin M, Pouzet B, Desnos M, Duboc D, et al. Myoblast transplantation for heart failure. Lancet 2001;357(9252):279–80.

[65] Dib N, Michler RE, Pagani FD, Wright S, Kerejakes DJ, Lengerich R, et al. Safety and feasibility of autologous myoblast transplantation in patients with ischemic cardiomyopathy: four-year follow-up. Circulation 2005;112(12):1748–55.

[66] de la Fuente LM, Stertzer SH, Argentieri J, Penaloza E, Miano J, Koziner B, et al. Transendocardial autologous bone marrow in chronic myocardial infarction using a helical needle catheter: 1-year follow-up in an open-label, nonrandomized, single-center pilot study (the TABMMI study). Am Heart J 2007;154(1). p. 79 e1–7.

[67] Amado LC, Saliaris AP, Schuleri KH, St John M, Xie JS, Catteneo S, et al. Cardiac repair with intramyocardial injection of allogeneic mesenchymal stem cells after myocardial infarction. Proc Natl Acad Sci U S A 2005;102(32):11474–9.

[68] Vale PR, Losordo DW, Milliken CE, Maysky M, Esakof DD, Symes JF, et al. Left ventricular electromechanical mapping to assess efficacy of phVEGF(165) gene transfer for therapeutic angiogenesis in chronic myocardial ischemia. Circulation 2000;102(9):965–74.

[69] Kornowski R, Fuchs S, Tio FO, Pierre A, Epstein SE, Leon MB. Evaluation of the acute and chronic safety of the biosense injection catheter system in porcine hearts. Catheter Cardiovasc Interv 1999;48(4):447–53.

[70] Kornowski R, Hong MK, Gepstein L, Goldstein S, Ellahham S, Ben-Haim SA, et al. Preliminary animal and clinical experiences using an electromechanical endocardial mapping procedure to distinguish infarcted from healthy myocardium. Circulation 1998;98(11):1116–24.

[71] Psaltis PJ, Zannettino AC, Gronthos S, Worthley SG. Intramyocardial navigation and mapping for stem cell delivery. J Cardiovasc Transl Res 2010;3(2):135–46.

[72] Tomkowiak MT, Klein AJ, Vigen KK, Hacker TA, Speidel MA, VanLysel MS, et al. Targeted transendocardial therapeutic delivery guided by MRI-x-ray image fusion. Catheterization Cardiovasc Interv 2011;78(3):468–78.

[73] de Silva R, Gutierrez LF, Raval AN, McVeigh ER, Ozturk C, Lederman RJ. X-ray fused with magnetic resonance imaging (XFM) to target endomyocardial injections: validation in a swine model of myocardial infarction. Circulation 2006;114(22):2342–50.

[74] Schmuck EG, Koch JM, Hacker TA, Hatt CR, Tomkowjak MT, Vigen KK, et al. Intravenous followed by X-ray fused with MRI-guided transendocardial mesenchymal stem cell injection Improves Contractility Reserve in a Swine Model of Myocardial Infarction. J Cardiovasc Transl Res 2015;8(7):438–48.

[75] Klein A, et al. Real-time, 3-D Trans-esophageal Echo-Guided Endomyocardial Treatment Delivery Targeting Myocardial Infarction Borders[Abstract] Trascatheter Therapeutics, San Francisco, CA, 2009.

[76] Dick AJ, Guttman MA, Raman VK, Peters DC, Pessanha BS, Hill JM, et al. Magnetic resonance fluoroscopy allows targeted delivery of mesenchymal stem cells to infarct borders in Swine. Circulation 2003;108(23):2899–904.
[77] Hatt CR, Jain AK, Parthasarathy V, Lang A, Raval AN. MRI-3D ultrasound-X-ray image fusion with electromagnetic tracking for transendocardial therapeutic injections: in-vitro validation and in-vivo feasibility. Comput Med Imaging Graph 2013;37(2):162–73.
[78] Raval AN, Schmuck EG, Tefera G, Leitzke C, Ark CV, Hei D, et al. Bilateral administration of autologous CD133+ cells in ambulatory patients with refractory critical limb ischemia: lessons learned from a pilot randomized, double-blind, placebo-controlled trial. Cytotherapy 2014;16(12):1720–32.
[79] Feng M, Li Y, Han Q, Bao X, Yang M, Zhu H, et al. Preclinical safety evaluation of human mesenchymal stem cell transplantation in cerebrum of nonhuman primates. Int J Toxicol 2014;33(5):403–11.
[80] Kim HY, Kim H, Oh KW, Oh SI, Koh SH, Baik W, et al. Biological markers of mesenchymal stromal cells as predictors of response to autologous stem cell transplantation in patients with amyotrophic lateral sclerosis: an investigator-initiated trial and in vivo study. Stem Cells 2014;32(10):2724–31.
[81] Karussis D, Karageorgiou C, Vaknin-Dembrinsky A, Gowda-Kurkalli B, Gomori JM, Kassis I, et al. Safety and immunological effects of mesenchymal stem cell transplantation in patients with multiple sclerosis and amyotrophic lateral sclerosis. Arch Neurol 2010;67(10):1187–94.
[82] Tian C, Wang X, Wang X, Wang L, Wang X, Wu S, et al. Autologous bone marrow mesenchymal stem cell therapy in the subacute stage of traumatic brain injury by lumbar puncture. Exp Clin Transplant 2013;11(2):176–81.
[83] Schmuck EG, Mulligan JF, Ertel RL, Kouris NA, Ogle BM, Raval AN, et al. Cardiac fibroblast-derived 3D extracellular matrix seeded with mesenchymal stem cells as a novel device to transfer cells to the ischemic myocardium. Cardiovasc Eng Technol 2014;5(1):119–31.
[84] Reis LA, Chiu LLY, Feric N, Fu L, Radisic M, et al. Biomaterials in myocardial tissue engineering. J Tissue Eng Regen Med 2016;10(1):11–28.
[85] Godier-Furnemont AF, Martens TP, Koeckert MS, Wan L, Parks J, Arai K, et al. Composite scaffold provides a cell delivery platform for cardiovascular repair. Proc Natl Acad Sci U S A 2011;108(19):7974–9.
[86] Johnson TD, Dequach JA, Gaetani R, Ungerleider J, Elhag D, Nigam V, et al. Human versus porcine tissue sourcing for an injectable myocardial matrix hydrogel. Biomater Sci 2014;2014:60283D.
[87] Oberwallner B, Brodarac A, Choi YH, Saric T, Anic P, Morawietz L, et al. Preparation of cardiac extracellular matrix scaffolds by decellularization of human myocardium. J Biomed Mater Res A 2014;102(9):3263–72.
[88] Schmuck EG, Mulligan JD, Ertel RL, Kouris NA, Ogle BM, Raval AN, et al. Cardiac fibroblast-derived 3D extracellular matrix seeded with mesenchymal stem cells as a novel device to transfer cells to the ischemic myocardium. Cardiovasc Eng Technol 2014;5(1):119–31.
[89] Wei HJ, Chen SC, Chang Y, Hwang SM, Lin WW, Lai PH, et al. Porous acellular bovine pericardia seeded with mesenchymal stem cells as a patch to repair a myocardial defect in a syngeneic rat model. Biomaterials 2006;27(31):5409–19.

[90] Wei HJ, Liang HC, Lee MH, Huang YC, Chang Y, Sung HW. Construction of varying porous structures in acellular bovine pericardia as a tissue-engineering extracellular matrix. Biomaterials 2005;26(14):1905–13.

[91] Chang Y, Lai PH, Wei HJ, Lin WW, Chen CH, Hwang SM, et al. Tissue regeneration observed in a basic fibroblast growth factor-loaded porous acellular bovine pericardium populated with mesenchymal stem cells. J Thorac Cardiovasc Surg 2007;134(1):65–73.

[92] Vashi AV, White JF, McLean KM, Neethling WM, Rhodes DI, Ramshaw JA, et al. Evaluation of an established pericardium patch for delivery of mesenchymal stem cells to cardiac tissue. J Biomed Mater Res A 2015;103(6):1999–2005.

[93] Rajabi-Zeleti S, Jalili-Firoozinezhad S, Azarnia M, Khayyatan F, Vahdat S, Nikeghbalian S, et al. The behavior of cardiac progenitor cells on macroporous pericardium-derived scaffolds. Biomaterials 2014;35(3):970–82.

[94] Badylak S, Liang A, Record R, Rullius R, Hodde J. Biomaterials 1999;20(23–24): 2257–63.

[95] Ferrand BK, Kokini K, Badylak SF, Geddes LA, Hiles MC, Morff RJ. Directional porosity of porcine small-intestinal submucosa. J Biomed Mater Res 1993;27(10):1235–41.

[96] Tan MY, Zhi W, Wei RQ, Huang YC, Zhou KP, Tan B, et al. Repair of infarcted myocardium using mesenchymal stem cell seeded small intestinal submucosa in rabbits. Biomaterials 2009;30(19):3234–40.

[97] Zhang X, Wang H, Ma X, Adila A, Wang B, Liu F, et al. Preservation of the cardiac function in infarcted rat hearts by the transplantation of adipose-derived stem cells with injectable fibrin scaffolds. Exp Biol Med (Maywood) 2010;235(12):1505–15.

[98] Segers VF, Lee RT. Biomaterials to enhance stem cell function in the heart. Circ Res 2011;109(8):910–22.

[99] Danoviz ME, Nakamuta JS, Marques FL, dos Santos L, Alvarenga EC, dos Santos AA, et al. Rat adipose tissue-derived stem cells transplantation attenuates cardiac dysfunction post infarction and biopolymers enhance cell retention. PLoS One 2010;5(8):pe12077.

[100] Ye L, Chang YH, Xiong Q, Zhang P, Zhang L, Somasundaram P, et al. Cardiac repair in a porcine model of acute myocardial infarction with human induced pluripotent stem cell-derived cardiovascular cells. Cell Stem Cell 2014;15(6):750–61.

[101] Dai W, Hale SL, Kay GL, Jyrala AJ, Kloner RA. Delivering stem cells to the heart in a collagen matrix reduces relocation of cells to other organs as assessed by nanoparticle technology. Regen Med 2009;4(3):387–95.

[102] Frederick JR, Fitzpatrick JR 3rd, McCormick RC, Harris DA, Kim AY, Muenzer JR, et al. Stromal cell-derived factor-1alpha activation of tissue-engineered endothelial progenitor cell matrix enhances ventricular function after myocardial infarction by inducing neovasculogenesis. Circulation 2010;122(Suppl. 11): S107–17.

[103] Simpson D, Liu H, Fan TH, Nerem R, Dudley SC Jr. A tissue engineering approach to progenitor cell delivery results in significant cell engraftment and improved myocardial remodeling. Stem Cells 2007;25(9):2350–7.

[104] Simpson DL, Dudley SC Jr. Modulation of human mesenchymal stem cell function in a three-dimensional matrix promotes attenuation of adverse remodelling after myocardial infarction. J Tissue Eng Regen Med 2013;7(3):192–202.

[105] Giraud MN, Ayuni E, Cook S, Siepe M, Carrel TP, Tevaearai HT, et al. Hydrogel-based engineered skeletal muscle grafts normalize heart function early after myocardial infarction. Artif Organs 2008;32(9):692–700.
[106] Arana M, Gavira JJ, Pena E, Gonzalez A, Abizanda G, Cilla M, et al. Epicardial delivery of collagen patches with adipose-derived stem cells in rat and minipig models of chronic myocardial infarction. Biomaterials 2014;35(1):143–51.
[107] Holladay CA, Duffy AM, Chen X, Sefton MV, O'Brien TD, Pandit AS. Recovery of cardiac function mediated by MSC and interleukin-10 plasmid functionalised scaffold. Biomaterials 2012;33(5):1303–14.
[108] Xiang Z, Liao R, Kelly MS, Spector M. Collagen-GAG scaffolds grafted onto myocardial infarcts in a rat model: a delivery vehicle for mesenchymal stem cells. Tissue Eng 2006;12(9):2467–78.
[109] Chachques JC, Trainini JC, Lago N, Cortes-Morichetti M, Schussler O, Carpentier A, et al. Myocardial assistance by grafting a new bioartificial upgraded myocardium (MAGNUM trial): clinical feasibility study. Ann Thorac Surg 2008;85(3):901–8.
[110] Chachques JC, Trainini JC, Lago N, Masoli OH, Barisani JL, Cortes-Morichetti M, et al. Myocardial assistance by grafting a new bioartificial upgraded myocardium (MAGNUM clinical trial): one year follow-up. Cell Transplant 2007;16(9):927–34.
[111] Blazquez R, Sanchez-Margallo FM, Crisostomo V, Baez C, Maestre J, Garcia-Lindo M, et al. Intrapericardial administration of mesenchymal stem cells in a large animal model: a bio-distribution analysis. PLoS One 2015;10(3):pe0122377.
[112] Li YH, Feng L, Zhang GX, Ma CG, et al. Intranasal delivery of stem cells as therapy for central nervous system disease. Exp Mol Pathol 2015;98(2):145–51.
[113] Wei N, Yu SP, Gu X, Taylor TM, Song D, Liu XF, et al. Delayed intranasal delivery of hypoxic-preconditioned bone marrow mesenchymal stem cells enhanced cell homing and therapeutic benefits after ischemic stroke in mice. Cell Transplant 2013;22(6):977–91.
[114] Azene N, Fu Y, Maurer J, Kraitchman DL. Tracking of stem cells in vivo for cardiovascular applications. J Cardiovasc Magn Reson 2014;16:7.
[115] Nguyen PK, Riegler J, Wu JC. Stem cell imaging: from bench to bedside. Cell Stem Cell 2014;14(4):431–44.
[116] Frangioni JV, Hajjar RJ. In vivo tracking of stem cells for clinical trials in cardiovascular disease. Circulation 2004;110(21):3378–83.
[117] Wu C, Ma G, Li J, Zheng K, Dang Y, Shi X, et al. In vivo cell tracking via (1)(8)F-fluorodeoxyglucose labeling: a review of the preclinical and clinical applications in cell-based diagnosis and therapy. Clin Imaging 2013;37(1):28–36.
[118] Fu Y, Kraitchman DL. Stem cell labeling for noninvasive delivery and tracking in cardiovascular regenerative therapy. Expert Rev Cardiovasc Ther 2010;8(8):1149–60.
[119] Wolfs E, Verfaillie CM, Van Laere K, Deroose CM. Radiolabeling strategies for radionuclide imaging of stem cells. Stem Cell Rev 2015;11(2):254–74.
[120] Musialek P, Tekieli L, Kostkiewicz M, Miszalski-Jamka T, Klimeczek P, Mazur W, et al. Infarct size determines myocardial uptake of CD34+ cells in the peri-infarct zone: results from a study of (99m)Tc-extametazime-labeled cell visualization integrated with cardiac magnetic resonance infarct imaging. Circ Cardiovasc Imaging 2013;6(2):320–8.
[121] Hofmann M, Wollert KC, Meyer GP, Menke A, Arseniev L, Hertenstein B, et al. Monitoring of bone marrow cell homing into the infarcted human myocardium. Circulation 2005;111(17):2198–202.

[122] Kang WJ, Kang HJ, Kim HS, Chung JK, Lee MC, Lee DS. Tissue distribution of 18F-FDG-labeled peripheral hematopoietic stem cells after intracoronary administration in patients with myocardial infarction. J Nucl Med 2006;47(8):1295–301.

[123] Blocklet D, Toungouz M, Berkenboom G, Lambermont M, Unger P, Preumont N, et al. Myocardial homing of nonmobilized peripheral-blood CD34+ cells after intracoronary injection. Stem Cells 2006;24(2):333–6.

[124] Terrovitis J, Lautamaki R, Bonios M, Fox J, Engles JM, Yu J, et al. Noninvasive quantification and optimization of acute cell retention by in vivo positron emission tomography after intramyocardial cardiac-derived stem cell delivery. J Am Coll Cardiol 2009;54(17):1619–26.

[125] Zhang Y, Dasilva JN, Hadizad T, Thorn S, Kuraitis D, Renaud JM, et al. (18)F-FDG cell labeling may underestimate transplanted cell homing: more accurate, efficient, and stable cell labeling with hexadecyl-4-[(18)F]fluorobenzoate for in vivo tracking of transplanted human progenitor cells by positron emission tomography. Cell Transplant 2012;21(9):1821–35.

[126] Kim MH, Woo SK, Lee KC, An GI, Pandya D, Park NM, et al. Longitudinal monitoring adipose-derived stem cell survival by PET imaging hexadecyl-4-(1)(2)(4) I-iodobenzoate in rat myocardial infarction model. Biochem Biophys Res Commun 2015;456(1):13–9.

[127] Kim MH, Woo SK, Kim KI, Lee TS, Kim CW, Kang JH, et al. Simple methods for tracking stem cells with (64)Cu-labeled DOTA-hexadecyl-benzoate. ACS Med Chem Lett 2015;6(5):528–30.

[128] Rosado-de-Castro PH, Pimentel-Coelho PM, Gutfilen B, Lopes de Souza SA, de Freitas GR, Mendez-Otero R, et al. Radiopharmaceutical stem cell tracking for neurological diseases. Biomed Res Int 2014;2014:417091.

[129] Schachinger V, Aicher A, Dobert N, Rover R, Diener J, Fichtlscherer S, et al. Pilot trial on determinants of progenitor cell recruitment to the infarcted human myocardium. Circulation 2008;118(14):1425–32.

[130] Goussetis E, Manginas A, Koutelou M, Peristeri I, Theodosaki M, Kollaros N, et al. Intracoronary infusion of CD133+ and CD133-CD34+ selected autologous bone marrow progenitor cells in patients with chronic ischemic cardiomyopathy: cell isolation, adherence to the infarcted area, and body distribution. Stem Cells 2006;24(10):2279–83.

[131] Caveliers V, De Keulenaer G, Everaert H, Van Riet I, Van Camp G, Verheye S, et al. In vivo visualization of 111In labeled CD133+ peripheral blood stem cells after intracoronary administration in patients with chronic ischemic heart disease. Q J Nucl Med Mol Imaging 2007;51(1):61–6.

[132] Kurpisz M, Czepczynski R, Grygielska B, Majewski M, Fiszer D, Jerzykowska O, et al. Bone marrow stem cell imaging after intracoronary administration. Int J Cardiol 2007;121(2):194–5.

[133] Sabondjian E, Mitchell AJ, Wisenberg G, White J, Blackwood KJ, Sykes J, et al. Hybrid SPECT/cardiac-gated first-pass perfusion CT: locating transplanted cells relative to infarcted myocardial targets. Contrast Media Mol Imaging 2012;7(1):76–84.

[134] Mitchell AJ, Sabondjian E, Sykes J, Deans L, Zhu W, Lu X, et al. Comparison of initial cell retention and clearance kinetics after subendocardial or subepicardial injections of endothelial progenitor cells in a canine myocardial infarction model. J Nucl Med 2010;51(3):413–7.

[135] Templin C, Zweigerdt R, Schwanke K, Olmer R, Ghadri JR, Emmert MY, et al. Transplantation and tracking of human-induced pluripotent stem cells in a pig model of myocardial infarction: assessment of cell survival, engraftment, and distribution by hybrid single photon emission computed tomography/computed tomography of sodium iodide symporter transgene expression. Circulation 2012;126(4):430–9.
[136] de Almeida PE, van Rappard JR, Wu JC. In vivo bioluminescence for tracking cell fate and function. Am J Physiol Heart Circ Physiol 2011;301(3):H663–71.
[137] Terrovitis JV, Smith RR, Marban E. Assessment and optimization of cell engraftment after transplantation into the heart. Circ Res 2010;106(3):479–94.
[138] van der Bogt KE, Schrepfer S, Yu J, Sheikh AY, Hoyt G, Govaert JA, et al. Comparison of transplantation of adipose tissue- and bone marrow-derived mesenchymal stem cells in the infarcted heart. Transplantation 2009;87(5):642–52.
[139] Wang J, Zhang S, Rabinovich B, Bidaut L, Soghomonyan S, Alauddin MM, et al. Human CD34+ cells in experimental myocardial infarction: long-term survival, sustained functional improvement, and mechanism of action. Circ Res 2010;106(12):1904–11.
[140] van der Bogt KE, Hellingman AA, Lijkwan MA, Bos EJ, de Vries MR, van Rappard JR, et al. Molecular imaging of bone marrow mononuclear cell survival and homing in murine peripheral artery disease. JACC Cardiovasc Imaging 2012;5(1):46–55.
[141] Li Z, Hu X, Mao J, Liu X, Zhang L, Liu J, et al. Optimization of mesenchymal stem cells (MSCs) delivery dose and route in mice with acute liver injury by bioluminescence imaging. Mol Imaging Biol 2015;17(2):185–94.
[142] Progatzky F, Dallman MJ, Lo Celso C. From seeing to believing: labelling strategies for in vivo cell-tracking experiments. Interface Focus 2013;3(3):p20130001.
[143] Muller-Borer BJ, Collins MC, Gunst PR, Cascio WE, Kypson AP. Quantum dot labeling of mesenchymal stem cells. J Nanobiotechnol 2007;5:9.
[144] Shah B, Clark P, Stroscio M, Mao J. Labeling and imaging of human mesenchymal stem cells with quantum dot bioconjugates during proliferation and osteogenic differentiation in long term. Conf Proc IEEE Eng Med Biol Soc 2006;1:1470–3.
[145] Lin S, Xie X, Patel MR, Yang YH, Li Z, Cao F, et al. Quantum dot imaging for embryonic stem cells. BMC Biotechnol 2007;7:67.
[146] Danner S, Benzin H, Vollbrandt T, Oder J, Richter A, Kruse C. Quantum dots do not alter the differentiation potential of pancreatic stem cells and are distributed randomly among daughter cells. Int J Cell Biol 2013;2013:918242.
[147] Rosen AB, Kelly DJ, Schuldt AJ, Lu J, Potapova IA, Doronin SV, et al. Finding fluorescent needles in the cardiac haystack: tracking human mesenchymal stem cells labeled with quantum dots for quantitative in vivo three-dimensional fluorescence analysis. Stem Cells 2007;25(8):p2128–38.
[148] Rak-Raszewska A, Marcello M, Kenny S, Edgar D, See V, Murray P. Quantum dots do not affect the behaviour of mouse embryonic stem cells and kidney stem cells and are suitable for short-term tracking. PLoS One 2012;7(3):pe32650.
[149] Ding D, Mao D, Li K, Wang X, Qin W, Liu R, et al. Precise and long-term tracking of adipose-derived stem cells and their regenerative capacity via superb bright and stable organic nanodots. ACS Nano 2014;8(12):12620–31.
[150] Chen G, Tian F, Li C, Zhang Y, Weng Z, Zhang Y, Peng R, et al. In vivo real-time visualization of mesenchymal stem cells tropism for cutaneous regeneration using NIR-II fluorescence imaging. Biomaterials 2015;53:265–73.

[151] Allard J, Li K, Lopez XM, Blanchard S, Barbot P, Rorive S, et al. Immunohistochemical toolkit for tracking and quantifying xenotransplanted human stem cells. Regen Med 2014;9(4):437–52.

[152] Kajstura J, Rota M, Whang B, Cascapera S, Hosoda T, Bearzi C, et al. Bone marrow cells differentiate in cardiac cell lineages after infarction independently of cell fusion. Circ Res 2005;96(1):127–37.

[153] Krause DS, Theise ND, Collector MI, Henegariu O, Hwang S, Gardner R, et al. Multi-organ, multi-lineage engraftment by a single bone marrow-derived stem cell. Cell 2001;105(3):369–77.

[154] Gargesha M, Qutaish M, Roy D, Steyer G, Bartsch H, Wilson DL. Enhanced volume rendering techniques for high-resolution color cryo-imaging data. Proc SPIE Int Soc Opt Eng 2009;7262:72655V.

[155] Roy D, Steyer GJ, Gargesha M, Stone ME, Wilson DL. 3D cryo-imaging: a very high-resolution view of the whole mouse. Anat Rec (Hoboken) 2009;292(3):342–51.

[156] Steyer GJ, Roy D, Salvado O, Stone ME, Wilson DL. Cryo-imaging of fluorescently-labeled single cells in a mouse. Proc SPIE Int Soc Opt Eng 2009;7262:72620W–W172620.

[157] Steyer GJ, Roy D, Salvado O, Stone ME, Wilson DL. Removal of out-of-plane fluorescence for single cell visualization and quantification in cryo-imaging. Ann Biomed Eng 2009;37(8):1613–28.

[158] Auletta JJ, Eid SK, Wuttisarnwattana P, Silva I, Metheny L, Keller MD, et al. Human mesenchymal stromal cells attenuate graft-versus-host disease and maintain graft-versus-leukemia activity following experimental allogeneic bone marrow transplantation. Stem Cells 2015;33(2):601–14.

[159] Srivastava AK, Kadayakkara DK, Bar-Shir A, Gilad AA, McMahon MT, Bulte JWM. Advances in using MRI probes and sensors for in vivo cell tracking as applied to regenerative medicine. Dis Model Mech 2015;8(4):323–36.

[160] Huang Z, Li C, Yang S, Xu J, Shen Y, Xie X, et al. Magnetic resonance hypointensive signal primarily originates from extracellular iron particles in the long-term tracking of mesenchymal stem cells transplanted in the infarcted myocardium. Int J Nanomedicine 2015;10:1679–90.

[161] Scharf A, Holmes S, Thoresen M, Mumaw J, Stumpf A, Peroni J. Superparamagnetic iron oxide nanoparticles as a means to track mesenchymal stem cells in a large animal model of tendon injury. Contrast Media Mol Imaging 2015;10(5):388–97.

[162] Ma N, Cheng H, Lu M, Liu Q, Chen X, Yin G, et al. Magnetic resonance imaging with superparamagnetic iron oxide fails to track the long-term fate of mesenchymal stem cells transplanted into heart. Sci Rep 2015;5:9058.

[163] Srivastava AK, Bulte JW. Seeing stem cells at work in vivo. Stem Cell Rev 2014;10(1):127–44.

[164] Ahrens ET, Flores R, Xu H, Morel PA. In vivo imaging platform for tracking immunotherapeutic cells. Nat Biotechnol 2005;23(8):983–7.

[165] Verdijk P, Scheenen TW, Lesterhuis WJ, Gambarota G, Veltien AA, Walczak P, et al. Sensitivity of magnetic resonance imaging of dendritic cells for in vivo tracking of cellular cancer vaccines. Int J Cancer 2007;120(5):978–84.

[166] Srinivas M, Heerschap A, Ahrens ET, Figdor CG, de Vries IJM. (19)F MRI for quantitative in vivo cell tracking. Trends Biotechnol 2010;28(7):363–70.

[167] Karussis D, karageorgiou C, Vaknin-Dembinsky A, Gowda-Kurkalli B, Gomori JM, Kassis I, et al. Safety and immunological effects of mesenchymal stem cell transplantation in patients with multiple sclerosis and amyotrophic lateral sclerosis. Arch Neurol 2010;67(10):1187–94.
[168] Callera F, de Melo CM. Magnetic resonance tracking of magnetically labeled autologous bone marrow CD34+ cells transplanted into the spinal cord via lumbar puncture technique in patients with chronic spinal cord injury: CD34+ cells' migration into the injured site. Stem Cells Dev 2007;16(3):461–6.
[169] Toso C, Vallee JP, Morel P, Ris F, Demuylder-Mischler S, Lepetit-Coiffe M, et al. Clinical magnetic resonance imaging of pancreatic islet grafts after iron nanoparticle labeling. Am J Transplant 2008;8(3):701–6.
[170] Geng K, Yang ZX, Huang D, Yi M, Jia Y, Yan G, et al. Tracking of mesenchymal stem cells labeled with gadolinium diethylenetriamine pentaacetic acid by 7T magnetic resonance imaging in a model of cerebral ischemia. Mol Med Rep 2015;11(2):954–60.
[171] Tachibana Y, Enmi J, Agudelo CA, Iida H, Yamaoka T. Long-term/bioinert labeling of rat mesenchymal stem cells with PVA-Gd conjugates and MRI monitoring of the labeled cell survival after intramuscular transplantation. Bioconjug Chem 2014;25(7):1243–51.
[172] Ris F, Lepetit-Coiffe M, Meda P, Crowe LA, Toso C, Armanet M, et al. Assessment of human islet labeling with clinical grade iron nanoparticles prior to transplantation for graft monitoring by MRI. Cell Transplant 2010;19(12):1573–85.
[173] Tang H, Sha H, Sun H, Wu X, Xie L, Wang P, et al. Tracking induced pluripotent stem cells-derived neural stem cells in the central nervous system of rats and monkeys. Cell Reprogram 2013;15(5):435–42.
[174] Blocki A, Beyer S, Dewavrin JY, Goralczyk A, Wang Y, Peh P, et al. Microcapsules engineered to support mesenchymal stem cell (MSC) survival and proliferation enable long-term retention of MSCs in infarcted myocardium. Biomaterials 2015;53:12–24.
[175] Bible E, Dell'Acqua F, Solanky B, Balducci A, Crapo PM, Badylak SF, et al. Non-invasive imaging of transplanted human neural stem cells and ECM scaffold remodeling in the stroke-damaged rat brain by (19)F- and diffusion-MRI. Biomaterials 2012;33(10):2858–71.
[176] Gaudet JM, Ribot EJ, Chen Y, Gilbert KM, Foster PJ. Tracking the fate of stem cell implants with fluorine-19 MRI. PLoS One 2015;10(3):pe0118544.

Chapter 4

Allogeneic Versus Autologous Mesenchymal Stromal Cells and Donor-to-Donor Variability

M. Qayed*,**, I. Copland*,**,†,1 and J. Galipeau*,**,†

*Department of Pediatrics, Emory University, Atlanta, Georgia, United States; **Aflac Cancer and Blood Disorders Center, Children's Healthcare of Atlanta, Atlanta, Georgia, United States; †Department of Hematology and Oncology, Winship Cancer Institute, Emory University, Atlanta, Georgia, United States

CHAPTER OUTLINE

4.1 INTRODUCTION

In this chapter, we will discuss the in vitro studies investigating the immunogenicity of mesenchymal stromal cells (MSCs), and how that may impact the choice of donor (autologous, HLA-matched, haploidentical, or third

[1]Posthumous

Mesenchymal Stromal Cells. http://dx.doi.org/10.1016/B978-0-12-802826-1.00004-0

party—derived from neither recipient nor donor) in hematopoietic stem cell transplantation (HSCT) and solid organ transplantation (SOT). We will review the influence of donors and disease on the quality and characteristics of expanded MSCs. MSCs from different tissue sources share many characteristics and fulfill the common definition of MSCs [1]. However, they have differing gene expression profile and varying levels of cytokine production, and their ability to differentiate into different tissues varies according to their tissue of origin [2]. Since bone marrow (BM)-derived MSCs are the best characterized population in a clinical setting, the discussion in this chapter will be focused on these cells.

4.2 IMMUNOGENICITY OF MSCs

MSCs having been considered immunoprivileged, mainly due to their default low-level expression of major histocompatibility complex (MHC) class II and lack of expression of costimulatory molecules CD40, CD80, and CD86. However, they have intermediate level expression of MHC class I [3–6]. In addition, in vitro assays consistently show that MSCs can suppress T-cell proliferation independent of MHC matching [7,8]. These considerations were essential for the potential to develop allogeneic MSCs as an "off the shelf" cellular therapy for immune-mediated disorders, transplantation medicine, and tissue engineering. The consequences of MSC immunogenic properties are twofold; allogeneic MSCs may be rejected, thus not persisting to exert their therapeutic properties, or they could elicit an immune response that may be detrimental to the host, and precipitate or increase the risk of graft rejection.

The evidence supporting the immunoprivileged status of MSC is controversial. While in vitro studies have demonstrated that MSCs have a low immunogenic profile [3,9–11], this has not been consistently reflected in vivo. Initial observations indicating an immune response were described by Djouad et al. [9]. They showed that allogeneic MSCs could engraft and result in new bone formation in immunocompetent mice. However, lymphocytic infiltrates were seen lining and invading the newly formed tissue. Since then, several groups have been able to reproduce in vivo experiments supporting the immunogenicity of allogeneic MSCs (Table 4.1) [12].

The addition of human MSCs expanded from BM in mixed lymphocyte cultures (MLC) suppresses T-cell proliferation regardless of the donor source [8]. However, in that same set of experiments by Le Blanc et al., the addition of a lower number of MSCs (10–1,000 cells) led to marked lymphocyte proliferation ranging from 40% to 190% of that observed in controls. This lymphocyte proliferation occurred with autologous and allogeneic MSCs, suggesting that it was independent of MHC matching.

Table 4.1 Summary of Animal Studies Examining Antidonor Immune Response to Allogeneic MSC

Author (Reference)	Disease Model	Inflammatory Response	Donor-Specific T-Cell Response	Antidonor Antibody Response
Djouad et al. [9]	Healthy mice	Lymphocytic infiltrates lining and invading newly formed bone tissue	–	–
Prigozhina et al. [19]	Healthy mice	Lymphocytic infiltration at the site of implantation of allo[a]-MSC[b]	–	–
Camp et al.[c]	Rat-brain injury	Infiltration of immune-reactive cells at site of injection. Increased $CD4^+$ and $CD8^+$ T cells at site of injection	–	–
Chen et al.[d]	Mouse-wound healing	No inflammatory response, reduced neutrophils and T cells at wound site	–	–
Eliopoulos et al. [19]	Healthy mouse	Increased proportion of host-derived lymphoid $CD8^+$, natural killer T, and NK infiltrating cells at site of allo-MSC implants	Increased antidonor T-cell response and accelerated rejection of repeat allo-MSC	–
Rafei et al.[e]	Mouse—autoimmune encephalitis	–	Pretreatment of MSC with IFN-γ[f] let to upregulation of MHC[g] I and MHC II and loss of MSC effect, consistent with immune rejection	–
Seifert et al. [31]	Rat—kidney transplant	Intragraft B cell infiltration and complement factor C4d deposits.	Antidonor T-cell response enhanced by donor-specific MSC pretreatment	Antidonor antibody enhanced by donor-specific MSC pretreatment
Isakova et al. [23,24]	Healthy Rhesus Macaques	Increased neutrophil and T-cell numbers, expansion of $CD8^+$, $CD16^+$, and $CD8^+/CD16^+$ lymphocytes	–	Detected, reexposure reveals allo-antigen recognition by B cells
Zangi et al. [7]	Mouse BMT[h]	–	Increased antidonor $CD4^+$ and $CD8^+$ T cells. Accelerated rejection of repeat allo-MSC	–
Poncelet et al. [21]	Pig—myocardial ischemia	–	Increased antidonor T cell response	Antidonor IgM and IgG detected and result in complement-mediated lysis. Antibody response increased with MSC rechallenge

(Continued)

Table 4.1 Summary of Animal Studies Examining Antidonor Immune Response to Allogeneic MSC (*cont.*)

Author (Reference)	Disease Model	Inflammatory Response	Donor-Specific T-Cell Response	Antidonor Antibody Response
Badillo et al. [20]	Healthy mouse	–	Increased $CD4^+$ T cell response	Detected at high titers
Nauta et al. [17]	Mouse BMT	–	Antidonor memory T cell response, and increased rejection of allogeneic donor bone marrow cells.	–
Beggs et al. [22]	Healthy baboon	–	Decreased host T cell responses to allo-antigen in majority of recipients	Antibodies reacting with donor PBMC[i] in all, alloantibodies reacting with MSC in some recipients
Schu et al.[j]	Healthy rat	–	–	Detected, and capable of complement- mediated lysis
Cho et al.[k]	Healthy pig	–	Detected after multiple exposures or allo-MSC pretreated with IFN-γ	Detected after multiple injections or allo-MSC pretreated with IFN-γ
Huang at al.[l]	Rat myocardial infarction	$CD4^+$ and $CD8^+$ T cell infiltrate at the site of differentiated allo MSC injection	Differentiated MSC induced T-cell proliferation in vitro	IgG specific to differentiated MSC detected

[a] *Allo, allogeneic.*
[b] *MSC, mesenchymal stromal cells.*
[c] *Camp DM, Loeffler DA, Farrah DM, Borneman JN, LeWitt PA. Cellular immune response to intrastriatally implanted allogeneic bone marrow stromal cells in a rat model of Parkinson's disease. J Neuroinflammation 2009;6:17.*
[d] *Chen L, Tredget EE, Liu C, Wu Y. Analysis of allogenicity of mesenchymal stem cells in engraftment and wound healing in mice. PLoS One 2009;4(9):e7119.*
[e] *Rafei M, Birman E, Forner K, Galipeau J. Allogeneic mesenchymal stem cells for treatment of experimental autoimmune encephalomyelitis. Mol Ther 2009;17(10):1799–803.*
[f] *IFN-γ, interferon gamma.*
[g] *MHC, major histocompatibility complex.*
[h] *BMT, bone marrow transplantation.*
[i] *PBMC, peripheral blood mononuclear cells.*
[j] *Schu S, Nosov M, O'Flynn L, et al. Immunogenicity of allogeneic mesenchymal stem cells. J Cell Mol Med 2012;16(9):2094–103.*
[k] *Cho PS, Messina DJ, Hirsh EL, et al. Immunogenicity of umbilical cord tissue derived cells. Blood 2008;111(1):430–8.*
[l] *Huang XP, Sun Z, Miyagi Y, et al. Differentiation of allogeneic mesenchymal stem cells induces immunogenicity and limits their long-term benefits for myocardial repair. Circulation 2010;122(23):2419–29.*

Similarly, in a set of experiments using a BM transplant murine model of Balb/c recipient and C3H donor mice, Mukonoweshuro et al. showed that undifferentiated allogeneic MSCs were immunogenic. In lymphocyte transformation assay (LTA) incorporating recipient mononuclear cells, cocultured 1:1 with donor (allogeneic) MSCs, resulted in significantly higher lymphocyte proliferation compared to recipient (syngeneic) MSCs [13]. The ratio of MSCs to lymphocytes used in these experiments was 1:1, which is lower than in other published work, and may explain the findings, as MSC-mediated inhibition of lymphocyte proliferation is dose-dependent, with testing started at a 2:1 ratio in some experiments, increasing as high as 100:1 [14,15].

In addition, in vivo experiments suggest that MSCs can evoke an immune response, and a memory T-cell response [16]. Nauta et al. examined the immunogenicity of MSCs in vivo in a murine allogeneic BMT model and showed that expanded MSCs induced T-cell memory in allogeneic mice that caused rejection of lymphocytes sharing transplantation antigens with MSCs. The infusion of host MSCs into recipients promoted engraftment of allogeneic BM cells and was associated with tolerance to donor and recipient antigens, while cotransplantation of allogeneic donor MSCs resulted in decreased engraftment, and a strong proliferative response of host splenocytes against donor cells. In addition, a memory response was triggered as suggested by decreased survival of allogeneic splenocytes in immunocompetent mice after injection of allogeneic MSCs [17]. Similarly, MSCs expanded from B6 mice BM and mixed with an osteoconductive matrix, were transplanted under the kidney capsule in syngeneic B6 mice as controls, as well as in allogeneic BALB/c recipients [18]. Evaluation after 3 months revealed ectopic osteohematopoietic complex formation in syngeneic transplants only, and lymphocyte infiltration was seen at the site of implantation in allogeneic mice, showing that allogeneic MSCs did not survive, suggesting rejection. Similarly, subcutaneous implantation of murine MSCs derived from C57Bl/6 mice and engineered to release erythropoietin in MHC mismatched Balb/c immunocompetent recipients resulted in a transient rise followed by a drop in hematocrit back to baseline value, while implantation in syngeneic mice produced a rapid rise in hematocrit that was maintained for over 200 days [19]. In this study allogeneic mice became refractory to repeat implantation, with a second implantation resulting in a lower and less sustained rise in hematocrit, and a third implantation leading to no rise in hematocrit consistent with alloimmunization. Immune responses to allogeneic MSCs were also demonstrated by increased proportion of host $CD8^+$ lymphoid and Natural Killer T cells in the allogeneic compared to the syngeneic implants. Splenocytes obtained from Balb/c recipient mice displayed a significant response with an increase in interferon-gamma (INF-γ) levels when exposed to C57Bl/6 MSC in vitro.

In a set of in vivo experiments, luciferase-labeled mouse MSCs were used and survival was assessed by imaging. MSCs had a significantly longer survival in syngeneic or immune-deficient NOD-SCID recipients compared to immunocompetent Balb/c mice. Intraperitoneal injection of MSCs resulted in cell survival for longer than 120 days in immune-deficient mice, compared to less than 40 days in immunocompetent mice. Following intravenous infusion, these cells lasted more than 40 days in syngeneic mice, whereas they were reduced at 20 days and eliminated by 40 days in allogeneic recipients. Balb/c mice that had previously rejected MSCs were reinfused with fibroblasts and these were rejected more rapidly than naive mice. The median time for rejection in these mice was 2 days, compared to 14 days in naive mice. Using T-cell receptor transgenic mice, the investigators were able to demonstrate the expansion of a TCR transgenic population with a memory phenotype ($CD4^+$ or $CD8^+$ with $CD122^+$, $CD44^+$ and $CD62L^{low}$) 40 days after MSC reinfusion [7].

In another set of murine experiments, Balb/c or B6 MSCs were injected intraperitoneally into B6 recipients. In vitro, B6 responder splenocyte proliferation was inhibited in MLC, this was observed for $CD4^+$ AND $CD8^+$ T-cells in response to Balb/c splenocytes and polyclonal mitogen stimulation. This suppression was similar when using B6 MSC and Balb/c MSCs, reproducing the observation that MSC inhibition in vitro is independent of MHC. Following intraperitoneal injection, analysis of T-cell proliferation showed an accelerated CD4 response to alloantigen compared to syngeneic recipients. Allogeneic recipients developed IgM and IgG antibodies with a high titer alloantibody response. When the serum from allogeneic MSC primed mice was incubated with third-party MSC, the IgG titers were lower, suggesting that the antibody response was mounted against specific MSC targets. To assess induction of tolerance, mice received allogeneic Balb/c skin grafts at 6 weeks, following MSC injection at 0 and 4 weeks. Injection of donor (Balb/c) or host (B6) matched MSCs failed to induce tolerance and all mice experience graft rejection [20]. While this data is derived from mouse experiments, using murine MSCs transferred to immune-competent species-matched recipients, the induction of an immune response demonstrated in multiple settings, and in different strains should raise caution. In particular, the assumptions related to MSC immune privilege when administered repeatedly in MHC mismatched immune replete recipients may be erroneous. Despite the inherent differences between mouse MSCs and human MSC cell physiology, the allogeneic immune response may be a barrier to the persistent effectiveness of a universal donor MSC source.

In addition to inducing a memory T-cell response, MSCs may be subject to clearance by the immune system through other mechanisms. In a pig experiment, allogeneic MSCs were introduced by intracardiac or subcutaneous transplantation. Antibody response was compared to peripheral blood mononuclear cell (PBMC) injection. In the subcutaneous group, injection of MSCs was followed by an IgM antidonor specific antibody response, though weaker than that following PBMC injection, and antidonor IgG was also detectable, though for a shorter duration of time. Although this humoral response did not result in a strong cytolytic activity as measured by flow cytometry and complement-dependent cytotoxicity assay (FCCA), upon rechallenge with subcutaneous injection, they elicited a cytotoxic antibody-mediated response. This was consistent with injections into an ischemic myocardium that elicited a strong IgM and IgG response and triggered a strong cytolytic activity [21]. This exposure was associated with increased donor-specific responses in MLC. This response was observed again upon rechallenge, with an earlier onset. It may be that the inflammatory environment in the ischemic myocardium, upregulated MSC expression of MHC class II, rendering the MSCs more immunogenic, and explaining the differences in the strength of the humoral immune response observed between intracardiac and subcutaneous injection routes. Similar observations with induction of antibodies against donor lymphocytes and MSCs have been seen in baboon experiments [22].

Intracranial injection of allogeneic MSCs into immunocompetent rhesus macaques resulted in significant increase in circulating white blood cells and the detection of MSC-specific alloantibodies. Expansion of $CD8^+$ T lymphocytes was correlated with the dose of MSCs, and degree of mismatch between donor and recipient [23]. Injection of allogeneic MSCs resulted in an increase in circulating neutrophil, lymphocyte, and eosinophil count as well as natural killer (NK) cell subsets, which was not observed in the autologous group. The alloimmune response was influenced by the haplotypes of the donor and the recipient, consistent with what is known for the immunogenicity of these haplotypes. Immunoglobulin levels were significantly higher 10 days postinjection in the allogeneic group when the recipients were rechallenged with subcutaneous donor MSCs injections. These antibodies were shown to be MSC-specific by flow cytometry of donor MSCs that were incubated with sera of transplant recipients and stained with rhesus-specific FITC-conjugated antibody against the immunoglobulin light chain [24].

MSCs express ligands that can activate NK cell receptors. This along with low expression of HLA class I may render them susceptible to NK-cell-mediated lysis. In vitro experiments have shown that while resting NK cells had no effect on MSC, the addition of NK cells cultures with IL-2 results in strong cytolytic activity. This result is seen with autologous and allogeneic MSCs. However, IFN-γ-treated autologous MSCs were less susceptible to lysis, and blocking of HLA class I by monoclonal antibodies resulted in an increase in cytotoxicity. This suggests that upregulation of HLA class I may protect autologous MSC from NK cell-mediated lysis [16,25]. It is not clear whether this explains why NK cells do not kill naturally occurring MSCs in vivo. Preincubation of MSC or NK cells with tacrolimus or rapamycin, drugs commonly used in the transplant setting, had no effect on MSC lysis [26].

Though resting MSCs do not express MHC class II, it can be demonstrated intracellularly by western blot, and undifferentiated MSCs can upregulate expression of MHC class II upon stimulation with IFN-γ [27]. Paradoxically, while IFN-γ is associated with MSC-mediated immunosuppression, it can stimulate MSCs to become antigen-presenting cells, able to present soluble exogenous peptides on MHC Class II molecules and activate antigen-specific $CD4^+$ T-cells [28,29]. In addition, MSCs have been shown to present extracellular antigen through MHC class I, and induce the proliferation of antigen-specific $CD8^+$ T-cells [30].

The concerns regarding MSC immunogenicity extend beyond their rejection and clearance by the immune system. The potential for allogeneic MSCs to trigger specific immune responses against MSC-associated allo-antigens may negatively impact their potential role in SOT and HSCT. It has been shown in multiple animal models (rat, mouse, pig) that the immune response mounted against allogeneic MSCs can result in accelerated rejection in kidney, skin, and BM transplant models [17,20,21,31]. The MSC donor source may not be critical in severely immunocompromised recipients, possibly due to the host inability to mount an immune response. Preclinical experiments in NOD/SCIB mice, mice undergoing total body irradiation, and preimmune fetal sheep, reveal similar effects in enhancing engraftment for autologous and allogeneic MSCs [32–39].

In vitro studies show that autologous and allogeneic MSCs are equally efficacious at suppressing T-cell proliferation. However, one study characterizing MSCs isolated from patients with Crohn's disease, MSCs derived from patients with Crohn's disease inhibited autologous PBMC (isolated from the same MSC donor) at the lowest concentration of 1:1,000 MSC/PBMC, and to a greater extent than healthy donor MSCs [40]. This effect was not observed in assessing autologous healthy donor MSC and healthy donor PBMC, so this effect may be unique to the underlying condition, and requires further evaluation.

4.3 CLINICAL CONSIDERATIONS FOR THE USE OF MSCs IN TRANSPLANTATION

MSCs have been considered therapy in conditions requiring immune suppression, and for tissue regeneration. The majority of trials use allogeneic MSCs, whether donor or third party, reflect a persistence of the scientific community's belief that MSCs are immune privileged. As more information on the immunogenicity of MSCs in vivo emerges, special considerations should be given to the MSC source depending on the clinical indication and recipient immune status. The interpretation of published results using these variable approaches is further complicated by the differences in MSC isolation and culture expansion conditions, which is likely to influence MSC behavior. For example, the use of fetal bovine serum (FBS) for MSC expansion may render the cells more immunogenic [19], and the number of passages may reduce their immune-suppressive qualities [41]. In addition, in most trials, MSCs are infused shortly after thawing, which markedly alters their biology [42,43]. We will review the clinical experience using MSC cotransplantation in allogeneic HSCT to address clinical challenges in BM graft rejection, as well as their use in SOT. MSC immunogenicity in these clinical conditions may not simply mitigate their immune-suppressive properties and result in lack of efficacy, but the risk of allosensitization may increase the risk of graft rejection.

4.3.1 MSC Cotransplantation in Allogeneic Hematopoietic Stem Cell Transplantation

The initial interest in the use of MSC in HSCT was for the treatment of graft versus host disease (GVHD). Le Blanc et al. published the first clinical use of third-party haploidentical MSCs for the treatment of a patient with steroid refractory GVHD [10], and this was followed by many others. Trials have also been conducted for the use of MSC for GVHD prophylaxis. Finally, MSCs are investigated in HSCT to facilitate engraftment of hematopoietic stem cells and to promote healing of regimen-related toxicities. The question of the timing of MSC infusion is under investigation, and mainly depends on which of the above complications is being targeted by this therapy. Completed and ongoing trials have used a variety of MSC sources; matched allogeneic, haploidentical, and third-party donor MSCs have all been reported.

4.3.1.1 Clinical trials of MSCs to promote engraftment

Infusion of MSCs around the time of HSCT may serve a dual role of promoting engraftment and preventing GVHD. A number of phase I–II trials have been completed and are summarized in Table 4.2 [44–55]. Most of these are single-arm trials with limited numbers of patients, though a few are randomized. There are wide variations in MSC donor source and HSCT-related variables (disease indication, HSCT graft, and intensity of preparative regimen). In a clinical trial of HLA-matched sibling donor HSCT using peripheral blood stem cells (PBSC) or BM grafts in

Table 4.2 Clinical Studies of Cotransplantation of MSC and Allogeneic HSCT in Humans

Author (Reference)	Patients Number/Age	Disease/ Conditioning	Graft	MSC Source	Outcomes
Lazarus et al. [55]	46/adults	Malignancy/MA[a]	BM[b]/PBSC[c] MSD[d]	Donor	Graft failure 0 AGVHD[e] grade II–IV 28% CGVHD[f] extensive 22%
MacMillan et al. [47]	8/pediatric	Malignancy/MA	Unrelated CBT[g]	Haploidentical parent	Graft failure 0 AGVHD grade II–IV 3 CGVHD extensive 0
Ball et al. [44]	14/pediatric Historical controls	Malignant and nonmalignant conditions/MA and NMA	HLA[h] MM[i] PBSC-CD34+ selected	Donor	Graft failure 0 (7/47 historical control, $p = 0.14$) AGVHD grade I-II 2 CGVHD limited 1
Bernardo et al. [46]	13/pediatric Historical controls	Malignancy/MA	Unrelated CBT	Haploidentical parent	Graft failure 2/13–15% (3% control, $p = \text{NS}$) AGVHD grade II 8% (13% in control), grade III–IV 0% (26% in control, $p = 0.05$) CGVHD extensive 0% (11% in control, $p = \text{NS}$[j])
Ning et al. [48]	25/adults 10 received MSC Randomized	Malignancy/MA	BM +/– PBSC MSD	Donor	Graft failure 0 AGVHD grade II 1 (8/15 in control) CGVHD 1 (4/15 in control) Relapse 6/10 (compared with 3/15 in controls, $p = .02$)
Baron et al. [45]	20/adults Historical controls	Malignancy/NMA	HLA MM PBSC	Third party	Graft failure 1 (0/16 in control) GVHD grade II–IV 35% CGVHD moderate-severe 65%
Zhang et al. [49]	12/adults	Malignant and nonmalignant conditions/MA and NMA (for patients with SAA)	PBSC	Matched sibling	Graft failure 0 AGVHD grade II–IV 2 CGVHD extensive 2
Kharbanda et al. [51]	4/pediatric	Nonmalignant/ NMA	MM BM (2) CBT (2)	Haploidentical (1) Third party (3)	Graft failure 4 (2 primary, 2 secondary)
Le Blanc et al. [52]	7/pediatric and adult	Malignant and nonmalignant conditions/ MA and NMA (3 patients with primary graft failure)	BM/PBSC MSD (3), MUD (2) MMUD (1) CBT	Matched sibling (3), Haploidentical (4)	Graft failure 0 AGVHD grade II 2 CGVHD 1

Table 4.2 Clinical Studies of Cotransplantation of MSC and Allogeneic HSCT in Humans (*cont.*)

Author (Reference)	Patients Number/Age	Disease/ Conditioning	Graft	MSC Source	Outcomes
Gonzalo-Daganzo et al. [50]	9/adult	Malignancy/MA	CBT	Haploidentical (7), third-party donor (2)	Graft failure 0 AGVHD grade II 4 CGVHD limited 1
Wu et al. [54]	20/pediatric 8 received MSC Randomized	Malignancy/MA	CBT	Third-party umbilical cord	Graft failure 0 AGVHD grade II 1 CGVHD limited 1
Li et al. [53]	17/pediatric and adult	SAA[k]	BM/PBSC Haploidentical	Third-party umbilical cord	Graft failure 1 AGVHD grade III-IV 23.5% CGVHD moderate-severe 14.2%

[a] *MA, myeloablative.*
[b] *BM, bone marrow.*
[c] *PBSC, peripheral blood stem cell.*
[d] *MSD, matched sibling donor.*
[e] *AGVHD, acute graft versus host disease.*
[f] *CGVHD, chronic graft versus host disease.*
[g] *CBT, cord blood transplantation.*
[h] *HLA, human leukocyte antigen.*
[i] *MM, mismatched.*
[j] *NS, not significant.*
[k] *SAA, severe aplastic anemia.*

adult patients with hematologic malignancies, 43 patients underwent coinfusion with MSC expanded from BM of their HLA-matched siblings. All patients received myeloablative conditioning regimens. MSCs were cryopreserved and thawed at the bedside prior to infusion. Thirteen patients (28%) developed grade II–IV acute GVHD, 22 (61%) developed chronic GVHD, and 11 (24%) had disease relapse. There was no graft rejection in this group. The authors concluded that the coinfusion of donor-derived MSCs was safe, but that the optimal dose and schedule for infusion required further study [55]. In a nonmyeloablative approach, 20 adult patients received third-party unrelated donor MSC within 2 h prior to graft HLA-mismatched unrelated PBSC infusion. MSCs were expanded in FBS and infused shortly post thaw. One patient had graft rejection, and the incidence of grade II–IV acute GVHD and moderate–severe chronic GVHD was 35% and 65% respectively. Outcomes were compared to a historical control group, with a significantly decreased 1-year nonrelapse mortality in the MSC group (10% vs 37%, $p = 0.02$). The data did not suggest that MSC infusion abrogated the graft vs leukemia effect (GVL), as the relapse rates in the two groups were similar [45].

Examining the use of MSCs in haploidentical transplantation, where the risk of graft rejection is augmented, Ball et al. infused donor MSCs in 14 children. All patients received $CD34^+$ selected BM or PBSC grafts, and all achieved hematopoietic engraftment. This was lower than the historical rate of graft failure in historic controls (15%) [44]. These trials, among others, have established the safety of using MSC in the early post-HSCT setting, with a signal of efficacy, particularly in haploidentical transplant, that warrants investigation in larger randomized trials.

4.3.1.2 Considerations for MSC donor source in Hematopoietic Stem Cell Transplantation

The formation of antidonor immune responses following the administration of allogeneic MSCs has been investigated in a limited number of studies. In one retrospective analysis of patients receiving matched-sibling donor, haploidentical, or third-party unrelated MSCs to promote engraftment or for treatment of hemorrhagic cystitis and GVHD, some patients developed anti-FBS antibodies [56]. This was also reported in a patient receiving MSCs for treatment of osteogenesis imperfecta [57]. This is consistent with the data that MSCs internalize and have FBS components present on the cell surface, and can be circumvented by the use of alternative culture media [58]. However, in another study, infusion of third-party MSCs did not result in HLA alloimmunization in patients treated for complications of HSCT [59]. Sundin et al. investigated the immune reaction of patients undergoing HSCT to third-party MSCs. Eighteen patients were included. Patient lymphocytes collected prior to MSC infusion responded to lymphocytes from the MSC donor or other healthy volunteer, but did not respond to MSCs from the same donor. Postinfusion proliferation assays repeated 1 week to 6 months postinfusion did not show an alloresponse against donated MSCs, but continued to show a response to lymphocytes, indicating that MSC donor tolerance was not induced. In vitro stimulation of recipient lymphocytes with MSCs resulted in a weak response at high MSC:lymphocyte ratio [56].

In SOT, it has been suggested that donor-derived MSCs are superior to recipient or unrelated third-party MSCs for inducing long-term tolerance [60]. In contrast to SOT, in allogeneic HSCT, GVHD is mediated by donor lymphocytes reacting against disparate recipient antigens, thus recipient, autologous, MSCs may be superior at inducing tolerance compared to allogeneic MSC. This concept is supported by a murine HSCT model, where the use of autologous MSCs significantly enhanced engraftment, and that was associated with tolerance to host and donor antigens [17]. The infusion of allogeneic donor MSCs may enhance the response of the host against donor antigens, resulting in increased risk of rejection. It is possible that third-party MSCs may regulate GVHD indirectly, by altering the cytokine milieu, as clinical responses are seen in clinical trials. Whether they are not immune-rejected due to the immune-suppression of the host, has not been assessed in clinical trials. It is also possible that allo-MSCs work to redirect the immune response of the donor T-cells toward the clearance of the infused MSCs, but this has not been assessed in patients.

There are only two reported trials using autologous MSCs in a patient population that shares some characteristics of patients undergoing allogeneic HSCT. In a phase I trial, the feasibility of collecting and expanding MSC was studied in patients in hematologic malignancies in complete remission. Twenty-three adult patients were enrolled, 12 of whom had undergone a previous autologous or syngeneic HSCT. Fifteen patients completed the expansion and underwent infusion without adverse events, with a median time of 28–49 days after collection [61]. The feasibility of infusing autologous MSCs in combination with autologous PBSCs was also investigated in 32 patients with breast cancer receiving autologous HSCT. Twenty-eight patients received MSC infusion. The median time to neutrophil and platelet recovery appeared to be shortened (median 8, and 8.5 days respectively). There were no adverse events related to MSC infusion [62].

The feasibility of expanding MSCs from patients undergoing HSCT has been investigated in children and adults with acute and chronic GVHD, and their properties compared to HSCT recipients without chronic GVHD, as well as to healthy donors. The majority of patients underwent HSCT for an underlying hematologic malignancy. There was no difference in morphology, immunophenotype, differentiation, and migratory potential. In one study, MSCs exhibited an initial lag in growth that resolved with continuous expansion. In vitro immunomodulatory properties, as exhibited by suppression of T-cell proliferation, upregulation of indoleamine 2,3-dioxygenase expression, and induction of regulatory T-cells, were similar to healthy donor MSCs [63,64]. In both studies, since MSC expansion was performed post-HSCT, chimerism analysis was performed, and revealed MSCs to remain of recipient origin, consistent with the majority of published reports [65,66]. Results from these studies disagree with previous reports, where MSCs isolated from patients with AML had changes in morphology, impaired differentiation, and did not suppress T-cell proliferation stimulated in a mixed lymphocyte reaction [67,68]. In a study of MSC expanded from patients with hematologic malignancies, with exposure to low-dose therapy, higher dose chemotherapy, or no therapy exposure, MSCs were morphologically similar between the groups, but there was a significant difference in expansion between patients receiving high-dose therapy and patients not receiving therapy [69]. This discrepancy may arise from the differences in the patient populations (receiving chemotherapy for hematologic malignancy, compared with post-HSCT status, where presumably there is no residual malignancy in the patients studied).

The reproduction of similar results of MSC expansion in patients with GVHD by two separate groups of investigators indicates that autologous MSCs from this patient population are a plausible option for use in treatment of GVHD [63,64], and a clinical trial investigating the safety and feasibility of this autologous approach is ongoing (NCT02359929—A Phase I Study of Mesenchymal Stromal Cells for the Treatment of Acute and Chronic Graft Versus Host Disease).

4.3.2 Considerations for MSC Donor Source in Solid Organ Transplantation

A number of studies have shown that syngeneic and allogeneic MSCs have different effects in animal organ transplantation models. MSCs that are of donor origin, or share HLA antigens with the donor, may be better at inducing tolerance, possibly through desensitization to donor antigens [60,70,71]. This is in contrast to HSCT, where the GVHD response is directed against the recipient, hence the argument for potential use of autologous MSCs. In a semiallogeneic heart transplant mouse model of B6C3 donor hearts into B6 recipients, pretransplant intraportal infusion of donor-derived MSCs induced prolonged cardiac allograft survival compared to untreated mice, which rejected the allograft within 13 days. This tolerance was associated with expansion of regulatory T-cells (Tregs) and the investigators also showed that Tregs generated were donor specific. Splenocytes obtained from B6 mice 7 days after MSC infusion, and >100 days in another set of experiments, were given to B6 naive mice 1 day before undergoing B6C3 heart transplantation; these mice had a significantly prolonged allograft survival compared to controls, with indefinite survival in 60% of recipients. Furthermore, mice receiving splenocytes had improved survival of second allografts from donor-specific C3 mice, but all rejected third-party (Balb/c) allografts [72]. The use of recipient-derived MSCs was associated with a weaker tolerogenic effect than donor MSCs, as transfer of a similar number of splenocytes had no effect on allograft survival, but that was overcome by increasing the number of infused splenocytes. The percentage of Tregs in the spleen and the expression of Foxp3 in the allografts were lower compared to levels in B6 mice receiving B6C3 hearts with donor MSC infusions.

4.4 MSC DONOR-TO-DONOR VARIABILITY

Autologous BM-derived MSCs have been used in clinical trials for cardiovascular, neurological, and autoimmune disorders, and represent a promising source of stem cells for regenerative medicine. Within an MSC population, there is heterogeneity with regards to proliferation, morphology, differentiation potential, and surface marker expression [2,73–76]. A variety of factors may affect the functional capabilities of autologous MSCs, and should be considered as these cells are evaluated for therapeutic indications.

4.4.1 Donor Age and Sex

The number of MSCs obtained by BM aspiration declines with age [77,78]. In one study, the isolation and expansion of MSCs by BM aspirate or from the iliac crest had a higher failure rate for donors older than 60 years of age [79]. Males and females had a similar failure rate (27.3 and 28.6% respectively). In a study of 53 donors (28 males and 25 females) with ages ranging from 13 to 80 years, phenotypical differences were noted; MSCs isolated from younger donors were smaller and had a higher proliferative rate [80]. In that study, mesodermal differentiation was not affected by donor age or sex. Migratory potential was found to be decreased in MSCs from older donors [77,81].

In vitro studies show a higher number of proliferative precursor cells in younger donors, and an age-related decrease in the fibroblast colony forming units and their differentiation potential to endothelial cells, osteoblasts, and adipocytes [77]. The gene expression profile of MSCs also varies with age, with MSCs from older donors showing decrease in Wnt-related transcripts (important in osteogenesis) and decreased expression of leptin receptor (leptin plays a role in controlling the balance between bone and fat in the BM) [82]. MSCs from older donors also exhibit shorter telomeres [83,84]. They have increased indices of oxidative damage and senescence [78].

Experiments in mouse models show lower numbers of isolated MSCs in females compared to males [85,86]. Functionally, female MSCs may offer an advantage, as estradiol has been shown to improve MSC function [87,88]. In humans, no differences in numbers of isolated MSCs were noted between males and females [77,79,80]; however, in one study, MSC derived from female donors inhibited T-cell proliferation more potently [80].

4.4.2 Underlying Disease

The characteristics of MSCs have been described in multiple diseases. This is an important step before considering the use of autologous MSCs for treatment because underlying disease, as well as treatment exposure, may affect MSC morphology, proliferative and differentiation capability, and immunomodulatory function. In hematologic diseases, abnormalities in MSCs have been described in multiple myeloma [89,90], hematologic malignancies [67,69], myelodysplastic syndrome [91], and aplastic anemia [92]. The characteristics of MSC harvested post HSCT are described in

Section 4.3.1.2. In autoimmune disease, abnormalities of MSC proliferative capabilities have been described in rheumatoid arthritis (RA), immune thrombocytopenic purpura (ITP), and systemic lupus erythematosus (SLE) [93–95]. And while MSC isolated in RA and SLE showed normal differentiation and immunosuppressive properties, in ITP, immunosuppressive properties were impaired. In contrast, MSCs isolated from patients with Crohn's disease are indistinguishable from healthy donor-derived MSC [40,96].

4.4.3 Inherent Donor Differences

MSCs isolated from different donors exhibit variable proliferative rates, and clonogenicity in vitro. In one study of 17 healthy donors, BM-derived MSC displayed a 12-fold difference in growth rate at first passage, and that did not correlate with age or sex [97]. Sampling bias related to the method of harvest may affect the composition of the MSC population. MSCs harvested simultaneously from two different sites from the same donor can have varying growth rates [97]. MSCs obtained from BM aspirates and cancellous bone chips had similar morphology and genetic characteristics, though culture and expansion was more successful from the BM aspirates (90.4% from 21 BM donors and 62.5% from cancellous bone specimens, $p = 0.02$) [79].

MSC characteristics may change with the number of passages, and that change varies by donor. In one study comparing properties of MSCs obtained from two donors who were similar in age, the percentage of STRO-1^+ MSC declined with increasing passages in one donor, but remained constant in the other. This correlated with the adipogenic differentiation potential of the MSCs [73]. While proliferative capacity was inversely related to passage, the extent decrease was different between the two donors. The size distribution of the MSCs and differentiation potential also varies with passage. In this same study, cells from one donor significantly increased in size, which correlated with the sharp drop in colony-forming units and adipogenic precursor frequency.

4.5 CONCLUSION

There is enthusiasm within the scientific medical community for the use of MSCs as a therapy for a broad range of immune-mediated disorders. However, the numerous published and ongoing trials have wide variations in MSCs used. While there is agreement that suppression of T-cell proliferation appears to be independent of MHC matching between the MSC and T-cells in vitro, multiple studies have now confirmed that allogeneic MSCs are subject to immune rejection, and can trigger donor-specific cellular and humoral immune responses in vivo. The suppressive culture conditions in vitro are associated with high cell and inhibitory cytokine concentration which do not reflect physiologic conditions in vivo, as systemic administration of MSCs results in host immune system exposure to much lower cell concentrations. In vivo, MSCs are lyzed by activated host NK cells, can activate both cellular and humoral immunity, induce a memory T-cell response, and result in formation of donor-specific antibodies. Thus, a possible explanation for the conflicting results of published clinical trials using MSCs is the rejection of allogeneic MSCs by the immune system. The possibility of a donor-specific immune response may adversely affect outcomes in allogeneic SOT and HSCT. Such immune responses may have a detrimental effect for allogeneic MSC therapeutic strategies at large. Notwithstanding, the use of third-party universal donor MSCs is certainly a convenient approach, especially for acute tissue injury syndromes where the tissue-sparing effects of MSCs take place before an anti-MSC immune takes effect. However, in light of the compelling data on allo MSC immunogenicity, the same may not be true for indications for repeated MSC infusion over a period of time where an adaptive immune response is triggered and sustained. Furthermore, inherent differences in MSCs functionality among donors, and the potential for accelerated clearance of MSCs in the setting of alloimmunization will certainly impact their potency in human clinical trials where efficacy is the primary endpoint. A robust appreciation of factors affecting potency of MSCs based on their immunological source, handling, and administration can inform the design of clinical trials optimized for meeting objective metrics of efficacy.

ABBREVIATIONS

BM Bone marrow
FBS Fetal bovine serum
FCCA Complement dependent cytotoxicity assay
GVHD Graft versus host disease
HSCT Hematopoietic stem cell transplantation
INF-γ Interferon-gamma
ITP Immune thrombocytopenic purpura
LTA Lymphocyte transformation assay
MHC Major histocompatibility complex
MLC Mixed lymphocyte cultures
MSCs Mesenchymal stromal cells
PBMC Peripheral blood mononuclear cell
PBSC Peripheral blood stem cell
RA Rheumatoid arthritis
SLE Systemic lupus erythematosus
SOT Solid organ transplantation
Tregs Regulatory T-cells

REFERENCES

[1] Horwitz EM, Le Blanc K, Dominici M, Mueller I, Slaper-Cortenbach I, Marini FC, et al. Clarification of the nomenclature for MSC: The International Society for Cellular Therapy position statement. Cytotherapy 2005;7(5):393–5.

[2] Wagner W, Wein F, Seckinger A, Frankhauser M, Wirkner U, Krause U, et al. Comparative characteristics of mesenchymal stem cells from human bone marrow, adipose tissue, and umbilical cord blood. Exp Hematol 2005;33(11):1402–16.

[3] Tse WT, Pendleton JD, Beyer WM, Egalka MC, Guinan EC. Suppression of allogeneic T-cell proliferation by human marrow stromal cells: implications in transplantation. Transplantation 2003;75(3):389–97.

[4] Pittenger MF, Mackay AM, Beck SC, Jaiswal RK, Douglas R, Mosca JD, et al. Multilineage potential of adult human mesenchymal stem cells. Science 1999;284(5411):143–7.

[5] Le Blanc K, Tammik C, Rosendahl K, Zetterberg E, Ringdén O. HLA expression and immunologic properties of differentiated and undifferentiated mesenchymal stem cells. Exp Hematol 2003;31(10):890–6.

[6] Deans RJ, Moseley AB. Mesenchymal stem cells: biology and potential clinical uses. Exp Hematol 2000;28(8):875–84.

[7] Zangi L, Margalit R, Reich-Zeliger S, Bachar-Lustig E, Beilhack A, Negrin R, et al. Direct imaging of immune rejection and memory induction by allogeneic mesenchymal stromal cells. Stem Cells 2009;27(11):2865–74.

[8] Le Blanc K, Tammik L, Sundberg B, Haynesworth SE, Ringden O. Mesenchymal stem cells inhibit and stimulate mixed lymphocyte cultures and mitogenic responses independently of the major histocompatibility complex. Scandinavian J Immunol 2003;57(1):11–20.

[9] Djouad F, Plence P, Bony C, Tropel P, Apparailly F, Sany J, et al. Immunosuppressive effect of mesenchymal stem cells favors tumor growth in allogeneic animals. Blood 2003;102(10):3837–44.

[10] Le Blanc K, Rasmusson I, Sundberg B, Gotherstrom C, Hassan M, Uzunel M, et al. Treatment of severe acute graft-versus-host disease with third party haploidentical mesenchymal stem cells. Lancet 2004;363(9419):1439–41.

[11] Di Nicola M, Carlo-Stella C, Magni M, Milanesi M, Longoni PD, Matteucci P, et al. Human bone marrow stromal cells suppress T-lymphocyte proliferation induced by cellular or nonspecific mitogenic stimuli. Blood 2002;99(10):3838–43.

[12] Griffin MD, Ryan AE, Alagesan S, Lohan P, Treacy O, Ritter T. Anti-donor immune responses elicited by allogeneic mesenchymal stem cells: what have we learned so far? Immunol Cell Biol 2013;91(1):40–51.

[13] Mukonoweshuro B, Brown CJ, Fisher J, Ingham E. Immunogenicity of undifferentiated and differentiated allogeneic mouse mesenchymal stem cells. J Tissue Eng 2014;5. 2041731414534255.

[14] Sudres M, Norol F, Trenado A, Gregoire S, Charlotte F, Levacher B, et al. Bone marrow mesenchymal stem cells suppress lymphocyte proliferation in vitro but fail to prevent graft-versus-host disease in mice. J Immunol 2006;176(12):7761–7.

[15] Jones S, Horwood N, Cope A, Dazzi F. The antiproliferative effect of mesenchymal stem cells is a fundamental property shared by all stromal cells. J Immunol 2007;179(5):2824–31.

[16] Crop MJ, Korevaar SS, de Kuiper RJN IJ, van Besouw NM, Baan CC, et al. Human mesenchymal stem cells are susceptible to lysis by CD8(+) T cells and NK cells. Cell Transplant 2011;20(10):1547–59.

[17] Nauta AJ, Westerhuis G, Kruisselbrink AB, Lurvink EG, Willemze R, Fibbe WE. Donor-derived mesenchymal stem cells are immunogenic in an allogeneic host and stimulate donor graft rejection in a nonmyeloablative setting. Blood 2006;108(6):2114–20.

[18] Prigozhina TB, Khitrin S, Elkin G, Eizik O, Morecki S, Slavin S. Mesenchymal stromal cells lose their immunosuppressive potential after allotransplantation. Exp Hematol 2008;36(10):1370–6.

[19] Eliopoulos N, Stagg J, Lejeune L, Pommey S, Galipeau J. Allogeneic marrow stromal cells are immune rejected by MHC class I- and class II-mismatched recipient mice. Blood 2005;106(13):4057–65.

[20] Badillo AT, Beggs KJ, Javazon EH, Tebbets JC, Flake AW. Murine bone marrow stromal progenitor cells elicit an in vivo cellular and humoral alloimmune response. Biol Blood Marrow Transplant 2007;13(4):412–22.

[21] Poncelet AJ, Vercruysse J, Saliez A, Gianello P. Although pig allogeneic mesenchymal stem cells are not immunogenic in vitro, intracardiac injection elicits an immune response in vivo. Transplantation 2007;83(6):783–90.

[22] Beggs KJ, Lyubimov A, Borneman JN, Bartholomew A, Moseley A, Dodds R, et al. Immunologic consequences of multiple, high-dose administration of allogeneic mesenchymal stem cells to baboons. Cell Transplant 2006;15(8–9):711–21.

[23] Isakova IA, Dufour J, Lanclos C, Bruhn J, Phinney DG. Cell-dose-dependent increases in circulating levels of immune effector cells in rhesus macaques following intracranial injection of allogeneic MSCs. Exp Hematol 2010;38(10). 957-67 e1.

[24] Isakova IA, Lanclos C, Bruhn J, Kuroda MJ, Baker KC, Krishnappa V, et al. Allo-reactivity of mesenchymal stem cells in rhesus macaques is dose and haplotype dependent and limits durable cell engraftment in vivo. PLoS One 2014;9(1):e87238.

[25] Spaggiari GM, Capobianco A, Becchetti S, Mingari MC, Moretta L. Mesenchymal stem cell-natural killer cell interactions: evidence that activated NK cells are capable of killing MSCs, whereas MSCs can inhibit IL-2-induced NK-cell proliferation. Blood 2006;107(4):1484–90.

[26] Hoogduijn MJ, Roemeling-van Rhijn M, Korevaar SS, Engela AU, Weimar W, Baan CC. Immunological aspects of allogeneic and autologous mesenchymal stem cell therapies. Hum Gene Ther 2011;22(12):1587–91.

[27] Le Blanc K, Tammik C, Rosendahl K, Zetterberg E, Ringden O. HLA expression and immunologic properties of differentiated and undifferentiated mesenchymal stem cells. Exp Hematol 2003;31(10):890–6.

[28] Stagg J, Pommey S, Eliopoulos N, Galipeau J. Interferon-gamma-stimulated marrow stromal cells: a new type of nonhematopoietic antigen-presenting cell. Blood 2006;107(6):2570–7.

[29] Chan JL, Tang KC, Patel AP, Bonilla LM, Pierobon N, Ponzio NM, et al. Antigen-presenting property of mesenchymal stem cells occurs during a narrow window at low levels of interferon-gamma. Blood 2006;107(12):4817–24.

[30] Francois M, Romieu-Mourez R, Stock-Martineau S, Boivin MN, Bramson JL, Galipeau J. Mesenchymal stromal cells cross-present soluble exogenous antigens as part of their antigen-presenting cell properties. Blood 2009;114(13):2632–8.

[31] Seifert M, Stolk M, Polenz D, Volk HD. Detrimental effects of rat mesenchymal stromal cell pre-treatment in a model of acute kidney rejection. Front Immunol 2012;3:202.

[32] Almeida-Porada G, Porada CD, Tran N, Zanjani ED. Cotransplantation of human stromal cell progenitors into preimmune fetal sheep results in early appearance of human donor cells in circulation and boosts cell levels in bone marrow at later time points after transplantation. Blood 2000;95(11):3620–7.

[33] Angelopoulou M, Novelli E, Grove JE, Rinder HM, Civin C, Cheng L, et al. Cotransplantation of human mesenchymal stem cells enhances human myelopoiesis and megakaryocytopoiesis in NOD/SCID mice. Exp Hematol 2003;31(5):413–20.

[34] Bensidhoum M, Chapel A, Francois S, Demarquay C, Mazurier C, Fouillard L, et al. Homing of in vitro expanded Stro-1- or Stro-1+ human mesenchymal stem cells into the NOD/SCID mouse and their role in supporting human CD34 cell engraftment. Blood 2004;103(9):3313–9.

[35] Hiwase SD, Dyson PG, To LB, Lewis ID. Cotransplantation of placental mesenchymal stromal cells enhances single and double cord blood engraftment in nonobese diabetic/severe combined immune deficient mice. Stem Cells 2009;27(9):2293–300.

[36] in't Anker PS, Noort WA, Scherjon SA, Kleijburg-van der Keur C, Kruisselbrink AB, van Bezooijen RL, et al. Mesenchymal stem cells in human second-trimester bone marrow, liver, lung, and spleen exhibit a similar immunophenotype but a heterogeneous multilineage differentiation potential. Haematologica 2003;88(8):845–52.

[37] Kim DW, Chung YJ, Kim TG, Kim YL, Oh IH. Cotransplantation of third-party mesenchymal stromal cells can alleviate single-donor predominance and increase engraftment from double cord transplantation. Blood 2004;103(5):1941–8.

[38] Kuci S, Kuci Z, Kreyenberg H, Deak E, Putsch K, Huenecke S, et al. CD271 antigen defines a subset of multipotent stromal cells with immunosuppressive and lymphohematopoietic engraftment-promoting properties. Haematologica 2010;95(4):651–9.

[39] Noort WA, Kruisselbrink AB, in't Anker PS, Kruger M, van Bezooijen RL, de Paus RA, et al. Mesenchymal stem cells promote engraftment of human umbilical cord blood-derived CD34(+) cells in NOD/SCID mice. Exp Hematol 2002;30(8):870–8.

[40] Bernardo ME, Avanzini MA, Ciccocioppo R, Perotti C, Cometa AM, Moretta A, et al. Phenotypical/functional characterization of in vitro-expanded mesenchymal stromal cells from patients with Crohn's disease. Cytotherapy 2009;11(7):825–36.

[41] von Bahr L, Sundberg B, Lonnies L, Sander B, Karbach H, Hagglund H, et al. Long-term complications, immunologic effects, and role of passage for outcome in mesenchymal stromal cell therapy. Biol Blood Marrow Transplant 2012;18(4):557–64.

[42] Chinnadurai R, Garcia MA, Sakurai Y, Lam WA, Kirk AD, Galipeau J, et al. Actin cytoskeletal disruption following cryopreservation alters the biodistribution of human mesenchymal stromal cells in vivo. Stem Cell Rep 2014;3(1):60–72.

[43] Galipeau J. The mesenchymal stromal cells dilemma—does a negative phase III trial of random donor mesenchymal stromal cells in steroid-resistant graft-versus-host disease represent a death knell or a bump in the road? Cytotherapy 2013;15(1):2–8.

[44] Ball LM, Bernardo ME, Roelofs H, Lankester A, Cometa A, Egeler RM, et al. Cotransplantation of ex vivo expanded mesenchymal stem cells accelerates lymphocyte recovery and may reduce the risk of graft failure in haploidentical hematopoietic stem-cell transplantation. Blood 2007;110(7):2764–7.

[45] Baron F, Lechanteur C, Willems E, Bruck F, Baudoux E, Seidel L, et al. Cotransplantation of mesenchymal stem cells might prevent death from graft-versus-host disease

(GVHD) without abrogating graft-versus-tumor effects after HLA-mismatched allogeneic transplantation following nonmyeloablative conditioning. Biol Blood Marrow Transplant 2010;16(6):838–47.

[46] Bernardo ME, Ball LM, Cometa AM, Roelofs H, Zecca M, Avanzini MA, et al. Co-infusion of ex vivo-expanded, parental MSCs prevents life-threatening acute GVHD, but does not reduce the risk of graft failure in pediatric patients undergoing allogeneic umbilical cord blood transplantation. Bone Marrow Transplant 2011;46(2):200–7.

[47] Macmillan ML, Blazar BR, DeFor TE, Wagner JE. Transplantation of ex-vivo culture-expanded parental haploidentical mesenchymal stem cells to promote engraftment in pediatric recipients of unrelated donor umbilical cord blood: results of a phase I-II clinical trial. Bone Marrow Transplant 2009;43(6):447–54.

[48] Ning H, Yang F, Jiang M, Hu L, Feng K, Zhang J, et al. The correlation between cotransplantation of mesenchymal stem cells and higher recurrence rate in hematologic malignancy patients: outcome of a pilot clinical study. Leukemia 2008;22(3):593–9.

[49] Zhang X, Li JY, Cao K, Lu H, Hong M, Qian S, et al. Cotransplantation of HLA-identical mesenchymal stem cells and hematopoietic stem cells in Chinese patients with hematologic diseases. Int J Lab Hematol 2010;32(2):256–64.

[50] Gonzalo-Daganzo R, Regidor C, Martin-Donaire T, Rico MA, Bautista G, Krsnik I, et al. Results of a pilot study on the use of third-party donor mesenchymal stromal cells in cord blood transplantation in adults. Cytotherapy 2009;11(3):278–88.

[51] Kharbanda S, Smith AR, Hutchinson SK, McKenna DH, Ball JB, Lamb LS Jr, et al. Unrelated donor allogeneic hematopoietic stem cell transplantation for patients with hemoglobinopathies using a reduced-intensity conditioning regimen and third-party mesenchymal stromal cells. Biol Blood Marrow Transplant 2014;20(4):581–6.

[52] Le Blanc K, Samuelsson H, Gustafsson B, Remberger M, Sundberg B, Arvidson J, et al. Transplantation of mesenchymal stem cells to enhance engraftment of hematopoietic stem cells. Leukemia 2007;21(8):1733–8.

[53] Li XH, Gao CJ, Da WM, Cao YB, Wang ZH, Xu LX, et al. Reduced intensity conditioning, combined transplantation of haploidentical hematopoietic stem cells and mesenchymal stem cells in patients with severe aplastic anemia. PLoS One 2014;9(3):e89666.

[54] Wu KH, Tsai C, Wu HP, Sieber M, Peng CT, Chao YH. Human application of ex vivo expanded umbilical cord-derived mesenchymal stem cells: enhance hematopoiesis after cord blood transplantation. Cell Transplant 2013;22(11):2041–51.

[55] Lazarus HM, Koc ON, Devine SM, Curtin P, Maziarz RT, Holland HK, et al. Cotransplantation of HLA-identical sibling culture-expanded mesenchymal stem cells and hematopoietic stem cells in hematologic malignancy patients. Biol Blood Marrow Transplant 2005;11(5):389–98.

[56] Sundin M, Ringden O, Sundberg B, Nava S, Gotherstrom C, Le Blanc K. No alloantibodies against mesenchymal stromal cells, but presence of anti-fetal calf serum antibodies, after transplantation in allogeneic hematopoietic stem cell recipients. Haematologica 2007;92(9):1208–15.

[57] Horwitz EM, Gordon PL, Koo WK, Marx JC, Neel MD, McNall RY, et al. Isolated allogeneic bone marrow-derived mesenchymal cells engraft and stimulate growth in children with osteogenesis imperfecta: Implications for cell therapy of bone. Proc Natl Acad Sci USA 2002;99(13):8932–7.

[58] Spees JL, Gregory CA, Singh H, Tucker HA, Peister A, Lynch PJ, et al. Internalized antigens must be removed to prepare hypoimmunogenic mesenchymal stem cells for cell and gene therapy. Mol Ther 2004;9(5):747–56.
[59] Yin F, Battiwalla M, Ito S, Feng X, Chinian F, Melenhorst JJ, et al. Bone marrow mesenchymal stromal cells to treat tissue damage in allogeneic stem cell transplant recipients: correlation of biological markers with clinical responses. Stem Cells 2014;32(5):1278–88.
[60] Hoogduijn MJ, Popp FC, Grohnert A, Crop MJ, van Rhijn M, Rowshani AT, et al. Advancement of mesenchymal stem cell therapy in solid organ transplantation (MISOT). Transplantation 2010;90(2):124–6.
[61] Lazarus HM, Haynesworth SE, Gerson SL, Rosenthal NS, Caplan AI. Ex vivo expansion and subsequent infusion of human bone marrow-derived stromal progenitor cells (mesenchymal progenitor cells): implications for therapeutic use. Bone Marrow Transplant 1995;16(4):557–64.
[62] Koc ON, Gerson SL, Cooper BW, Dyhouse SM, Haynesworth SE, Caplan AI, et al. Rapid hematopoietic recovery after coinfusion of autologous-blood stem cells and culture-expanded marrow mesenchymal stem cells in advanced breast cancer patients receiving high-dose chemotherapy. J Clin Oncol 2000;18(2):307–16.
[63] Copland IB, Qayed M, Garcia MA, Galipeau J, Waller EK. Bone marrow mesenchymal stromal cells from patients with acute and chronic graft-versus-host disease deploy normal phenotype, differentiation plasticity, and immune-suppressive activity. Biol Blood Marrow Transplant 2015;21(5):934–40.
[64] Wang B, Hu Y, Liu L, Hu K, Tie R, He Y, et al. Phenotypical and functional characterization of bone marrow mesenchymal stem cells in patients with chronic graft-versus-host disease. Biol Blood Marrow Transplant 2015;21(6):1020–8.
[65] Bartsch K, Al-Ali H, Reinhardt A, Franke C, Hudecek M, Kamprad M, et al. Mesenchymal stem cells remain host-derived independent of the source of the stem-cell graft and conditioning regimen used. Transplantation 2009;87(2):217–21.
[66] Rieger K, Marinets O, Fietz T, Korper S, Sommer D, Mucke C, et al. Mesenchymal stem cells remain of host origin even a long time after allogeneic peripheral blood stem cell or bone marrow transplantation. Exp Hematol 2005;33(5):605–11.
[67] Zhao ZG, Liang Y, Li K, Li WM, Li QB, Chen ZC, et al. Phenotypic and functional comparison of mesenchymal stem cells derived from the bone marrow of normal adults and patients with hematologic malignant diseases. Stem Cells Dev 2007;16(4):637–48.
[68] Zhi-Gang Z, Wei-Ming L, Zhi-Chao C, Yong Y, Ping Z. Immunosuppressive properties of mesenchymal stem cells derived from bone marrow of patient with hematological malignant diseases. Leukemia Lymphoma 2008;49(11):2187–95.
[69] Kemp K, Morse R, Wexler S, Cox C, Mallam E, Hows J, et al. Chemotherapy-induced mesenchymal stem cell damage in patients with hematological malignancy. Ann Hematol 2010;89(7):701–13.
[70] Inoue S, Popp FC, Koehl GE, Piso P, Schlitt HJ, Geissler EK, et al. Immunomodulatory effects of mesenchymal stem cells in a rat organ transplant model. Transplantation 2006;81(11):1589–95.
[71] Popp FC, Eggenhofer E, Renner P, Slowik P, Lang SA, Kaspar H, et al. Mesenchymal stem cells can induce long-term acceptance of solid organ allografts in synergy with low-dose mycophenolate. Transplant Immunol 2008;20(1–2):55–60.

[72] Casiraghi F, Azzollini N, Cassis P, Imberti B, Morigi M, Cugini D, et al. Pretransplant infusion of mesenchymal stem cells prolongs the survival of a semiallogeneic heart transplant through the generation of regulatory T cells. J Immunol 2008;181(6):3933–46.

[73] Lo Surdo J, Bauer SR. Quantitative approaches to detect donor and passage differences in adipogenic potential and clonogenicity in human bone marrow-derived mesenchymal stem cells. Tissue Eng Part C Methods 2012;18(11):877–89.

[74] Pevsner-Fischer M, Levin S, Zipori D. The origins of mesenchymal stromal cell heterogeneity. Stem Cell Rev 2011;7(3):560–8.

[75] Russell KC, Phinney DG, Lacey MR, Barrilleaux BL, Meyertholen KE, O'Connor KC. In vitro high-capacity assay to quantify the clonal heterogeneity in trilineage potential of mesenchymal stem cells reveals a complex hierarchy of lineage commitment. Stem Cells 2010;28(4):788–98.

[76] Schafer R, Dominici M, Muller I, Horwitz E, Asahara T, Bulte JW, et al. Basic research and clinical applications of non-hematopoietic stem cells, 4-5 April 2008, Tubingen, Germany. Cytotherapy 2009;11(2):245–55.

[77] Xin Y, Wang YM, Zhang H, Li J, Wang W, Wei YJ, et al. Aging adversely impacts biological properties of human bone marrow-derived mesenchymal stem cells: implications for tissue engineering heart valve construction. Artif Organs 2010;34(3):215–22.

[78] Stolzing A, Jones E, McGonagle D, Scutt A. Age-related changes in human bone marrow-derived mesenchymal stem cells: consequences for cell therapies. Mech Ageing Dev 2008;129(3):163–73.

[79] Bertram H, Mayer H, Schliephake H. Effect of donor characteristics, technique of harvesting and in vitro processing on culturing of human marrow stroma cells for tissue engineered growth of bone. Clin Oral Implants Res 2005;16(5):524–31.

[80] Siegel G, Kluba T, Hermanutz-Klein U, Bieback K, Northoff H, Schafer R. Phenotype, donor age and gender affect function of human bone marrow-derived mesenchymal stromal cells. BMC Med 2013;11:146.

[81] Yu JM, Wu X, Gimble JM, Guan X, Freitas MA, Bunnell BA. Age-related changes in mesenchymal stem cells derived from rhesus macaque bone marrow. Aging Cell 2011;10(1):66–79.

[82] Churchman SM, Ponchel F, Boxall SA, Cuthbert R, Kouroupis D, Roshdy T, et al. Transcriptional profile of native CD271+ multipotential stromal cells: evidence for multiple fates, with prominent osteogenic and Wnt pathway signaling activity. Arthritis Rheumatism 2012;64(8):2632–43.

[83] Baxter MA, Wynn RF, Jowitt SN, Wraith JE, Fairbairn LJ, Bellantuono I. Study of telomere length reveals rapid aging of human marrow stromal cells following in vitro expansion. Stem Cells 2004;22(5):675–82.

[84] Jones E, English A, Churchman SM, Kouroupis D, Boxall SA, Kinsey S, et al. Large-scale extraction and characterization of CD271+ multipotential stromal cells from trabecular bone in health and osteoarthritis: implications for bone regeneration strategies based on uncultured or minimally cultured multipotential stromal cells. Arthritis Rheumatism 2010;62(7):1944–54.

[85] Katsara O, Mahaira LG, Iliopoulou EG, Moustaki A, Antsaklis A, Loutradis D, et al. Effects of donor age, gender, and in vitro cellular aging on the phenotypic, functional, and molecular characteristics of mouse bone marrow-derived mesenchymal stem cells. Stem Cells Dev 2011;20(9):1549–61.

[86] Strube P, Mehta M, Baerenwaldt A, Trippens J, Wilson CJ, Ode A, et al. Sex-specific compromised bone healing in female rats might be associated with a decrease in mesenchymal stem cell quantity. Bone 2009;45(6):1065–72.

[87] Erwin GS, Crisostomo PR, Wang Y, Wang M, Markel TA, Guzman M, et al. Estradiol-treated mesenchymal stem cells improve myocardial recovery after ischemia. J Surg Res 2009;152(2):319–24.

[88] Hong L, Sultana H, Paulius K, Zhang G. Steroid regulation of proliferation and osteogenic differentiation of bone marrow stromal cells: a gender difference. J Steroid Biochem Mol Biol 2009;114(3–5):180–5.

[89] Garderet L, Mazurier C, Chapel A, Ernou I, Boutin L, Holy X, et al. Mesenchymal stem cell abnormalities in patients with multiple myeloma. Leukemia Lymphoma 2007;48(10):2032–41.

[90] Corre J, Mahtouk K, Attal M, Gadelorge M, Huynh A, Fleury-Cappellesso S, et al. Bone marrow mesenchymal stem cells are abnormal in multiple myeloma. Leukemia 2007;21(5):1079–88.

[91] Klaus M, Stavroulaki E, Kastrinaki MC, Fragioudaki P, Giannikou K, Psyllaki M, et al. Reserves, functional, immunoregulatory, and cytogenetic properties of bone marrow mesenchymal stem cells in patients with myelodysplastic syndromes. Stem Cells Dev 2010;19(7):1043–54.

[92] Chao YH, Peng CT, Harn HJ, Chan CK, Wu KH. Poor potential of proliferation and differentiation in bone marrow mesenchymal stem cells derived from children with severe aplastic anemia. Ann Hematol 2010;89(7):715–23.

[93] Kastrinaki MC, Sidiropoulos P, Roche S, Ringe J, Lehmann S, Kritikos H, et al. Functional, molecular and proteomic characterisation of bone marrow mesenchymal stem cells in rheumatoid arthritis. Ann Rheum Dis 2008;67(6):741–9.

[94] Sun LY, Zhang HY, Feng XB, Hou YY, Lu LW, Fan LM. Abnormality of bone marrow-derived mesenchymal stem cells in patients with systemic lupus erythematosus. Lupus 2007;16(2):121–8.

[95] Perez-Simon JA, Tabera S, Sarasquete ME, Diez-Campelo M, Canchado J, Sanchez-Abarca LI, et al. Mesenchymal stem cells are functionally abnormal in patients with immune thrombocytopenic purpura. Cytotherapy 2009;11(6):698–705.

[96] Chinnadurai R, Copland IB, Ng S, Garcia M, Prasad M, Arafat D, et al. Mesenchymal stromal cells derived from Crohn's patients deploy indoleamine 2,3-dioxygenase-mediated immune suppression, independent of autophagy. Mol Ther 2015;23(7):1248–61.

[97] Phinney DG, Kopen G, Righter W, Webster S, Tremain N, Prockop DJ. Donor variation in the growth properties and osteogenic potential of human marrow stromal cells. J Cell Biochem 1999;75(3):424–36.

Chapter 5

Mesenchymal Stromal Cell Production in Academic Centers: Challenges and Opportunities

D.H. McKenna

Department of Laboratory Medicine and Pathology, Division of Transfusion Medicine, University of Minnesota Medical School, Minneapolis, Minnesota, United States

CHAPTER OUTLINE

5.1 INTRODUCTION

In general, the successful development and testing of any novel cellular therapy product requires a cohesive, interactive relationship between the basic research laboratory, the translational development laboratory, the clinical/cGMP laboratory, and the quality assurance/regulatory and principal investigator's teams. As academic centers have begun to see value in the investment required for such an enterprise and as mesenchymal stromal cells (MSCs) gain momentum in the clinical arena, an increasing number of academic centers are becoming involved in MSC manufacturing. While MSC production itself is fairly straightforward, there are methodological,

Mesenchymal Stromal Cells. http://dx.doi.org/10.1016/B978-0-12-802826-1.00005-2

logistical, scientific, and regulatory considerations surrounding cell manufacturing and implementation in a clinical trial. Only with a team approach utilizing individuals from a wide array of backgrounds, including various specialties in medicine, basic and applied science, technology, quality assurance and regulatory affairs, can an academic center effectively traverse the path from the research laboratory to clinical manufacturing and human clinical trials.

5.2 APPROACH TO MANUFACTURING

5.2.1 General Considerations

There are fundamentally three stages or tiers in the process of moving a cellular therapy from the "bench to the bedside": (1) Evaluation of Research and Preclinical Studies; (2) Scale-up and Optimization; and (3) Methods Validation [1]. A prospective cellular therapy emanating from discoveries in the research laboratory must first be tested for "proof-of-concept" in relevant in vitro and in vivo model systems. With sufficient preclinical data demonstrating safety and efficacy in relevant animal models, clinically appropriate product manufacturing methods are developed, and ultimately optimized and validated prior to initiation of the clinical trial. The three general stages of development are further discussed below.

5.2.1.1 Evaluation of research and preclinical studies

At this stage, the research and preclinical data is evaluated for feasibility to move forward into the clinical arena. This evaluation includes an assessment from several perspectives, including those from medical, scientific, technical, and regulatory angles. The results of "proof-of-concept" studies, which may include survival kinetics and distribution, and toxicology studies are reviewed. The overall goal is to define the desired characteristics of the cellular product, establish possible methods of cell manufacturing and product testing, and identify appropriate instrumentation and optimal reagents. Often, instrumentation used in early studies is not optimized for large/clinical-scale cell production, and reagents may not be suitable for human use for a variety of reasons. Notably, this stage provides opportunities for collaboration within the larger academic setting. Engineering and/or biotechnology departments or institutes may have valuable input on new devices, cellular scaffolds, and optimization of preservation or stability testing. The technology transfer is initiated with methods being transitioned to the translational development laboratory, which in many cases may be part of the clinical/cGMP laboratory. Project management commences with timelines and milestones established, and a pre-IND meeting/call may occur at this stage as well.

5.2.1.2 Scale-up and optimization

Prior to the initiation of the clinical trial, development of a reproducible, large-scale/clinically relevant methodology is required. Frequently, research-based methods are not refined, lacking standard operating procedures and details that would allow smooth transition into cGMP manufacturing. The translational development/clinical/cGMP laboratory will have developed various tools and processes to facilitate scale-up and optimization of manufacturing. Examples may include forms for facilitating technology transfer (such as a process map; see Fig. 5.1) or a calculation tool, for example, to determine optimal seeding of a culture vessel based upon input of seeding density and media depth requirements. See Fig. 5.2. A more well-established laboratory will have overarching processes for product development in place. An outline of such a process is shown in Fig. 5.3. Finally, it is at this phase that the standard operating procedures (SOPs) and batch production record (BPR) are initially developed.

5.2.1.3 Methods validation

The validation phase begins when a clinically appropriate and optimized method has been defined. The SOPs and BPR are established, and a validation plan is written. The validation plan is then executed. Typically in the cell therapy field, a validation consists of three successive runs which must pass preestablished specifications. The chemistry, manufacturing, and controls (CMC) section of the investigational new drug (IND) application and often the validation results are provided to FDA in support of the IND application and, in some cases, the institutional review board (IRB) submission.

While the ultimate goal of the three-tiered approach described above is to provide for a consistent safe, pure, and potent product, it should also improve upon efficiency and laboratory logistics, as well as cost containment. There are often additional issues, such as contractual considerations, intellectual property, and risk assessment and management. The Product Development Process (PDP, shown in Fig. 5.3) outlines these issues as well as others. The PDP integrates effectively into the three general stages of development discussed above. Tier one (Evaluation of Research and Preclinical Studies) corresponds roughly to the Proposal and Assessment of the PDP; Tier two (Scale-up and Optimization) is equivalent to Development of the PDP, and; Tier three (Methods Validation) parallels Validation of the PDP, leading to Clinical Use.

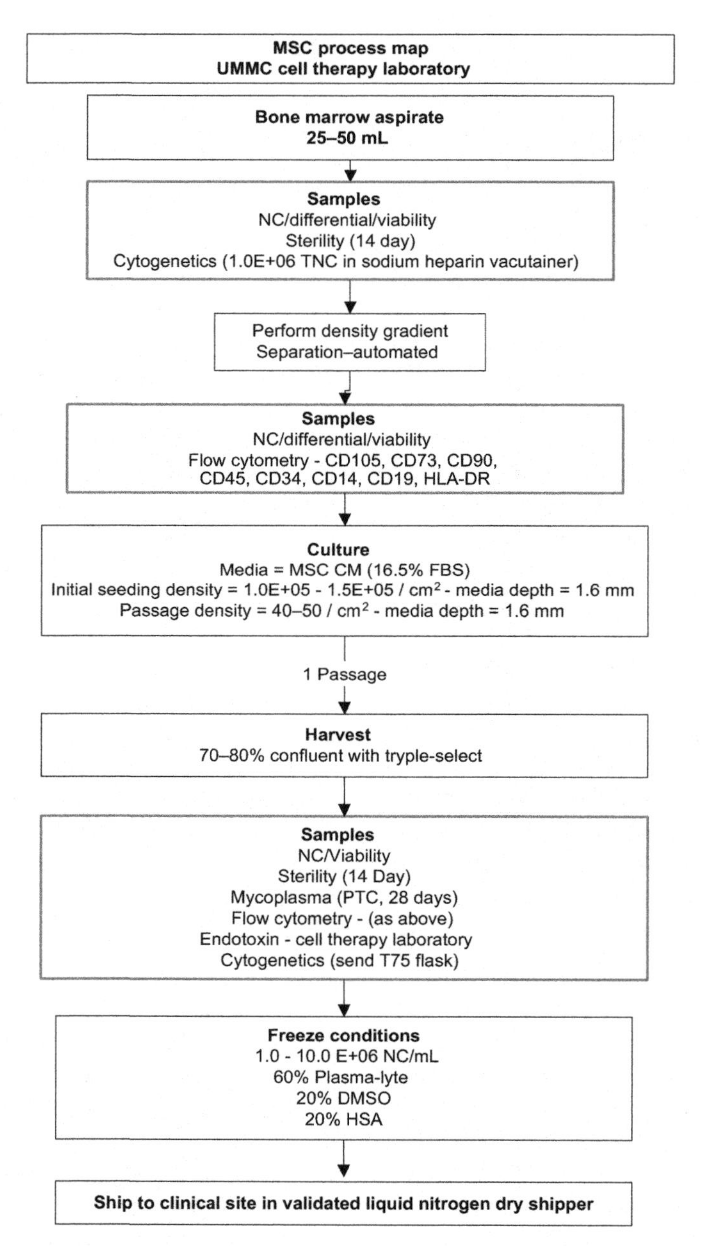

■ **FIGURE 5.1 MSC process map.** Developing a rudimentary process map early in development can guide the successful technology transfer, scale-up, and optimization of a product. As the process is fine-tuned, the process map is updated, and eventually finalized prior to validation. *MSC-CM*, MSC culture medium; *NC*, nucleated cells; *PTC*, points to consider; *TNC*, total nucleated cells.

Flask seeding calculator

Enter the following information			
	Seeding density	*1.60E + 05*	/cm2
	Media depth	*1.6*	mm

Culture dishes			
Dish size	**Surface area (cm2)**	**Volume/dish(ml)**	**TNC/dish**
100 mm × 20 mm	58	9.3	9.28E + 06
Culture plates			
Plate size	**Surface area (cm2/well)**	**Volume/well(ml)**	**TNC/Well**
96-Well	0.32	0.051	5.12E + 04
48-Well	1.1	0.176	1.76E + 05
24-Well	1.9	0.304	3.04E + 05
12-Well	3.5	0.6	5.60E + 05
6-Well	9.6	1.5	1.54E + 06
Culture flasks			
Flask size	**Surface area (cm2)**	**Volume/flask (ml)**	**TNC/flask**
T-25 upright	12.5	2	2.00E + 06
T-25	25	4	4.00E + 06
T-75 upright	24	4	3.84E + 06
T-75	75	12	1.20E + 07
T-150	150	24	2.40E + 07
T-175 upright	44	7	7.04E + 06
T-175	175	28	2.80E + 07
T-185	185	30	2.96E + 07
Cell factories			
Factory size	**Surface area (cm2)**	**Volume/factory (ml)**	**TNC/factory**
CF-1	632	101	1.01E + 08
CF-2	1264	202	2.02E + 08
CF-4	2528	404	4.04E + 08
CF-5	3160	506	5.06E + 08
CF-10	6320	1,011	1.01E + 09
CF-40	25280	4,045	4.04E + 09

■ **FIGURE 5.2 Culture vessel seeding calculator.** Simple tools such as this seeding calculator can assist with technology transfer, scale-up, and optimization. By entering seeding density and media depth requirements based upon review of preclinical studies, the translational laboratory can determine the amount of cells to add to culture vessels. (Developed by Darin Sumstad, University of Minnesota Medical Center)

5.2.2 MSC-Specific Considerations

MSCs were first described by Friedenstein et al. roughly 50 years ago when his group isolated them from bone marrow [2]. Since that time, investigators have further defined marrow-derived MSCs [3], isolated them from several other sources [4,5], and moved MSCs into a number of clinical trials for several applications [6]. While there is a remarkable level of activity of MSCs in the clinical arena, there are substantial differences in the manufacturing

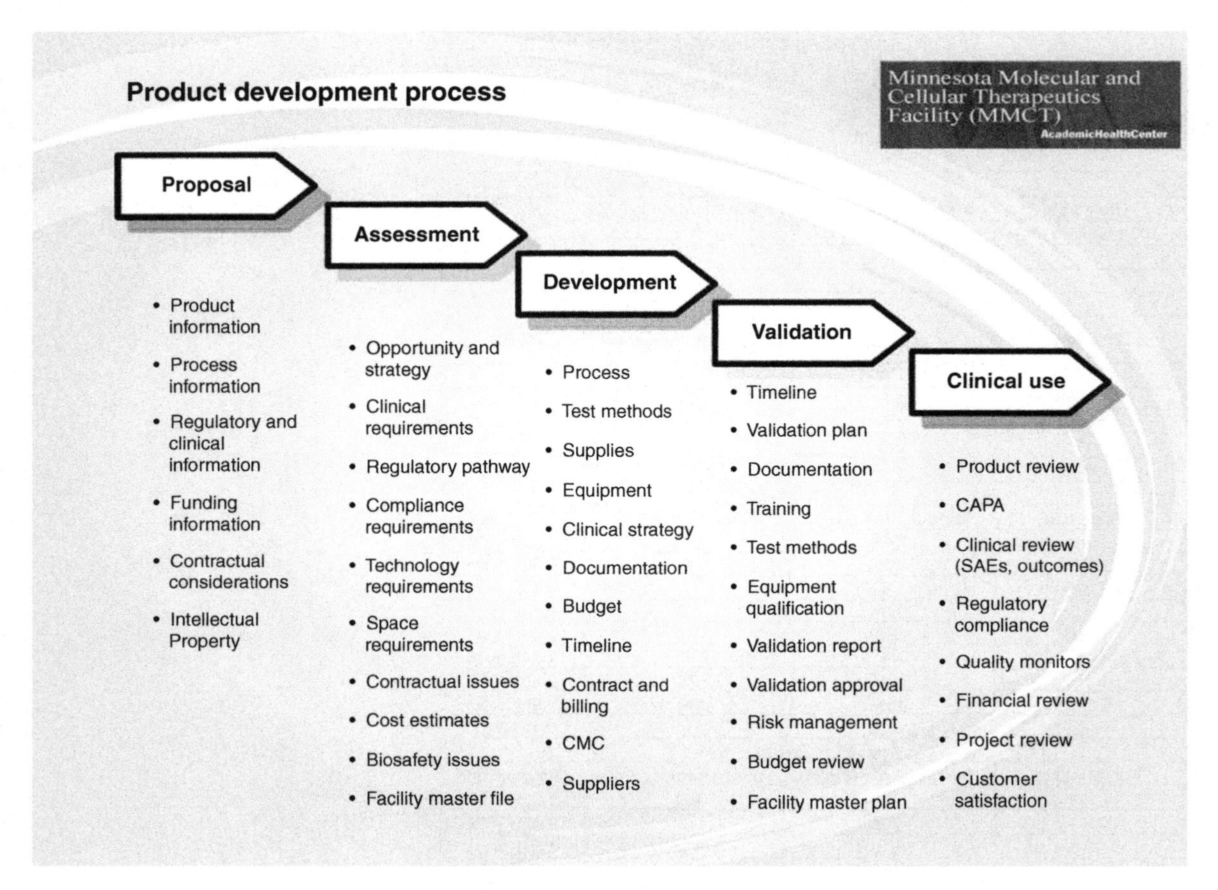

FIGURE 5.3 Product development process. This example of a product development process highlights the critical steps and considerations in the process of late translational development of cellular therapies. (Developed by Diane Kadidlo, Molecular and Cellular Therapeutics, University of Minnesota)

of these cells. Some examples of variation in methods include source material, selection of donor, culture container/device, base medium, medium supplements, seeding density, passage number, cryopreservation, and quality control/lot release testing [6]. While these methods-based variations are specific topics of chapters in this text, a brief discussion of each follows.

5.2.2.1 *Source material and selection of donor*

As mentioned above, MSCs have been isolated from several different tissues. In a review of over 200 clinical trials, Sharma et al. [6] noted the majority to involve bone marrow (56%), followed by umbilical cord blood (UCB, 17%), adipose (12%), and other tissues (eg, placenta, endometrium, trademarked, and unknown, cumulatively 15%). Although it remains to be determined if MSCs from a given source are optimal for certain applications, UCB does have a higher frequency of MSCs as compared to marrow [7] and thus may be a logical source of MSCs for this reason, if they are shown to be functionally unique or as effective as marrow-derived MSCs. Adipose-derived MSCs also appear to have higher frequencies, 100–1000-fold as compared to marrow. MSCs manufactured using adipose tissue have been shown to have a high potential for angiogenesis/vasculogenesis, making adipose tissue a potential choice for this application [8].

MSCs may be autologous or allogeneic (related or unrelated). The screening and testing of allogeneic donors of MSCs (eg, health questionnaire, viral testing) is similar to that of other cell- or tissue-based products, and it very closely mirrors that of a blood donor. As per FDA (21CFR1271, *Human Cells, Tissues, and Cellular and Tissue-Based Products, or HCT/Ps*), donors of MSCs require testing for HIV-1 (FDA-licensed screening test either for anti-HIV-1 or combination test for anti-HIV-1 and anti-HIV-2; and FDA-licensed screening NAT test for HIV-1, or combination NAT that includes HIV-1), HIV-2 (FDA-licensed screening test either for anti-HIV-2 or combination test for anti-HIV-1 and anti-HIV-2), HBV [FDA-licensed screening test for Hepatitis B surface antigen (HBsAg) and for total antibody to Hepatitis B core antigen (anti-HBc, IgG, and IgM)], HCV (FDA-licensed screening test for anti-HCV; and FDA-licensed screening NAT test for HCV, or combination NAT that includes HCV), and *Treponema pallidum* (FDA-cleared screening test for syphilis or FDA-cleared diagnostic serologic test for syphilis). Donors of viable, leukocyte-rich HCT/Ps require additional testing for HTLV-1 and -2 and CMV. MSC donors will often be tested for both HTLV and CMV, as most cell therapy laboratories in the academic setting manufacture leukocyte-rich products and prefer to adhere to a single, comprehensive standard operating procedure for donor infectious disease testing. Donor infectious disease testing is summarized in Table 5.1. More details of the specific requirements can be found on the US Food and Drug Administration website (http://www.fda.gov/BiologicsBlood Vaccines/SafetyAvailability/TissueSafety/ucm151757.htm).

The age of the donor appears to be important, with marrow from children containing a higher concentration of CFU-Fs than that from adults [9]. Moreover, increased donor age has been shown to directly correlate with detrimental effects in terms of proliferation and potency of MSCs [10]. Thus, if a choice of allogeneic donor is possible, the evidence supports derivation of cells from marrow of a healthy, younger donor [9,10].

Table 5.1 Summary of Donor Infectious Disease Testing for MSCs

Infectious Agent	Required Testing
HIV, type 1[a]	FDA-licensed screening test for anti-HIV-1 and FDA-licensed screening test for NAT
HIV, type 2	FDA-licensed screening test for anti-HIV-1
HBV	FDA-licensed screening test for HBsAg and for total antibody (IgG and IgM) to core antigen (anti-HBc)
HCV	FDA-licensed screening test for anti-HCV and FDA-licensed screening NAT for HCV
Treponema pallidum	FDA-cleared screening test for syphilis or FDA-cleared diagnostic serologic test for syphilis
HTLV I/II [b]	FDA-licensed screening test for anti-HTLV I/II
CMV [b]	FDA-cleared screening test for anti-CMV (total IgG and IgM)

[a] *Establishments not utilizing an FDA-licensed screening test that tests for group O antibodies must evaluate donors for risk associated with HIV group O infection.*
[b] *Testing only required for donors of viable, leukocyte-rich HCT/Ps.*

5.2.2.2 *Culture container/device*

MSCs grow as adherent cells, and as they approach confluence on the culture surface they may be serially passaged for further expansion. Because cell yield directly correlates to surface area of the culture platform, multilayered cell factories have been used [11]. Depending upon the size of culture, which can be quite large for clinical applications, this approach may be fairly labor intensive and quite costly. Alternatively, it is possible to expand MSCs by using a bioreactor. Bioreactors newer to the market are typically closed system setups, and as such, may be preferred for cGMP manufacturing [12].

5.2.2.3 *Culture medium and supplements*

The most commonly used culture medium is a basal medium, such as alpha-MEM or DMEM, with a serum source for nutrients and growth factors [13]. A recent FDA perspective confirmed that most clinical trials in the United States utilize MSCs manufactured with fetal bovine serum (FBS) as the serum source. Over 80% of the submissions used FBS in the culture in a concentration range of 2–20% FBS with 10% being the most common [14]. While investigators would be using FBS only from qualified herds, concerns for the use of FBS do remain and include risk of transfer of immunogenic xenoproteins and infectious disease, such as transmissible spongiform encephalopathy (TSE). This has led investigators to consider alternatives to FBS, including human plasma or serum, human UCB serum, and human platelet derivatives such as platelet lysate [11] in lieu of a completely defined, serum-free medium [15]. Human platelet lysate holds much promise as an alternative to FBS. While much remains to be learned, there have been interesting discoveries, including that of Lohmann et al. [16] which noted increased proliferation and decreased senescence in MSC cultures using platelet lysate from younger donors. Finally, various growth factors have been investigated in MSC culture. Basic fibroblast growth factor (bFGF), in particular, has shown encouraging results with positive effects, including maintenance of "stemness" and reduction of senescence [17,18].

5.2.2.4 Culture schema

Seeding density is an important variable for culture in general, and while this is certainly the case with MSCs, both initial mononuclear cell (MNC) and subsequent MSC seeding densities vary substantially among investigators. Published clinical trials describe initial MNC seeding at concentrations ranging from higher densities (eg, 1.70×10^5 MNC/cm^2) to lower densities (eg, 5.0×10^4 MNC/cm^2) [6]. An initial seeding density as low as 4×10^4 MNC/cm^2 has been described in support of rapid, large-scale expansion of MSCs [11]. Seeding densities are typically much lower on subsequent passages, with low (eg, $1.0–5.0 \times 10^3$ MSC/cm^2) or very low (eg, $1.0–5.0 \times 10^1$ MSC/cm^2) plating densities not uncommon.

It is apparent that both expansion rate and yield of more primitive, multipotent progenitors are inversely related to the seeding density and culture time of each passage. Sekiya et al. provide a compelling argument for compromise between conditions that provide the highest overall yield and those that provide the greatest amount of early progenitor cells [19]. It is important to understand the concept of proliferative age of MSCs. MSCs have a restricted lifespan and ultimately reach senescence (ie, cellular functions become diminished and the risk for acquiring mutations and inflammatory phenotype increases) rendering them less than ideal for clinical applications [20]. Interestingly, a recent study by von Bahr et al. in GVHD patients showed 1-year survival of 75% in patients receiving early-passage MSCs (passages 1–2) in contrast to 21% in patients getting later passage MSCs (passages 3–4) [21]. Similar analyses of additional clinical trials of MSCs may shed further light on the importance of the approach to culture, and studies such as that of Bellayr et al. [22] will soon lead to a better understanding of the impact of culture design and optimization of culture schema.

5.2.2.5 Cryopreservation

MSCs for clinical use are most commonly frozen in 10% DMSO within an electrolyte solution (eg, PlasmaLyteA) and a protein source (eg, human serum albumin) using a controlled rate freezer. This approach is based upon hematopoietic stem cell cryopreservation and is not optimized for MSCs [23]. MSCs are then stored in liquid nitrogen (liquid or vapor phase) to maintain temperatures below −150°C, preserving cell viability and function upon thaw. Frozen MSCs can be transported in liquid nitrogen dry shippers to a clinical site where they can be thawed for final preparation and patient administration.

Cryopreservation of MSCs has become a focus of intense debate following the failed Osiris-sponsored phase III clinical trial of MSCs for GVHD in 2009. This trial failed to meet its primary endpoint, and speculation surrounding the etiology of the presumed functional deficits of the cells ensued [24]. Cryopreservation was one of the possibilities, and several investigators subsequently demonstrated impaired function of MSCs when tested in vitro immediately upon thawing [25,26]. Interestingly, there have also recently been studies refuting this consequence of cryopreservation [27,28]. It may be that the impact of cryopreservation is dependent upon intended function of the cells or the way by which that function is tested.

5.2.2.6 ***Quality control and lot release testing***

Quality control (QC) and lot release testing for MSCs is fairly standard and includes the typical testing for novel cell therapies—viability, sterility, Gram stain, endotoxin, mycoplasma, and immunophenotype—and often includes cytogenetics testing. If a cell banking approach is used, as is the case in 35% of IND submissions to FDA [14], additional testing may be required, as determined through discussion with FDA. A potency assay must, of course, be included and varies depending upon clinical application and/or presumed mechanism of action. The potency assay is by far the most difficult QC/lot release assay to develop. However, in the United States it is required by FDA to be established prior to initiating phase III clinical trials and validated before biologics licensing application (BLA) submission. FDA defines potency as "the specific ability or capacity of the product, as indicated by appropriate laboratory tests or by adequately controlled clinical data obtained through the administration of the product in the manner intended, to effect a given result." [21 CFR 600.3(s)]. The chapter on MSC characterization covers the potency assay and related topics in much greater depth. An example of QC/lot release testing for MSCs is shown in Table 5.2.

5.2.2.7 ***Summary***

As discussed above, there are multiple variables in the manufacturing and testing of MSCs. Overall, clinical manufacturing should be based upon methods used in preclinical studies. This approach would both meet FDA/

Table 5.2 Summary of Quality Control Testing of MSCs

Assay	Example Test Methods	Specification
Sterility (14 days)	Per 21 CFR 610.12 /USP, automated system	No growth
Mycoplasma	Per FDA Points to Consider; PCR-based	Negative
Endotoxin	LAL/USP: chromogenic, turbidometric, etc.	<5 EU/kg/total product
Gram stain[a]	Gram's method	No organisms seen
Immunophenotype	Flow cytometry:	
	positive markers—CD73, CD90, CD105	≥85% expression
	Negative markers—CD14, CD19, CD34, CD45, HLA-DR	≤15% expression
Viability	Trypan blue, acridine orange/propidium iodide (microscopy); 7-AAD (flow cytometry)	≥70%
Cytogenetics	Giemsa-banding (karyotype)	No clinically significant abnormality
Functional/potency	Trilineage assay, MLR-based, cytokine measurement, PCR-based	Assay and indication-dependent

[a] Gram stain would not likely be required with cryopreserved MSCs and finalized sterility testing showing no growth. Freshly administered MSCs would, however, require Gram stain.

regulator expectations and provide the greatest likelihood of achieving results seen in the proof-of-principle/preclinical studies. Substantial changes to the preclinical approach (eg, cell source, culture device, medium/supplements) will likely require additional preclinical studies. Common considerations for alteration to manufacturing may include changing from FBS to human platelet lysate or moving from cell factories to a hollow fiber bioreactor. These changes may provide clear benefits for meeting cGMPs (ie, eliminating xenogeneic components and perhaps achieving closed-system culture, respectively); however, the alterations may also substantially modify the inherent qualities or characteristics of the cells. Stroncek et al. used global gene expression analysis to compare MSCs manufactured at eight sites using varying methods of processing and noted considerable intercenter variability. Variables in manufacturing included use of FBS versus platelet lysate, concentration of FBS, inclusion of bFGF, and culture in cell factories/flasks versus bioreactor [29]. A functional assay is planned in conjunction with this study and may shed light on the impact of variables in manufacturing on cellular function. Clearly, it is critical to initiate modifications to manufacturing as early as possible in the development process, then finalize methods and move forward. When modifications are substantial, comparability must be demonstrated before proceeding further in the development process.

5.2.3 Challenges

There are certainly challenges for enabling cell therapies in the academic setting. Many of the challenges are highlighted in the preceding sections on considerations for manufacturing, and several of these would undoubtedly overlap with the experience in the biopharmaceutical realm. Below are a few additional challenges that may be more specific to academia where innovation is abundant but resources may be limited.

5.2.3.1 *Assembly of the team*

The criticality of the team approach in cell therapy translational development cannot be under emphasized. A wide range of skillsets are necessary, and multiple team members with backgrounds in medicine, basic and applied science, technology, quality assurance, and regulatory affairs are required for success. A typical academic clinical research center will have strong representation in most, and possibly all, of these areas. Expertise in regulatory affairs may be the most difficult resource to locate in an academic setting, and when available to the team, these individuals may be most experienced with pharmaceuticals or devices, and not biologics. If a suitable regulatory expert is not available, a quality assurance expert centered in cellular therapy may have a solid working knowledge of the regulations and may be able to help the team with regulatory challenges.

Once the team is assembled, it is important to utilize the skills of each member and to remember that the leader may be different depending upon the particular project and/or the stage of development. Regular meetings will move the project along, and the frequency of meetings may change, again depending upon the stage of development. Every institution will likely function somewhat differently with regard to roles and responsibilities and their overlap. With this in mind, it is very important to include those involved with the cGMP facility as early as possible. This group may or may not have much to contribute to the science early in development, but their insight into the optimal development pathway is invaluable. The cGMP team will facilitate the transition of the cell therapy product into the clinical arena. This process often involves studies that are typically not of interest to the basic research laboratory, but without which therapies would never reach clinical trials [30–32].

5.2.3.2 Funding

Historically, investigators have both underestimated the cost of development and cGMP production of cell therapies and failed to include sufficient support for cell therapy production in grant applications. The appreciation for the complexity and cost of such material has improved, and mechanisms such as the NIH-sponsored Production Assistance for Cellular Therapies (PACT) contract have accelerated progress in the field by defraying costs of manufacturing [33]. However, with NIH grant support leveling out or actually dropping based on constant (2003) dollars (see http://officeofbudget.od.nih.gov/approp_hist.html), academic institutions are increasingly having to search for other sources of funding for cell therapies. Partnering with biopharmaceutical companies has become more common in recent years, leveraging resources and expertise from both academia and industry [34].

5.2.3.3 Balancing workload

As a cell therapy program grows, balancing workload becomes an ongoing effort. Particularly with projects at different levels of development, this can be a challenge. Activity level for projects in the clinical trial stage will be dictated by patient enrollment. Most phase I clinical trials have few patients, and enrollment is often gradual, especially with a dose escalation schema. However, as trials progress to phases II and III, enrollment picks up, scheduling becomes more complex, and it can be difficult to work other projects at earlier stages of development into the workflow. This may be unique to the academic center, in that many support multiple types of cell therapies [1], whereas most biopharmaceutical companies limit their portfolio in cell therapy to one or two products [34]. The issue of workload balance may be less significant for MSCs at the clinical trial stage depending upon the approach to manufacturing. That is, third-party cryopreserved MSCs thawed on the day of administration, for example, may be stockpiled and held in liquid nitrogen storage awaiting enrollment of patients.

5.2.3.4 Other challenges

There are many potential other challenges that may fit into the categories above but are worth special mention here. A cell therapy lab/cGMP manufacturing facility may be asked to support more than one trial with MSCs. Ideally, it would be optimal or certainly most straightforward to use just one standardized method, set of reagents, assays, etc. However, depending upon the investigator's research, preclinical studies, and intended effect this may not be the best scientific approach. All of these factors must be considered in determining the route in support of more than one MSC trial.

Supporting more than one trial or even a single multicenter trial in MSCs may also cause challenges for maintaining inventory and demand for cells. As an example a phase II randomized, double-blind placebo-controlled study of MSCs for acute lung injury is ongoing at four academic institutions with centralized MSC manufacturing [35]. This trial and others of similar design in acutely ill patients highlight several inventory considerations. Issues related to MSC products often must be resolved by the manufacturing laboratory and receiving cell therapy laboratories without involvement of clinical investigators to maintain blinding. Also, to further efforts toward blinding, each site must continually stock enough cells for at least two cell-treated patients. A rapid enrollment of two successive patients receiving MSCs could unintentionally unblind an investigator if sufficient doses of MSCs are not available.

Another issue related to inventory is limiting number of third-party donors if that is the intent of an MSC study involving such cells. The literature suggests that donor age, gender, and health can influence the quality and/or characteristics of MSCs [9,10,36]. Restricting the number of MSC donors for a clinical trial may at least limit the impact of this variable in a given study. However, access to a small pool of regular donors may be challenging due to donor schedule/availability, IRB stipulations, etc.

A final challenge for multicenter trials worth mentioning is that of training and coordinating final cell processing for patient administration. Standard operating procedures must be transferred to the participating sites. Often, the lead site will host a training session with representatives from each cell therapy laboratory in attendance [37]. If funding allows, this is probably the best approach, assuring optimal recovery of cells and consistency of final MSC preparation.

5.2.4 Opportunities

Many of the challenges noted above may prove to be opportunities as well. While moving MSC or any cell therapy toward clinical trials can be challenging, over time the team improves and becomes more efficient with each successive project. Partnering with biopharmaceutical companies allows for opportunities to gain experience with different technologies and to learn alternative approaches to product development. Partnering with other academic laboratories or facilities in general provides the opportunity to leverage strengths and talents of participating groups and affords the chance to increase study number ("n") to maximize success and study statistical power. Both PACT and BEST (Biomedical Excellence for Safer transfusion) have been quite successful in this regard [38–41]. Having to attend to workload balance will undoubtedly improve workflow and laboratory efficiency. Finally, supporting the manufacturing of MSCs of different types (source, methods, etc.) expands expertise in the laboratory. Ultimately, the greatest opportunity afforded to those working in academic cell therapy is the chance to make a difference in the lives of people afflicted with a serious illness or disease. MSCs hold great potential for the treatment of several diverse medical conditions, and academic cell therapy programs remain focused working diligently to move the field forward to the clinic.

5.3 CONCLUSION

The possibilities for MSCs in the clinical arena appear vast. Much remains to be determined regarding their utility and efficacy; however, with preclinical studies supporting applications in the broad areas of immune modulation and regenerative medicine, several medical conditions may be treatable. The academic center is ideally positioned to participate, if not lead, the movement from the research laboratory into human clinical trials. The team approach is essential, as expertise from a broad array of specialties is needed. Institutional support, for example, in the form of expertise (eg, regulatory personnel), infrastructure (eg, establishment and maintenance of cGMP facility), and seed funds for investigators to use toward translational development is critical as well. Support through governmental funding pathways has been successful, and continued support is essential for moving MSCs and other cellular therapies along the translational pipeline [42]. There continue to be challenges, but many of these are actually also opportunities. Those challenges found not to be opportunities are overcome by the promise of great potential of MSCs.

ABBREVIATIONS

Alpha-MEM	Alpha-Minimum Essential Medium
BEST	Biomedical Excellence for Safer Transfusion
bFGF	Basic fibroblast growth factor
BPR	Batch production record
CFU-F	Colony forming unit-fibroblast
cGMP	Current good manufacturing practices
CMC	Chemistry, manufacturing, and controls
CMV	Cytomegalovirus
DMEM	Dulbecco's Modified Eagle Medium
DMSO	Dimethyl sulfoxide
FBS	Fetal bovine serum
FDA	Food and Drug Administration
GVHD	Graft-versus-host disease

HBc	Hepatitis B core
HBs Ag	Hepatitis B surface antigen
HBV	Hepatitis B virus
HCV	Hepatitis C virus
HCT/Ps	Human cells, tissues, and cellular and tissue-based products
HIV	Human immunodeficiency virus
HTLV	Human T-lymphotropic virus
IND	Investigational new drug
IRB	Institutional review board
MNC	Mononuclear cell
NAT	Nucleic acid testing
NIH	National Institutes of Health
PACT	Production Assistance for Cellular Therapies
QC	Quality control
SOP	Standard operating procedures
TSE	Transmissible spongiform encephalopathy
UCB	Umbilical cord blood

REFERENCES

[1] McKenna D, et al. The Minnesota molecular & cellular therapeutics facility: a state-of-the-art biotherapeutics engineering laboratory. Transfus Med Rev 2005;19(3): 217–28.

[2] Friedenstein AJ, et al. Heterotopic of bone marrow. Analysis of precursor cells for osteogenic and hematopoietic tissues. Transplantation 1968;6:230–47.

[3] Dominici M, et al. Minimal criteria for defining multipotent mesenchymal stromal cells. The International Society for Cellular Therapy position statement. Cytotherapy 2006;8:315–7.

[4] Zuk PA, et al. Human adipose tissue is a source of multipotent stem cells. Mol Biol Cell 2002;13:4279–95.

[5] In 't Anker PS, et al. Isolation of mesenchymal stem cells of fetal or maternal origin from human placenta. Stem Cells 2004;22:1338–45.

[6] Sharma RR, et al. Mesenchymal stem or stromal cells: a review of clinical applications and manufacturing practices. Transfusion 2014;54:1418–37.

[7] Lu L, et al. Isolation and characterization of human umbilical cord mesenchymal stem cells with hematopoiesis-supportive function and other potentials. Haematologica 2006;91(8):1017–26.

[8] Moseley TA, et al. Adipose-derived stem and progenitor cells as fillers in plastic and reconstructive surgery. Plast Reconstr Surg 2006;118(Suppl. 3):121S–8S.

[9] Stolzing A, et al. Age-related changes in human bone marrow-derived mesenchymal stem cells: consequences for cell therapies. Mech Ageing Dev 2008;129(3):163–73.

[10] Baxter MA, et al. Study of telomere length reveals rapid aging of human marrow stromal cells following in vitro expansion. Stem Cells 2004;22(5):675–82.

[11] Schallmoser K, et al. Rapid large-scale expansion of functional mesenchymal stem cells from unmanipulated bone marrow without animal serum. Tissue Eng Part C Methods 2008;14(3):185–96.

[12] Rojewski MT, et al. GMP-compliant isolation and expansion of bone marrow-derived MSCs in the closed, automated device Quantum Cell Expansion System. Cell Transplant 2013;22:1981–2000.

[13] Sotiropoulou PA, et al. Characterization of the optimal culture conditions for clinical scale production of human mesenchymal stem cells. Stem Cells 2006;24:462–71.

[14] Mendicino M, et al. MSC-based product characterization for clinical trials: an FDA perspective. Cell Stem Cell 2014;14(2):141–5.

[15] Kinzebach S, Bieback K. Expansion of mesenchymal stem/stromal cells under xenogenic-free culture conditions. Adv Biochem Eng Biotechnol 2013;129:33–57.

[16] Lohmann M, et al. Donor age of human platelet lysate affects proliferation and differentiation of mesenchymal stem cells. PLoS One 2012;7(5):e37839.

[17] Battula VL, et al. Human placenta and bone marrow derived MSC cultured in serum-free, b-FGF-containing medium express cell surface frizzled-9 and SSEA-4 and give rise to multilineage differentiation. Differentiation 2007;75(4):279–91.

[18] Eom YW, et al. The role of growth factors in maintenance of stemness in bone marrow-derived mesenchymal stem cells. Biochem Biophys Res Commun 2014;445(1):16–22.

[19] Sekiya I, et al. Expansion of human adult stem cells from bone marrow stroma: conditions that maximize the yields of early progenitors and evaluate their quality. Stem Cells 2002;20(6):530–41.

[20] Digirolamo CM, et al. Propagation and senescence of human marrow stromal cells in culture: a simple colony-forming assay identifies samples with the greatest potential to propagate and differentiate. Br J Haematol 1999;107(2):275–81.

[21] von Bahr L, et al. Long-term complications, immunologic effects, and role of passage for outcome in mesenchymal stromal cell therapy. Biol Blood Marrow Transplant 2012;18(4):557–64.

[22] Bellayr IH, et al. Gene markers of cellular aging in human multipotent stromal cells in culture. Stem Cell Res Ther 2014;5(2):59.

[23] Hanna J, Hubel A. Preservation of stem cells. Organogenesis 2009;5(3):134–7.

[24] Galipeau J. The mesenchymal stromal cells dilemma—does a negative phase III trial of random donor mesenchymal stromal cells in steroid-resistant graft-versus-host disease represent a death knell or a bump in the road? Cytotherapy 2013;15(1):2–8.

[25] François M, et al. Cryopreserved mesenchymal stromal cells display impaired immunosuppressive properties as a result of heat-shock response and impaired interferon-γ licensing. Cytotherapy 2012;14(2):147–52.

[26] Moll G, et al. Do cryopreserved mesenchymal stromal cells display impaired immunomodulatory and therapeutic properties? Stem Cells 2014;32:2430–42.

[27] Luetzkendorf J, et al. Cryopreservation does not alter main characteristics of Good Manufacturing Process-grade human multipotent mesenchymal stromal cells including immunomodulating potential and lack of malignant transformation. Cytotherapy 2015;17(2):186–98.

[28] Cruz FF, et al. Freshly thawed and continuously cultured human bone marrow-derived mesenchymal stromal cells comparably ameliorate allergic airways inflammation in immunocompetent mice. Stem Cells Transl Med 2015;4(6):615–24.

[29] Stroncek DF, et al. Comparison of bone marrow stromal cell (BMSC) production methods and products from multiple centers. Cytotherapy 2015;17(6):S51.

[30] Koepsell SA, et al. Successful "in-flight" activation of natural killer cells during long-distance shipping. Transfusion 2012;53(2):398–403.

[31] Sumstad D, et al. Reduction of non-clinical-/non-cGMP-grade culture reagents and measurement of residual ingredients in final early phase cellular therapy products. Cytotherapy 2009.

[32] Gee A, et al. A multi-center comparison study between the Endosafe® PTS™ rapid release testing system and conventional test methods for detecting endotoxin in cell therapy products. Cytotherapy 2008;10(4):427–35.

[33] Wood D, et al. An update from the United States National Heart, Lung, and Blood Institute-funded Production Assistance for Cellular Therapies (PACT) program: a decade of cell therapy. Clin Transl Sci 2014;7(2):93–9.

[34] French A, et al. Global strategic partnerships in regenerative medicine. Trends Biotechnol 2014;32(9):436–40.

[35] Liu KD, et al. Design and implementation of the START (STem cells for ARDS Treatment) trial, a phase 1/2 trial of human mesenchymal stem/stromal cells for the treatment of moderate-severe acute respiratory distress syndrome. Ann Intensive Care 2014;4:22.

[36] Siegel G, et al. Phenotype, donor age and gender affect function of human bone marrow-derived mesenchymal stromal cells. BMC Med 2013;11:146.

[37] Gee AP, et al. Multicenter cell processing for cardiovascular regenerative medicine applications: the Cardiovascular Cell Therapy Research Network (CCTRN) experience. Cytotherapy 2010;12(5):684–91.

[38] McKenna D, et al. CD34+ cell selection using small volume marrow aspirates: a platform for novel cell therapies and regenerative medicine. Cytotherapy 2010;12:170–7.

[39] Bloom DD, et al. A reproducible immunopotency assay to measure mesenchymal stromal cell-mediated T-cell suppression. Cytotherapy 2015;17(2):140–51.

[40] Pamphilon D, et al. Storage characteristics of cord blood progenitor cells: report of a multicentre study by the cellular therapies team of the Biomedical Excellence for Safer Transfusion (BEST) Collaborative. Transfusion 2011;51(6):1284–90.

[41] Maria Nawrot, et al. An inter-laboratory study to evaluate the performance of a novel seven day colony forming unit (CFU) assay. BEST Collaborative. Transfusion 2011;51:2001–5.

[42] Ouseph S, et al. Cellular therapies clinical research roadmap: lessons learned on how to move a cellular therapy into a clinical trial. Cytotherapy 2015;17(4):339–43.

Chapter 6

Bioreactor for Scale-Up: Process Control

E. Abraham, S. Gupta, S. Jung and E. McAfee
Lonza Walkersville Inc., Walkersville, MD, United States

CHAPTER OUTLINE

Mesenchymal Stromal Cells. http://dx.doi.org/10.1016/B978-0-12-802826-1.00006-4

6.1 INTRODUCTION

The question of which platform to choose for human MSC clinical manufacturing comes down to a few critical considerations: quality, cost, quantity, and comparability.

The roots of current-day MSC manufacturing are based in cell culture methods used in academic labs around the world, primarily consisting of two-dimensional (2D) cell culture in static plastic vessels such as tissue culture flasks or multilayered cell factories. Two-dimensional culture of cells dates back several decades. With the advent of plastic materials, plastic dishes, and flasks for cell culture became widely used. Indeed, one of the hallmarks of MSCs is their rapid adherence to 2D tissue culture vessels. This property is exploited by isolating these cells from a mixed starting population. Thus, flask-based 2D culture is the major starting point for MSC isolation and expansion [1]. This method of isolating and expanding MSCs has in turn been used for larger-scale expansion for clinical trials of several MSC-based cell therapies.

A central concept in increasing efficiencies of adherent, contact inhibited, cell culture is the concept of maximizing the growth surface area to vessel volume (SA/V) ratio of the culture platform. The more growth surface area that can be incorporated into a given volume, the more cells one can produce from that volume, thereby creating a more efficient process. Small culture flasks, such as T-75, T-175, and T-225, have a surface area which may not be sufficient for scaling up MSC culture to produce a large enough dosage even for relatively small clinical trials. For this reason, the 2D flask concept was expanded into 10-layer and 40-layer cell factories, which have growth surface areas much larger than T-175 flasks (36- and 144-fold larger surface area per vessel, respectively). However, the increased surface area provided by cell factories represents only a marginal increase in efficiency as cell factories are essentially expanded flasks. Furthermore, cell factories are bulky, difficult to manipulate, and take up large volumes of space in incubators. An additional improvement to the 2D culture method was the introduction of the Hyperstack (Corning, Tewksbury, MA), which incorporates growth surface with a gas-permeable membrane, eliminating the need for headspace for aeration and thus allowing an incremental improvement in efficiency by increasing the SA/V ratio.

The concept of maximizing SA/V has a fundamental impact on both the cost of manufacturing cells and the number of cells that can realistically be manufactured. Two of the main cost drivers in cell manufacturing are space and manpower. Clean-room space is expensive to build and maintain, and skilled manpower is clearly also a cost driver. The smaller the SA/V ratio, the more clean room space will be needed. Also, with 2D-based culture, which is manual and labor intensive, the manpower requirements could possibly drive up the cost of manufacturing to a point of unsustainability.

Another aspect of classical 2D culture of MSC is the fact that the culture environment is poorly controlled. The culture is conducted in incubators that maintain the temperature, CO_2 concentration, and humidity. However, online monitoring of critical culture parameters, such as dissolved oxygen, pH, and concentration of key energy sources (eg, glucose) and metabolites in media, is not possible. Moreover, the process of medium feeding is conducted outside of the incubators, and as usually 50–100% medium exchange is performed, the cells experience recurring spikes in levels of dissolved oxygen, glucose, pH, temperature, etc. This can have a detrimental impact on the quality of the cells and increase batch-to-batch variability of the final cell product.

Since planar (2D) systems can only offer incremental increase in the SA/V ratio, the use of microcarriers or other three-dimensional (3D) culture surfaces in bioreactors, either in suspension or in a packed-bed configuration, has been envisioned for late-stage clinical and large-scale commercial manufacture of MSCs. The use of bioreactor platforms for the manufacture of MSC presents some challenges, but to a large extent alleviates many of the major issues associated with 2D culture systems. Although solid and microporous microcarriers that provide outer surfaces for cell attachment and growth still conform to the principle of monolayer cell culture, the efficiencies of SA/V are dramatically increased in this type of culture. Some culture substrates such as macroporous microcarriers or woven and nonwoven fibers (eg, Fibra-Cel) are designed for high-density cell culture. They represent an additional increase in SA/V versus solid and microporous microcarriers. However, they present challenges, in terms of the removal of the cells from the substrates at the end of culture. As an example of the difference in SA/V ratio between 2D systems and 3D bioreactors, Fig. 6.1 shows the efficiency of microcarrier-mediated suspension culture using stirred-tank bioreactors, which is at least an order of magnitude more efficient than 2D-based culture.

The increased efficiency has a dramatic impact on the cost of manufacturing cells and creates the ability to manufacture the large cell numbers needed for many clinical indications. The dose of MSC used for therapies ranges from 100×10^6 to 1000×10^6 cells per dose. Following is an example, based on MSC therapy for real-world stroke treatment follows.

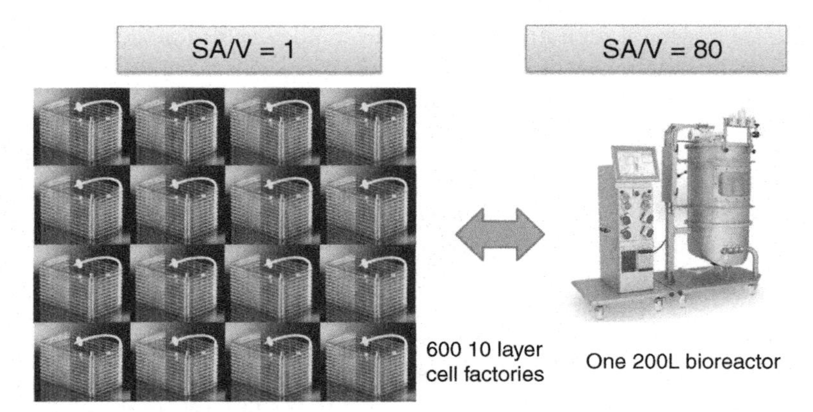

■ **FIGURE 6.1** Example of difference in SA/V ratio between 10-layer cell factories and microcarrier-based suspension culture in a 200 L stirred-tank bioreactor. Given that the SA/V of the cell factories is 1, the SA/V ratio of microcarrier bioreactor culture is 80-fold.

Stroke assumptions:

- Annual number of patients per year in the United States—800,000
- Percentage of patient population treated—30% (240,000)
- Dose per patient—300×10^6 cells
- 2D-based platform—80,000 cells/cm^2 at harvest in a 10-layer cell factory platform
- 3D-based platform—1×10^6 cells/mL cells at harvest in a 1000 L suspension microcarrier-based bioreactor
- Treatment consists of one dose/patient
- No adjustment for suboptimal harvest, cell viability, or cells retained for QC
- Number of 10 layer cell factories needed—142,405 per year or 2,738 per week
- Number of 1000 L bioreactors needed—72 per year or 1.4 per week

Using current methods, the size of a 2D-based manufacturing lot is capped at approximately one hundred 10-layer cell factories per batch, primarily due to clean room space and downstream processing time. Thus, large-scale manufacturing (needing more than one hundred 10-layer cell factories per batch) will require bioreactor platforms. Clearly, this hypothetical yet conservative calculation demonstrates the impossibility of using 2D platforms for larger or even medium-size cell therapy trials, while showing the necessity of using suspension-based bioreactor platforms.

Another consideration for bioreactor use is the ability to enable continuous monitoring of key process parameters and control of culture conditions. In bioreactors, dissolved oxygen, pH, and temperature levels are continuously monitored and maintained at predefined set points for the duration of the culture. This allows for optimized cell culture conditions throughout the culture period and the prevention of drastic changes in these critical parameters which occurs in 2D culture during medium exchanges. Moreover, with the advent of online monitoring for glucose and other medium components, media perfusion rates can be linked to levels of nutrients or other metabolites so as to optimize culture conditions and to minimize unnecessary media exchange, thus further reducing costs. Obtaining optimal cell quality is further enhanced by the fact that downstream processing of a bioreactor process is in most cases significantly faster than processing multiple cell factory units when controlling for cell yield and manpower. During the downstream process, MSCs are exposed to harsh conditions including shear stress and absence of anchorage, glucose, and other nutrients. Thus, minimizing the length of exposure to these conditions is one aspect of facilitating high cell quality in terms of viability and critical quality attributes.

Use of a bioreactor, however, may have its downsides which include potential cell damage or change to product critical attributes due to the introduction of shear stress from the impeller, and microcarrier bead-to-bead collisions during the culture process; the cost of operating and maintaining bioreactor systems; the time required for training to operate these systems; and the downstream separation of carriers from cells. However, it would seem that the advantages of bioreactor suspension culture in most cases outweigh the downsides. In addition, other viable alternatives currently do not exist for large-scale manufacturing of MSCs.

The effect of suspension culture on the biological activity of cells is a key question to consider before switching from 2D platforms. Depending on the carrier type, there are key differences between static 2D culture and 3D suspension culture, such as shear stress, altered culture conditions, and different cell–cell interactions. Due to these differences, MSCs cultured in these two platforms may have a somewhat divergent biological profile. Particularly for MSC therapies, in which the mechanism of action is largely based on the secretome, changes to culture conditions may change critical attributes of the product. An example would be increased secretion of VEGF in response to suspension culture conditions [2].

The relevance of this consideration depends largely on the stage of therapy development. In therapies that have not yet reached the clinic, this issue is of lesser importance. However, in therapies that are in clinical stages, switching from a 2D culture to suspension platform will require close examination of cell properties and potency as well as a comparability study to demonstrate to the regulators that product characteristics such as potency are unaltered.

6.2 SEED TRAIN

The seed train refers to a series of unit operations selected to expand cells after thawing of a cryopreserved working cell bank (WCB) vial, prior to being transferred to a final production step. On the upstream side, each expansion step consists of a defined quantity of culture vessels of a certain scale and the time spent at that step. For example, the first expansion step of an MSC culture from the WCB vial may occur in three 10-layer cell stacks for 3–4 days. Each expansion step is larger in scale than the previous one, leading up to the final expansion step, after which one enters the downstream phase of the process, where the unit operations focus on removing any undesired components from the product, concentrating the cells and formulating them in the appropriate buffer for cryopreservation.

Bioreactor single-use vessels are available at several scales, as discussed in Section 6.5. However, one must think about the seed train all the way from tissue acquisition to final dose preparation to determine the scale of the final expansion step and the needed seed train to facilitate this final step. The seed train is determined by the number of cells needed for the final expansion step (which in turn is determined by lot size needed to satisfy clinical need) as well as by size of the WCB and the desired number of cell population doublings. In addition, the scale of the final expansion should not just be the largest available, but some considerations should be given to the bottlenecks that may lie at the downstream end of the operation. For example, for an indication with an extremely large patient population such as stroke, one may calculate that a 2000 L working volume bioreactor is the most cost effective way to produce one dose of product. However, the visual inspection setup may not be able to handle the number of doses that such a bioreactor would produce in the short window of time that cells can spend in cryopreservation buffer. These downstream bottlenecks are detailed in Section 6.7. With these considerations in mind, one should consider the following questions when designing the seed train for an MSC manufacturing process, which is described using a flow chart in Fig. 6.2.

- What is the number of cells obtained from the primary tissue source?
- What is the number of cells needed for seeding for the final expansion step?
- What is the maximum number of population doubling level allowed for the final product?

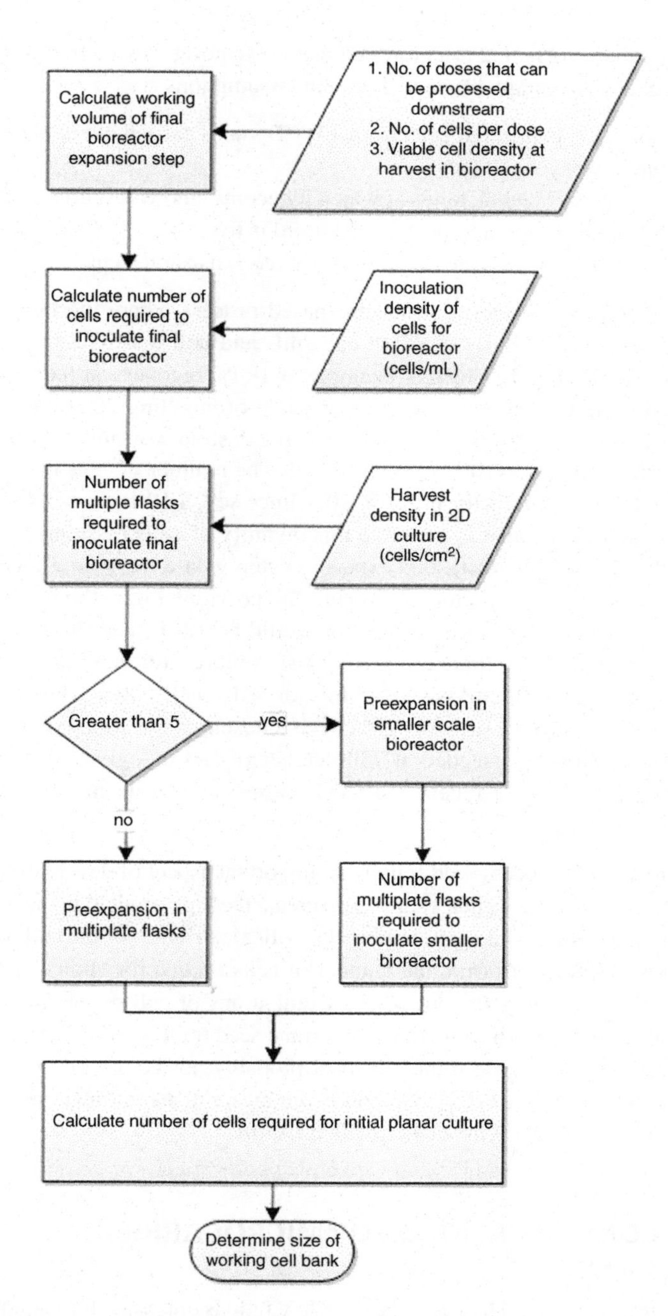

■ **FIGURE 6.2** A decision-making flowchart to design a seed train for an MSC therapeutic product.

The flow chart in Fig. 6.2 is a proposed method to design a seed train for an allogeneic MSC therapy product. The main assumptions are:

- The initial expansion of cells from a WCB vial is carried out in planar culture (10-layer cell factories).
- The number of cell factories at which it becomes more cost effective to move to bioreactor culture (for seed train) is five.
- The upstream phase consists of two or three expansion steps.

As an example, consider a bioreactor-based process where the bioreactor inoculation density is 2×10^4 cells/mL and cell density at harvest is 1.5×10^6 cells/mL. For this example, a 100% recovery at harvest is assumed. The downstream process is capable of handling 2000 doses (1×10^8 cells/dose). At this harvest density, the desired working volume to obtain the needed 2×10^{11} cells is ~135 L. The number of cells required to inoculate this bioreactor is 2.7×10^9. Since seven 10-layer cell factories (CF-10) are needed to generate this quantity of cells (assuming 4×10^8 cells/CF-10 at harvest), this expansion step would be more effective if carried out in a bioreactor. Following the previous logic, the working volume of the preexpansion bioreactor would be 1.8 L, and the number of cells needed to inoculate the preexpansion bioreactor is 3.6×10^7. A planar surface of 570 cm^2 is needed for 3.6×10^7 cells, so a 1-layer cell factory with a surface area of 632 cm^2 is chosen as the first expansion step. Planar culture is seeded at 7500 cells/cm^2 and this gives a desired WCB size of 4.75×10^6 cells/vial. This case is shown in the seed train described in Fig. 6.3.

In summary, the process seed train is an important aspect of developing an MSC expansion process that should help create the highest yield in the most cost-effective way, while maintaining the critical attributes of the cell. One should work backward from the number of cells needed for clinical doses and carefully consider the downstream limitations of cell processing and expansion. Using this information, an optimal seed train should be devised. Although this exact seed train will most probably not be used during the early phases of product development, being aware of the ultimate process and striving to mimic it in smaller scale is useful.

6.3 CONSIDERATIONS FOR MICROCARRIER SELECTION

As discussed previously, increasing the SA/V ratio is important for manufacturing MSCs at a commercially viable cost. This can mainly be achieved by using microcarriers in bioreactors or other fibrous scaffolds in packed/fixed

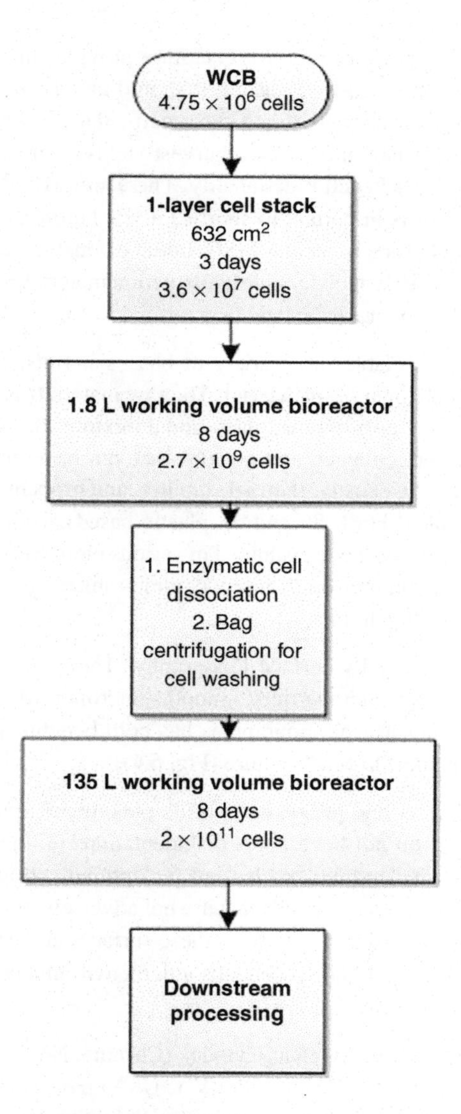

FIGURE 6.3 An example of a seed train of an MSC manufacturing process yielding 2,000 doses per batch (100 million cells per dose).

bed systems. Fixed- and packed-bed systems provide higher SA/V ratios than microcarrier-based platforms, but the latter is preferred due to the relative ease of detaching the cells from the microcarriers. Cell detachment from microcarriers takes less time and provides a higher efficiency of harvest per unit volume than harvesting cells from packed/fixed beds.

While the use of microcarriers in MSC culture provides increased surface area for cell attachment and expansion, it should also be noted that using a microcarrier that is not well suited for a particular process could result in cell aggregate formation, decreased harvest yields, decreased culture viability, and/or changes in cell functionality. There are several microcarriers in the market, and it is important to shortlist a few before proceeding with experimental testing for a particular MSC. Initial evaluation should focus on the physical properties, which include construction material, surface charge, surface coating, diameter, shape, and density.

Microcarriers are available in a variety of base materials, surface-treated or untreated, and charged or uncharged. The raw materials used to produce microcarriers include polystyrene, glass, and a mixture of peptide and protein components (eg, collagen and gelatin) that can have an effect on the toxicity, permeability, density, thermal stability, and propensity to generate unwanted particulates. For MSC culture, plastic-based carriers with or without protein coatings have been evaluated in various bioreactor processes [3]. Table 6.1 shows a comparison of a few common microcarrier types available for adherent cell culture.

Microcarriers also vary by surface topography. There are three common surfaces available for microcarriers: smooth, microporous, and macroporous. Each of these surface topographies has both benefits and challenges associated with respect to MSC culture (Fig. 6.4).

Smooth microcarriers [eg, plastic and plastic plus microcarriers (Pall, Port Washington, NY)] do not have pores or indentations on their surface and are often coated with proteins or charged for optimal cell attachment and growth. They are best used for cells that are not adversely affected by shear. Cells are typically removed easily from these surfaces at harvest, and there is reduced possibility of heterogeneous microenvironments as the cells grow as a monolayer.

Microporous microcarriers such as Cytodex (Corning, NY, USA) and GEM (Global Cell Solutions, Charlottesville, VA, USA) have surface pores that are too small for cells to infiltrate, yet provide slightly increased surface area and indentations for better cell anchorage. The pores of these carriers are also capable of allowing various nutrients to diffuse across the microcarriers' inner structure [3].

Macroporous microcarriers such as Cultispher (Percell, Åstorp, Sweden) and Cytopore (GE Healthcare, Pittsburgh, PA, USA) have internal pores with sizes ranging from 30 to 400 μm in diameter, providing a very large surface area for the growth of anchorage-dependent cells. MSCs have been

Table 6.1 Comparison of Commercially Available Microcarriers

Microcarrier Name	Manufacturer	Core Bead Material	Coating	Charged?	Relative Density (g/cm^3)	Diameter (μm)	Surface Area (cm^2/g)
Hillex CT	Pall/Solohill	Modified Polystyrene	n/a	Yes	1.090–1.150	160–200	257
Hillex II	Pall/Solohill	Modified Polystyrene	n/a	Yes	1.090–1.150	160–200	515
Plastic	Pall/Solohill	Polystyrene	n/a	No	1.022–1.030, 1.041–1.049	90–150, 125–212	480, 360
Plastic Plus	Pall/Solohill	Polystyrene	n/a	Yes	1.022–1.030, 1.041–1.049	90–150, 125–212	480, 360
Glass-Coated	Pall/Solohill	Polystyrene	High Silica Glass	Negatively	1.022–1.030	125–212	360
Pronectin F	Pall/Solohill	Polystyrene	rRDG-containing protein	Yes	1.022–1.030, 1.041–1.049	90–150, 125–212	480, 360
FACT III	Pall/Solohill	Polystyrene	Porcine Collagen Gelatin Protein	Yes	1.022–1.030, 1.034–1.046	90–150, 125–212	480, 360
Collagen Coated	Pall/Solohill	Polystyrene	Porcine Collagen Gelatin Protein	No	1.090–1.150, 1.022–1.030, 1.041–1.049	160–200, 125–212	515, 360
Synthemax II-polystyrene	Corning	Polystyrene	Synthemax II	No	1.022–1.030	125–212	360
Collagen Coated	Corning	Polystyrene	Collagen	No	1.022–1.030	125–212	360
Positive charge	Corning	Polystyrene	n/a	Yes	1.022–1.030	125–212	360
Nontreated	Corning	Polystyrene	n/a	No	1.022–1.030	125–212	360
Enhanced attachment polystyrene	Corning	Polystyrene	CellBIND	Yes	1.022–1.030	125–212	360
CultiSpher-G	PerCell	Porcine Gelatin	n/a	No	1.04	130–380	4000–7000
CultiSpher-S	PerCell	Porcine Gelatin	n/a	No	1.04	130–380	4000–7000
Cytodex 1	GE Healthcare	DEAE-dextran	n/a	Yes	1.03	131–220	440
Cytodex 3	GE Healthcare	DEAE-dextran	Denatured Collagen	No	1.04	133–215	270
GEM (Global Eukaryotic Microcarrier)	Global Cell Solutions	Alginate	Gelatin, Matrigel, Collagen, Fibronectin, Poly-D-lysine	Yes	Varies	75–150	Varies

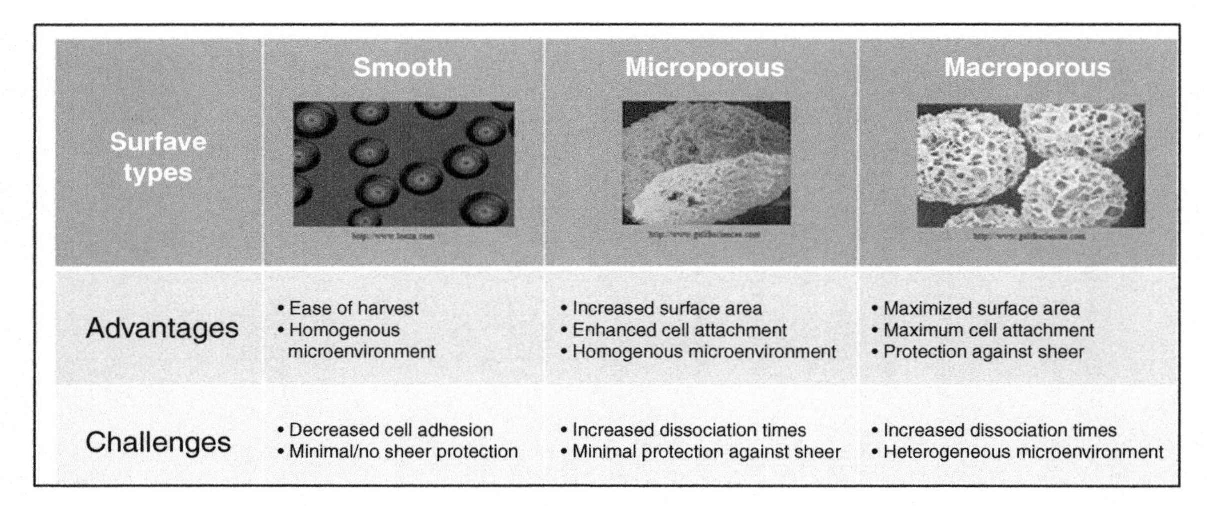

FIGURE 6.4 Advantages and challenges associated with microcarrier topography.

observed attaching to and expanding both on the surface of and within such microcarriers (Lonza Walkersville, Inc., Walkersville, MD, USA, unpublished). MSCs growing within these larger pores are provided with increased protection from shear forces created by agitation, microcarrier bead-to-bead collision, and gas sparging in bioreactors.

Deciding which microcarrier to use will depend on the individual process. For example, one could expect a relatively easy dissociation of MSCs from a smooth microcarrier as opposed to porous microcarriers, since the dissociation reagent would not have to penetrate the interior of the carrier. However, if the smooth microcarrier was coated or charged, it is possible that the cells would be more tightly attached to the microcarrier surface, resulting in increased difficulty to dissociate the cells. Also, porous microcarriers may provide better protection in highly dense, high agitation cultures when compared to smooth microcarriers and will usually provide more surface area for cell expansion. However, porous microcarriers may also have a higher probability of changing cell attributes due to the heterogeneous microenvironments within the carrier and the possible development of nutrient gradients as one moves from the surface of the carrier toward the center. Both solid and macroporous carriers should be characterized in terms of cell growth and ease of dissociation at the end of culture. Finally, the critical quality attributes of the cells should be examined to decide which type of carrier to move forward with.

Size and density of the microcarrier are also important considerations. It is possible to maintain homogenous suspensions under lower agitation speeds when using smaller, less dense microcarriers. However, if a process utilizes sedimentation procedures for media exchange, the use of larger, denser microcarriers can decrease the time for the microcarriers to settle to the bottom of the culture vessel.

Additional considerations are the challenges associated with the scale-up and commercialization of the process. For example, it is important to know what options are available for obtaining research and GMP-grade microcarriers from manufacturers. To prevent hurdles during clinical and commercial phases, it is important to investigate early in the development cycle of a microcarrier process whether a GMP-grade version of the chosen carrier exists.

Since microcarriers are in contact with the cells, microcarrier sterility is essential. If autoclaving microcarriers, one should evaluate the thermal and mechanical stability of the carriers after sterilization. If microcarrier stability is compromised by autoclaving, alternative sterilization options such as gamma irradiation or dry heat can be considered. While there may be cost savings associated with thermal sterilization, it may not be feasible to do so with large quantities of microcarriers at the commercial scale.

6.3.1 Experimental Strategies for Microcarrier Selection

Once the selection of carriers has been narrowed to three or four criteria, it is useful to perform a thorough screening and comparison of the candidates. Some of the experimental variables that must be investigated are:

- Microcarrier concentration
- Cell to microcarrier ratio for seeding
- Bioreactor agitation rate required for homogenous suspension of the carrier
- Cell attachment efficiency
- Aggregate formation
- Cell dissociation and separation protocol
- Critical quality attributes (CQAs) of MSCs grown on the carrier
- Trace residuals and particulate counts post-harvest

Utilizing small disposable spinner flasks at the earliest stages of microcarrier selection is recommended to observe the microcarrier–cell interactions at a relatively low cost. There are also automated, small-scale bioreactor systems that can be utilized for high-throughput screening of microcarrier–cell interactions in a controlled environment, but these require large upfront investment and operator training. Small-scale studies (15–500 mL culture volumes) can yield useful information regarding many of the mentioned variables. However, scaling-up presents its own challenges, which may result in the need for additional process optimization.

If a 2D process has already been developed, then much of that information can be used during the development of a microcarrier-based process. As an example, if the specific surface area of the carriers is known, then the seeding density of cells per unit area from the 2D process can be translated into the 3D process. If there is no established 2D protocol, it is advisable to begin with the microcarrier manufacturer's recommendations or consult the literature for use in a particular media or vessel first, then perform optimization studies during the process development stage of the project.

Another critical process parameter to control in stirred suspension cultures is agitation speed of the impeller. Differences in microcarrier density will affect the required agitation rate to maintain a homogenous suspension. As cells proliferate on the microcarrier surface, the cell–microcarrier complexes will require higher agitation rates to stay in suspension because of their increased mass. In addition, it is possible to decrease and/or control the size of microcarrier–cell aggregates with higher agitation. However, extremely high agitation rates could cause adverse effects on the cell viability and CQAs because of the shear stress. Moreover, higher agitation rates could lead to foaming in serum-containing media.

Proliferation of anchorage-dependent cells depends on efficient cell adhesion to the microcarrier. Choosing a carrier with either a porous structure or a collagen/fibronectin coating are two suggestions for increasing cell attachment efficiency in 3D culture. Attachment can be assessed by the presence of elongated and flattened cells on the carrier surface.

Low agitation speeds or static seeding conditions usually maximize cell attachment as well as help to ensure homogenous seeding, while static conditions allow cells time to flatten and elongate in a controlled environment without being challenged by shear forces. To maximize homogenous seeding it is also possible to inoculate MSC cultures under static conditions with the incorporation of intermittent, low-speed mixing steps.

When optimizing seeding parameters, culture samples should be observed at several time intervals to monitor cell attachment efficiency by either direct or indirect methods. Direct methods include performing visual observations under a microscope or obtaining cell counts of both the supernatant and cells dissociated from the microcarrier culture sample (for a quantitative assessment of cell attachment efficiency). For smooth and microporous microcarrier types, it may be feasible to determine optimal seeding conditions through light microscopy. However, for macroporous microcarriers, the use of light microscopy is difficult as many cells attach on the internal surface of pores. Therefore, fluorescent staining of cells and observation using fluorescent microscopy or obtaining cell counts post detachment of cells from carrier is recommended.

In microcarrier-based culture, MSCs may form complexes of several microcarriers with cells growing and bridging across discrete carriers. This is likely facilitated by the secretion of extracellular matrix by the cells. Because of the formation of concentration gradients, these aggregates can limit optimal oxygen and nutrient transfer to the cells within the microcarrier–cell complexes, and also prolong dissociation of cells from the microcarriers. Microcarrier aggregation is more likely to occur when using coated microcarriers in low agitation cultures, so increasing agitation speeds in some cases can reduce the number of observed aggregates. However, it is important to ensure that the increased shear forces produced by the higher agitation speeds do not adversely affect the cells.

Once a microcarrier is selected, it is important to determine the optimal protocol for dissociating the MSCs from the microcarriers. A successful microcarrier dissociation process should:

- Require minimal time
- Obtain a high yield (percentage of attached cells that are obtained after harvest)
- Produce cells of high viability in single cell suspension
- Maintain product CQAs

It is recommended to use harvesting methods from the literature or the microcarrier manufacturer as a starting point and optimize the protocol through experimentation, as needed. This could include comparing agitation speeds, concentrations and mixtures of proteolytic enzymes, process buffers, harvest pH, temperature, and preenzymatic washes and processing times. After dissociation, an additional step to separate the cells from the microcarriers is required as part of the downstream processing, which is discussed in more detail later in the chapter. Once the final product has been separated from microcarriers, characterization assays specific to the end product should be performed to ensure the culture and harvesting protocols have had no detrimental effect on product quality.

At more advanced stages of process optimization, it is also critical to evaluate the noncell particulate load and minimize the presence of particulates in the final product. Particulates can be generated from the carriers or from other single-use components that come into contact with cells and media during the process.

To summarize, the use of 3D culture for MSC manufacturing is a viable method for obtaining MSCs at high quality and acceptable cost. The choice of microcarrier or other 3D substrates used for cell culture needs to be carefully evaluated and compared to assure an optimal process and cells that meet all requirements.

6.4 IN-PROCESS CULTURE CONTROL: OPTIONS AND COMPONENTS

To maximize harvest yields and maintain consistent quality, MSC culture processes must ensure that key process parameters (KPP) such as dissolved oxygen, dissolved carbon dioxide (CO_2), ammonium ions, pH, temperature, glucose, and lactate are monitored in a real-time, semicontinuous or continuous basis. The ideal monitoring systems should be:

- Highly precise and accurate
- Single use or noncontact. If this is not possible, then it should be easy to sterilize
- Inert, with no impact on final product quality or culture conditions
- Scalable as the process moves through research to commercial stages
- GMP-compliant
- Easy to install, maintain, and operate
- Cost-effective

There are several options for monitoring key process parameters such as nutrient and waste concentrations. Offline methods require a sample to be removed from the culture and analyzed outside of the bioreactor. Online sampling methods are able to measure key process parameters within the bioreactor system without product loss. At a minimum, a bioreactor should have online measurement and control of temperature, pH, dissolved oxygen, and agitation rate.

6.4.1 Offline Measuring Methods

There are various types of stand-alone instruments used to measure key process parameters offline and validate online measurements. For example, while many bioreactors are equipped to monitor pH online once a bioreactor process starts, the pH probe must first be calibrated against standardized buffers. However, once the process is initiated, intermittent one-point standardization is conducted by removing a sample and measuring its pH using an offline pH meter. Periodic offline measurements of the glucose and lactate concentrations can help establish the glucose consumption rate as well as ratios of lactate generation to glucose consumption. These measurements can help establish if cell metabolism is aerobic or anaerobic. There are three main options for measuring culture nutrient and metabolite concentrations offline—enzyme-linked membrane technologies, ion-selective electrodes, or photometric technologies. Enzyme-linked membrane technologies, such as the NOVA BioProfile Flex (NOVA Biomedical, Waltham, MA) or YSI 2900 Biochemistry Analyzer (YSI, Inc., Yellow Springs, OH), measure the free flow of electrons when a particular analyte of interest comes in contact with the immobilized enzyme layer of a biosensor membrane. For example, with the YSI 2900 the culture supernatant is introduced to the instrument's fluidic flowpath and comes in contact with an immobilized enzyme layer of the biosensor membrane. Once the substrate touches the immobilized enzyme, the substrate is oxidized and produces hydrogen peroxide. The hydrogen peroxide is then oxidized on the platinum electrode of the biosensor and the free flow of electrons creates a current proportional to the substrate concentration [4]. Some of the compounds commonly measured using enzyme-linked membrane technologies are glucose, lactate, glutamine, and glutamate.

Ion-selective electrodes require a permeable, ion-selective membrane, reference electrode, and internal filling solution of a fixed concentration of the ion of interest (eg, sodium, potassium, ionized calcium, and ammonium). The charge created when the ions diffuse across the permeable membrane in the presence of the culture supernatant is then correlated to a concentration measurement of the particular analyte [5]. Ion-selective electrodes can be used to measure dissolved gas levels and ammonium ions.

Photometric Technologies, for example, Roche Cedex (Roche, Basel, Switzerland) and Optocell Cubian (Optocell, Bielefeld, Germany), incorporate absorbance measurements into the system to determine the concentration of a given substrate. Essentially, a sample of the culture supernatant is loaded into a specialized cuvette, the substrate reagent specific to the analyte of interest is added and mixed with sample, the mixture is incubated for 10–20 min at 37°C, and the nicotinamide adenine dinucleotide phosphate (NADPH) or chromogen formation is measured by UV absorbance and correlated to the analyte concentration [6].

When working with MSCs grown on microcarriers, cell density can be assessed via offline methods or online methods. For offline sampling, a representative sample must first be aseptically removed from the bioreactor, the cells dissociated and separated from the microcarriers, and the cells

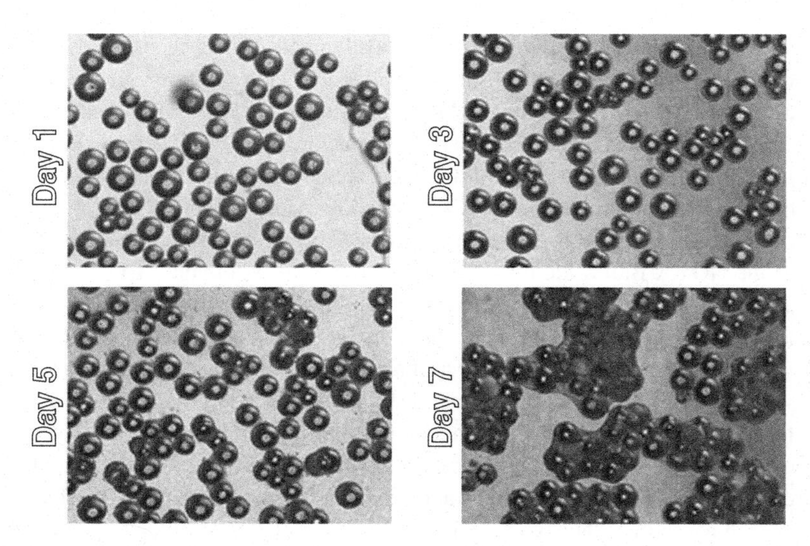

■ **FIGURE 6.5** BM-hMSC aggregation in microcarrier bioreactor culture days 1, 3, 5, and 7 postinoculation (*Source*: Lonza Walkersville, Inc., Walkersville, MD, USA, unpublished)

counted using automated or manual counting methods. However, culture sampling may become problematic if microcarrier aggregates form—this leads to nonrepresentative samples because larger aggregates do not pass through the small inner diameter sampling tube. Without optimal process control, aggregates are common and removing a representative sample from the reactor requires large sample volumes in the later stages of the culture (Fig. 6.5).

6.4.2 Automated Sampling Systems

There are several offline instruments, such as the previously discussed NOVA and YSI systems, which have the ability to aseptically remove and transfer a sample from the bioreactor culture to the instrument for analysis. This removes the need for a technician to remove a sample and also allows for sampling during off-work hours. However, the use of microcarriers may make sampling from MSC suspension cultures problematic, as a microcarrier separation step is recommended to keep the microfluidic pathways free of obstructions. As cultures become more confluent, the probability of aggregate formation increases, and obtaining representative, homogeneous samples for cell counting can become more challenging. It is possible to filter out microcarriers to allow for off-line supernatant analysis using these automated sampling instruments, but there is the risk of clogging and costs associated with initial development and testing of these types of sample removal devices.

While offline monitoring of bioreactor cultures is beneficial, it is labor intensive, has risk associated with operator-dependent errors, and can only supply a few data points per day. Ultimately, a better option for most MSC bioreactor processes is to incorporate online monitoring technologies.

6.4.3 Online Measuring Methods

Traditionally, planar tissue culture systems were incapable of real-time measurements of pH and dissolved gas tensions while bioreactors relied on invasive, product contact electrochemical probes. These probes must be autoclaved before each use and require extensive maintenance. More recently, single-use chemical optical sensors, also referred to as optodes, have enabled noninvasive measurements of microcarrier cultures [7]. A major manufacturer of optodes is PreSens (Precision Sensing GmbH, Regensburg, Germany). The PreSens microsensor optodes are referred to as "Sensor Spots," and are used to monitor pH, dissolved oxygen, or dissolved CO_2 of cultures in conjunction with an optical fiber fixed opposite the sensor spot in glass or transparent plastic vessels. Due to the smaller size, Sensor Spot technology can be incorporated into smaller culture vessels where installation of a probe is not feasible. For example, the SDR SensorDish system (PreSens, Precision Sensing GmbH, Regensburg, Germany) allows for the continuous, online monitoring of either pH or DO in 6- or 24-well plates manufactured with the PreSens optodes. It is also possible for end users to apply sensor spots to other commercial available culture vessels, such as 125 mL spinner flasks [7]. Furthermore, the pH and dissolved oxygen sensor spot technology has been incorporated in several bioreactor vessel types and sizes, including the Celligen BLU (Eppendorf, Hauppauge, NY) Sartorius AMBR and Sartorius Cultibag (Sartorius, Bohemia, NY), and Applikon BioSep (Applikon Biotechnology, Delft, The Netherlands) vessels.

Because of the labor intensiveness and inaccuracy of offline cell counting, a move to online monitoring of cell density is desirable. There are some probes currently on the market that measure the capacitance in a small region around the probe and correlate this to the viable cell density in the bioreactor [ABER Futura Biomass sensor (Applikon Biotechnology, Foster City, CA), Fogale Nanotech (Nimes, France)]. The intact plasma membranes of viable cells act as capacitors in the presence of an electric field. When a temporary electric field is generated by the probe, a charge builds near the nonconductive membrane of each cell and the resulting capacitance can be measured. This capacitance is proportional to the membrane bound volume of viable cells [8]. Lonza has performed preliminary studies to assess the usefulness of capacitance probes with MSC microcarrier cultures (data not published). From these studies, Lonza has concluded that a correlation between capacitance measurements and viable cell density for MSC microcarrier bioreactor cultures can be made when total viable cell density is greater than 5×10^5 viable cells/mL and the culture suspension is homogeneous.

Because of the tendency of MSCs to secrete extracellular matrix and create large microcarrier aggregates, it is recommended to keep the flow path within the bioreactor vessel as clear as possible; any obstructions or narrow spaces between tubes and walls are likely to be hotspots for aggregate formation. Therefore, adding unnecessary probes in the microcarrier culture is not recommended.

6.5 PLATFORM COMPONENTS

When considering the development of a bioreactor-based process for MSC expansion, it is helpful to take a platform-based approach and devote efforts to optimizing each component of this platform. A basic platform would consist of the following components:

- Growth medium
- Microcarrier
- Vessel
- Controller

The topics of medium optimization and microcarrier use are discussed in detail in other sections of this chapter and therefore will only be mentioned briefly in this section.

6.5.1 Growth Medium

Most studies of MSC expansion in both planar and bioreactor systems report using FBS at a range of 2–20% (v/v) in medium. It is desirable to make efforts to reduce the level of FBS used as much as the process allows, or to switch to serum-free medium. This is preferred to minimize process variations that are a result of using different lots of FBS, as well as to reduce the risks associated with animal-derived components and to remove supply chain constraints of these components.

Depending on the chosen vessel, scale, cell density, and agitation speeds, maintenance of key process parameters such as pH and dissolved oxygen may require gas sparging through the liquid medium (sparging refers to gas transfer by bubbling a gas through the liquid as opposed to passive transfer across the gas–liquid interface in the headspace). If the medium contains FBS at any concentration, gas sparging can lead to the development of a significant amount of foam at the surface of the culture medium, which increases the risk of compromised vessel integrity, decreased viable cell density, and reduces the surface area for cell growth as microcarriers become trapped in the foam. In case of foaming, the addition of an antifoaming agent such as pluronic into medium should be considered. It is also important to consider that FBS reduces shear stress in the bioreactor environment; therefore, when using serum-free medium, or when high shear stress is needed for culture, the addition of compounds that increase medium viscosity should be considered.

6.5.2 Vessel

The first consideration when choosing a vessel is selecting which material is best for the process: plastic (single-use vessel), glass, or a stainless steel. Each type has its advantages and disadvantages depending on the needed volume. Single-use vessels are the fastest to set up, do not require cleaning or sterilization, and are readily available in most formats from 0.5 to 50 L. However, it is not possible to introduce modifications if needed, which may be important especially during process development, and single-use vessels are less cost effective in some cases. Glass vessels are flexible in that modifications and changes to dip tubes can be done and then the complete vessel can be sterilized by autoclaving. But cleaning, assembling, and sterilizing glass vessels are time consuming and prone to human error. Stainless-steel vessels are usually used for larger volumes when single use is not available.

Depending on the source of the power input required to keep the medium mixed, bioreactors are classified into two main categories: mechanically driven and hydraulically driven.

6.5.2.1 *Mechanically driven bioreactors*

For a mechanically driven bioreactor, the power is supplied through an electrical motor. The motor is connected to an impeller in the vessel or to the platform supporting the vessel. The main categories of mechanically driven bioreactors are stirred-tank bioreactors, rocker bioreactors, and orbital shaken bioreactors.

6.5.2.1.1 Stirred-tank bioreactors

Stirred-tank bioreactors (STR) consist of one or more impellers in the center of the chamber. These impellers can be mounted on a vertical shaft or directly on the bottom of the vessel. In recent years, some novel impeller types have been developed where the axis of rotation is not strictly vertical (See Pall PadReactor in Table 6.2). Stirred-tank bioreactors are one of the

Table 6.2 Primary Manufacturers of Single-Use Bioreactors for Cell Therapy Manufacturing

Brand Name	Manufacturer	Available Working Volumes
ambr 15	Sartorius	15 mL
Univessel SU	Sartorius	0.6–2 L
BIOSTAT STR	Sartorius	12.5–1000 L
Allegro STR	Pall	60–200 L
PadReactor Mini	Pall	13 L
PadReactor	Pall	25–1200 L
Mobius	EMD Millipore	3–2000 L
MagDrive Vertical Wheel	PBS Biotech	2–15 L
Hyperforma	Thermo Fisher Scientific	30–2000 L
BioBLU	Eppendorf	0.25–50 L
Xcellerex	GE Life Sciences	4.5–2000 L

most widely understood bioreactor models and have been successfully used in the monoclonal antibody production industry for over two decades. It is also the most widely represented bioreactor type in the industry. Some of the main manufacturers of stirred-tank, single-use bioreactors along with the range of working volumes are listed in Table 6.2.

6.5.2.1.2 Rocker bioreactors

Rocker bioreactors are a class of reactors consisting of a platform on which the vessel is placed, which provides movement around one or more axes by an electrical motor. The main advantage of rocker bioreactors in many processes is the low shear environment for the cells, as they are not directly exposed to the fast moving tip of an impeller blade. In addition, it is easier to achieve good gas mixing at large working volumes in rocker bioreactors without the need for sparging bubbles through the culture medium. The main brands in this category of reactors are Sartorius BIOSTAT Cultibag (Goettingen, Germany), GE WAVE Bioreactors (Marlborough, MA), Pall XRS system (Port Washington, NY), and the Finesse SmartRocker (Santa Clara, CA).

6.5.2.2 *Hydraulically driven bioreactors*

The main distinguishing feature of hydraulically driven bioreactors is that they incorporate some of the functionality of both a 2D planar system and a stirred tank bioreactor. In these systems, the cells are fixed to a substrate while the medium is in motion (either recirculated or perfused). Within this category, there are several different vessel types, some of which are discussed as follows.

6.5.2.2.1 Pall Xpansion multiplate bioreactor system

Pall Xpansion Multiplate bioreactor system (Pall, Port Washington, NY) consists of multiple circular polystyrene plates (ranging from 10 to 200) spaced 1.6 mm apart. Each plate has a surface area of 612 cm^2. The vessel is completely filled with medium and the cells adhere to the plates. A series of gas-permeable tubings run through the center for gas transfer and the medium itself can be perfused for nutrient replacement. This platform is an attempt to conserve the traditional multiplate planar model, while providing an advanced level of process parameter control in a closed system along with automated medium feeding.

6.5.2.2.2 Fixed-bed bioreactors

Fixed-bed bioreactors refer to a class of bioreactors where the carrier particles are tightly packed and provide a 3D-like matrix for the cells to adhere to, while the medium is continuously passed through the packed bed. One example of this class is the Eppendorf BioBLU Packed-Bed vessel, which is a single-use 5L vessel preloaded with Fibra-Cel Disks. One hundred fifty grams of these dual layer nonwoven polyester and polypropylene disks provide a surface area of 0.12 m^2. Fixed-bed reactors are also available at a much larger scale, as represented by the Pall iCELLis platform. The iCELLis system provides up to 500 m^2 of surface area in a 25 L working volume. However, while fixed-bed reactors provide an extremely high surface area-to-volume ratio, their main disadvantage is that traditional enzymatic methods of cell detachment are inefficient because the cells are enmeshed in the substrate.

6.5.2.2.3 Hollow fiber bioreactors

Hollow fiber bioreactors consist of a bundle of polymeric fibers assembled in a cylindrical cartridge. The cells adhere to the inner surface of these hollow fibers while the medium circulates through the lumen. The two main examples of this technology today are the Quantum Cell Expansion system (Terumo BCT, Lakewood, CO) and the FiberCell System (FiberCell Systems, Frederick, MD). Both of these systems provide a surface area just over 2 m^2. As with the fixed-bed bioreactors, harvesting the cells is a challenge and no large-scale platform is available at the moment.

6.5.3 Controller

The bioreactor controller is the "brain" of the bioreactor platform. Its primary function is to *monitor* and control the critical process parameters within the vessel (eg, pH, culture temperature, dissolved oxygen tension, and media exchange). The reactor is able to provide an output to specified process

Table 6.3 Examples of Sensors and Control Strategies for Process Parameters in Bioreactor-Based Manufacturing

Process Parameter	Sensor	Control Element
Temperature	Resistance temperature detector	Heating Blanket/Water Jacket
pH	pH probe	Carbon dioxide mass flow controller to reduce pH pump for base addition to increase pH
Dissolved oxygen tension	Polarographic oxygen sensor	Oxygen mass flow controller
Glucose	Spectroscopic sensor	Pump for glucose solution feed

parameter control elements to *correct* for any deviation in the value of these parameters from the user-defined set point. The number of parameters that can be monitored and controlled is limited by the number of sensors and control elements incorporated into the bioreactor vessel, which is recognized by the bioreactor controller. Examples of sensors and control elements are provided for the most common process parameters in Table 6.3. Please note that this table is only exemplary and does not provide an exhaustive list of sensors, parameters, or control strategies.

6.6 SCALABILITY/COMPARABILITY

When approaching the culture of MSC in bioreactors, scalability is one of the most important considerations. As mentioned previously in the chapter, it is possible that for large indications, bioreactor vessels with working volumes 200–1000 L or larger will be needed to supply market demand. However, for R&D, process development, and preclinical/early clinical trials these scales are not required and not effective from a throughput and cost perspective. Therefore, understanding the capacity for scaling up bioreactor platforms from small volumes used in R&D and early clinical trials (eg, 15 mL to 5 L) to larger volumes used in commercial scale manufacturing (eg, 50 –1000 L) without significantly changing the culture dynamic or cell biology is critical. Differences in the scalability of culture parameters and platform components between small and large bioreactors should be considered when choosing a bioreactor platform and when scaling up or down. Examples of parameters that may change due to scaling are shear stress, oxygen transfer rates (moving from head space based gas supply to sparging), the necessity of using antifoaming compounds, length of downstream processing time, etc. However, these differences should be minimized and taken into consideration to the extent possible. Lack of attention to scalability can result in changes to the growth kinetics, cell biology, and the critical quality attributes of the cells, which may result in loss of cell efficacy and inability to scale the process.

The use of small-scale cell culture platforms to study cells and optimize manufacturing components and processes (eg, media composition, feeding rate, growth factor addition, cell–surface and cell–cell interaction, etc.) can be very effective and allow a high-throughput design of experiment (DOE) approach. In many cases, the correct R&D approach may be extremely beneficial in terms of improving cell quality and reducing cost of goods. Examples of small-scale bioreactors that facilitate this type of research and process development include the ambr15 system (Sartorius Stedim Biotech GmbH, Goettingen, Germany), which includes 24 or 48 independently controlled 15 mL vessels, each with an impeller, continuous measurement of pH, dissolved oxygen and temperature, and the option for automated sampling and media replacement. However the shape of the vessels is cuboidal, and thus this system is not identical in architecture to most single-use bioreactors used for larger-scale MSC expansion (which are typically cylindrical). The ambr15 system and its performance with regard to MSC culture is further discussed later in Section 6.8. Another small-scale option is the Pall Micro-24 MicroReactor (Pall, Port Washington, NY) in which the vessels are cylindrical, but not equipped with an impeller. They are agitated by an orbital shaking mechanism and are thus dissimilar to larger stirred-tank bioreactor platforms that have an impeller to agitate the liquid. Slightly larger small-scale bioreactor options are various 250 mL to 1 L vessels. These are offered by multiple companies and include the Applikon MiniBio (glass reactor, Applikon Biotechnology, Delft, The Netherlands), Eppendorf DASbox (250 mL working volume, single use), and Eppendorf Celligen BLU 1c (1 L working volume, single use, Eppendorf, Hauppauge, NY). While these options are valuable and an attempt is being made by device manufacturers to offer scalable options for each, it is important to perform thorough evaluations of the process and the cells when scaling up.

Process parameters that should be considered when choosing and evaluating both upscale and downscale systems for MSC processes include:

- mixing time
- shear stress
- impeller design
- impeller tip speed
- method of gas delivery (head space versus sparging)
- k_La (the volumetric mass-transfer coefficient describing the efficiency with which oxygen is delivered to the fluid within the bioreactor vessel)
- downstream processing methods

- processing time
- separation method of cells from carriers

Critically, differences in these parameters should be minimized between scales as much as possible, and the effect of changes to these parameters on cell biology and critical product attributes must be examined closely when up-scaling so as not to change product attributes and lose product potency. The FDA strongly recommends adherence to the Quality by Design (QbD) approach, wherein the design space and operating space for process parameters is established early in the process development phase. Any effect of deviations of the key process parameters and critical process parameters on the critical quality attributes of the product should be considered and understood.

In the best of cases, scaling up will require close examination of cell characteristics and biology as changes are made. Understanding the impact of process changes on cell biology and potency is based first and foremost on understanding the mechanism of action of each therapy in the context of the specific clinical indication. MSC therapies that are based on a secretome mode of action are especially sensitive to process changes due to the sensitivity of protein expression to culture environmental changes. Relevant environmental changes include a wide array of physiochemical states that compose the milieu of the cell environment. Examples would be media composition, osmolality, pH, dissolved gas composition, cell–substrate interactions, cell–cell interactions, shear stress, etc. However, environmental or process changes that are a result of up-scaling may be detrimental to a certain MSC type, while having no effect or a positive effect on others. Good understanding of the therapy-specific mechanism of action allows one to assess whether process changes have an effect on product-critical attributes and potency.

Comparability is usually discussed in the context of regulatory requirements. However, it is important to remember that being assured that cells perform comparably well throughout the clinical trial process and beyond is first and foremost in the interest of the manufacturer. The loss of critical cell attributes due to process changes and up-scaling can result in clinical trial failure. Even worse, losing critical product attributes and not knowing that this is the case will result in failure for reasons unknown and that cannot be remedied. This reinforces the need for a complete understanding of the mechanism of action and in vitro and in vivo assays directly tied to this mechanism.

These assays can in turn be used to continually assess the critical quality attributes of the cells as the process is scaled up. As soon as a significant change is detected, one can immediately go back to look at the process change, whether it indeed has resulted in detrimental effects to the cell product, and devise a solution to address this. The types of assays relevant to MSC are versatile and discussed elsewhere in this book. The design of a comparability study for regulators that is mandated by up-scaling of the process (which is most probably accompanied by various process changes) needs to be very carefully considered. Clearly, an emphasis must be made on looking at critical product attributes; however, additional assays to ensure comparability should be considered.

A theoretical example would be a MSC therapy that targets an ischemic condition. This theoretical product has recently undergone up-scaling to 200 L bioreactors from 5 L bioreactors in the context of moving from a successful phase II to Phase III. An in vitro cell-based potency assay has been established; however, ancillary assays such as an in vivo animal model, secretion of several critical proteins, and expression levels of several relevant genes have also been established. The comparability study would be composed of comparing cells cultured for Phase II (stock) versus using an identical process to cells manufactured using the upscaled process. Ideally, several lots would be run, using cells from several donors in appropriate numbers to ensure adequate statistical power. Final product would be tested using the above assays and results analyzed to assess comparability. While some differences may be detected, a careful evaluation of both statistical significance and relevance of changes to product potency and efficacy must be undertaken and presented to the regulators. Finally, a discussion with regulators regarding the approach to comparability and the interpretation of comparability results is performed.

Critical aspects of comparability and up-scaling include:

- Upscaling should be done carefully while attempting to minimize physiochemical changes to the cell environment.
- The "right" cell expansion platform given the intended population size to be treated and the expected dose range should be chosen and implemented as early as possible.
- Physiochemical changes due to up-scaling and process changes (including materials) must be understood and tracked carefully.
- Understanding the product mechanism of action as best as possible is imperative.
- Development of relevant and robust assays to assess critical product attributes, first and foremost a potency assay, is imperative.
- An array of relevant assays should be used continuously to assess the cell product throughout process changes and up-scaling, and for comparability studies between different manufacturing process changes to ensure the cell product maintains critical attributes.

In summary, up-scaling and other process changes are needed and unavoidable in cell therapy development. Subsequently, resulting comparability studies are important. However, if changes are minimized to the extent possible, done correctly, and if cell attributes are monitored throughout process development, and comparability is appropriately designed and powered, these hurdles can be overcome.

6.7 DOWNSTREAM PROCESSING

Downstream processing refers to all the unit operations that occur once final cell expansion has been completed for the final therapeutic product. Conventionally, the downstream processing in a MSC process begins with the detachment of cells from the substrate and ends with the cryopreservation of the final product. It may be argued that the downstream part of a cell therapy manufacturing operation has the larger share of challenges and bottlenecks compared to the upstream operations because of the short window of time available to transfer MSCs from their preferred adherent state in culture medium into a cryopreserved state. As a result, the potential of thousands of liters of working volume (and trillions of cells produced) cannot yet be realized, because the downstream bottlenecks do not allow us to process such large numbers of cells [9]. The sequential unit operations comprising the downstream processing and the options available to achieve those are detailed in the following sections.

6.7.1 Cell Detachment

As with planar systems, for bioreactor platforms MSCs must be detached from the substrate, which could be microcarriers for stirred tank systems, a stationary matrix in fixed-bed systems, or the inside surface in a hollow fiber bioreactor. Enzymatic methods are most commonly employed, the targets of the proteolytic enzymes being the adhesion proteins on the cell surface. Trypsin has been most commonly reported in studies that include the expansion of MSCs on solid microcarriers [10,11]. However, if digestible microcarriers are used, then the enzyme mixture selected should target not only the cell surface adhesion proteins but also completely digest the substrate itself. This phase of the downstream processing needs extensive optimization to determine the maximum amount of time that the cells can be exposed to the enzyme as well as the optimal pH and temperature for optimal enzymatic activity.

6.7.2 Microcarrier Removal

This unit operation is necessary if indigestible microcarriers are used, since the carriers and cells are suspended together after completion of the cell detachment phase. One way to achieve separation of cells from carriers is by filtering based on the size difference between cells and carriers; MSC diameter is typically less than 30 μm [12], while most commercially available microcarriers are 80 μm or above in diameter. At smaller scales, this filtration may be achieved by the use of Baxter blood filters that consist of two mesh screens with a pore size of 40 and 150 μm, respectively. Custom filters with a choice of pore size and material are also available from vendors such as Zenpure (Manassas, VA, USA). The Harvestainer BPC product (Thermo Fisher Scientific, Waltham, MA, USA) is a filtration option for large-scale separations. It consists of a bag-within-a-bag design, with the 25 L inner bag being constructed of a mesh that traps microcarriers while allowing cells to flow into the outer 200 L container. For any filtration-based unit operation, it is necessary to optimize the flow rates for best cell recovery. Filters are prone to clogging, and thus backwashing may be required to increase viable cell recovery. In addition, it is important to note that if the pumping procedure is used to transfer the suspension across the filter, this may lead to particulate generation from the microcarriers. If microcarrier debris is smaller than the filter pore size, it may contaminate the final product.

6.7.3 Cell Washing and Concentration

The cell washing unit operation is required to remove any unwanted components from the cell suspension. This includes any animal products (such as FBS) and proteoloytic enzymes used to detach cells from the culture substrate. The concentration of cells in the product dose is usually much higher than that in culture at the time of harvest; therefore, the concentration step is typically carried out in tandem with the washing unit operation. There are two main approaches for cell washing and concentration—centrifugation and tangential flow filtration.

6.7.3.1 *Tangential flow filtration*

Tangential flow filtration is a technology successfully employed in large-scale protein manufacturing processes, where the suspension containing the cells and product (secreted protein) is pumped along the surface of a membrane. The trans-membrane pressure forces any molecules smaller than the membrane pore size through to the filtrate side. In protein processing, the proteins on the filtrate side are the desired product, and the process can be carried out at large scales with the cells exposed to high shear and trans-membrane pressures. However, the same process can be adapted for cell therapy manufacturing by keeping the shear rate and pressures low enough such that the cell phenotype and viability are not affected during concentration and washing. This process has been used at Lonza (Walkersville, MD) to achieve 60–100-fold concentration while maintaining viability and recovery [9].

6.7.3.2 *Centrifugation*

Centrifugation can be further classified into conventional and fluidized-bed modes. In conventional centrifugation, the cells are compacted against a solid surface as a result of the centrifugal force. This may be detrimental to cell health and also decreases the washing efficiency. The most common method of conventional centrifugation is a traditional bench-top centrifuge; this is not appropriate for clinical or commercial scale manufacturing because it is an open, nonscalable process. However, some other technologies incorporate closed, single-use disposables along with automation to conduct multiple rounds of centrifugation and bring residual levels below the required specification. Examples of such equipment are the CellSaver (Haemonetics, Braintree, MA), the COBE blood processor (TerumoBCT, Lakewood, CO), the Cytomate system (Baxter, Deerfield, IL), the CARR Centritech Lab III and CARR Centritech Unifuge (Pneumatic Scale Angelus, Clearwater, FL). The UniFuge represents a large-scale conventional centrifugation system in which the cell suspension is continuously fed into the single-use cylindrical chamber, where the cells are compacted against the wall while the clear supernatant is continuously removed. Once the cylinder fills with cells, the solids are discharged and the cycle can be repeated until all cells have been processed.

In counterflow centrifugation, the cells are maintained in a fluidized bed as a result of the flow of cell suspension being parallel but opposite to the direction of centrifugal force. The cells remain in the developing bed while residuals and clear supernatant are continuously removed. Two examples of counterflow centrifugation equipment are the Elutra from Terumo BCT (mainly used at smaller scale for separation of peripheral blood into its constituents) and the kSep400 and kSep6000 systems from KBI Biopharma. The kSep systems are based on automated, closed, and single-use technology with chamber capacity ranging from 100 mL to 6 L. The main advantage of counterflow centrifugation is that cells are not compacted against a solid surface and that multiple washes can be achieved in a relatively short period of time.

6.7.4 Fill

Once the washed and concentrated cells are obtained, they must be formulated in the appropriate cryopreservation buffer. This buffer is usually dimethyl sulfoxide (DMSO)-based. Extensive studies must be performed to determine the maximum amount of time that the cells can be suspended

Table 6.4 Examples of Filling Equipment for Cell Therapy Manufacturing

Equipment	Vendor	Vials/h
Fill-It	Sartorius	1,440
M1 manual station	Aseptic Technologies	120
L1 Robot Line		600
PX Filling Line		10,800

The throughout columns assume a 4.5 mL fill in a vial.

in the cryopreservation buffer in a liquid state as this time window can be a major bottleneck in the downstream operation. At least two major unit operations (Fill and Visual Inspection) must be performed while the cells are suspended in DMSO-based buffer.

Some of the main examples of filling equipment and their throughput for a 4.5 mL fill are shown in Table 6.4.

6.7.5 Visual Inspection

One of the pressing current issues in cell therapy is how to minimize and deal with particulates in final product. These particulates can be generated or introduced during the manufacturing process from single-use components, welding, microcarriers, etc. Guidelines regarding acceptance criteria in terms of size and number of particles in injectables are in regulatory flux and vary from product to product. However, a 100% inspection of all vials of an injectable drug is recommended. Considerations for visual inspection include the methodology, verifying that vials with visible particles are identified and rejected, and from a time perspective, that the visual inspection process is done within a timeframe that is not detrimental to the cells. At present, visual inspection is carried out manually by trained operators against a well-lit background. The minimum size of particles that can be detected depends on the operator, color and opacity of vial and product, and the presence of protein or cell aggregates, but is considered to be approximately 50 μm.

6.8 PROCESS OPTIMIZATION

So far, it has been demonstrated that the use of scalable bioreactors with microcarriers represents a superior means of producing large quantities of MSCs over planar cell culture systems. However, it should be realized that employing bioreactors will also incorporate many more process parameters than a planar system. Therefore, to develop a robust manufacturing platform for MSC, it is crucial to carry out systematic process development and optimization studies by performing a detailed investigation of key process parameters for their effect on cell growth and other characteristics.

Developing and optimizing a culture process for a specific cell type requires understanding the characteristics of cells and their requirements for attachment to substrate and subsequent growth in culture. Investigating the effects of nutritional and physicochemical parameters, such as medium pH and oxygen tension, on cell growth and other characteristics are often sufficient for static planar culture systems. In dynamic stirred culture environment, however, fluid mechanics introduced by the stirring impeller of bioreactors can also significantly affect cells. Moreover, the use of microcarriers will make the culture system more complex. The microcarrier-based stirred culture environment induces shear forces through the interaction of microcarrier beads with small turbulent eddies and bead-to-bead collisions, which can influence significantly cell growth and other characteristics. Therefore, each of the critical parameters associated with bioreactors and microcarriers in addition to classical cell culture process variables should be identified and optimized. Table 6.5 shows fundamental parameters for adherent cell culture in microcarrier-mediated stirred-tank bioreactors, which were further grouped under different categories related to medium type, medium composition, medium feeding (particularly important for high cell density culture and discussed in Section 6.9), microcarrier, bioreactor, physicochemical variables, harvest, etc. Depending on cell types, some of these parameters may have a major impact on cell attachment/growth and other properties, while others may have minor effects. It has been demonstrated that the attachment and growth of MSC and other mammalian cells is significantly affected by medium type and formulation [13,14], microcarrier type [13,15–17], impeller type and speed [15,18,19], and seeding density of cells and microcarriers (unpublished data). Therefore, it would be rational to study these major factors particularly at the early phase of MSC process development and investigate the effect of other parameters later. Once optimized, it is important to standardize all the parameters to make the performance of the process consistent and reproducible.

In addition to identifying the individual effect of each of these parameters, it is also important to examine the effect of selected parameters in concert during the process optimization studies since these often interact with each other, resulting in synergic effects. Therefore, employing high-throughput technologies into the process development/optimization studies is necessary so that DOE-based cell culture experiments can be implemented. In this regard, the use of a microscale bioreactor system, such as ambr15

Table 6.5 Parameters Grouped Under Different Categories That Need to Be Considered for the Culture of Adherent Cells in Microcarrier-Based Bioreactors

Category	Parameter
Medium type	■ Serum-based ■ Serum-free/xeno-free/animal-free ■ Chemically defined
Medium composition	■ Serum[a] ■ Growth factors ■ Attachment factors ■ Shear protecting agents ■ Viscosity modulating agents
Medium feeding (for high-density culture)	■ Fed-batch ■ Perfusion
Microcarriers	■ Selection of microcarriers (chemical, physical, and geometrical properties, eg, solid, microporous, versus macroporous microcarriers) ■ Microcarrier-coating material (particularly for serum-free culture) ■ Microcarrier loading density
Bioreactors	■ Impeller type and size ■ Stirring protocol[a] ❑ impeller tip speed ❑ stirring direction (upward versus downward) ❑ continuous versus intermittent stirring ❑ manipulation of stirring speed according to cell growth ■ Aeration protocol ❑ headspace aeration ❑ sparging (together with the use of antiforming agents) ■ Material of the inner surface of vessel and impeller (to avoid the attachment of cells and microcarriers) ■ Bioreactor geometry and size ■ Location of impeller and other instruments submerged in culture
Physicochemical variables	■ pH[a] ■ dO_2 ■ dCO_2 ■ Temperature ■ Viscosity
Harvest	■ Proteolytic enzyme ■ Cell wash ■ Mechanical force (intermittent or continuous) ■ Cell separation from microcarriers
Others	■ Cell inoculation density ■ Cell-to-bead ratio ■ Initial working volume ■ Preparation of microcarriers and cell inocula ■ Bead-to-bead transfer ■ Addition of antiforming agent[b] ■ Addition of base[b]

[a] *Parameters that may have different optimal values for initial cell attachment phase and for growth phase.*
[b] *Parameters that are highly related to high-density culture.*

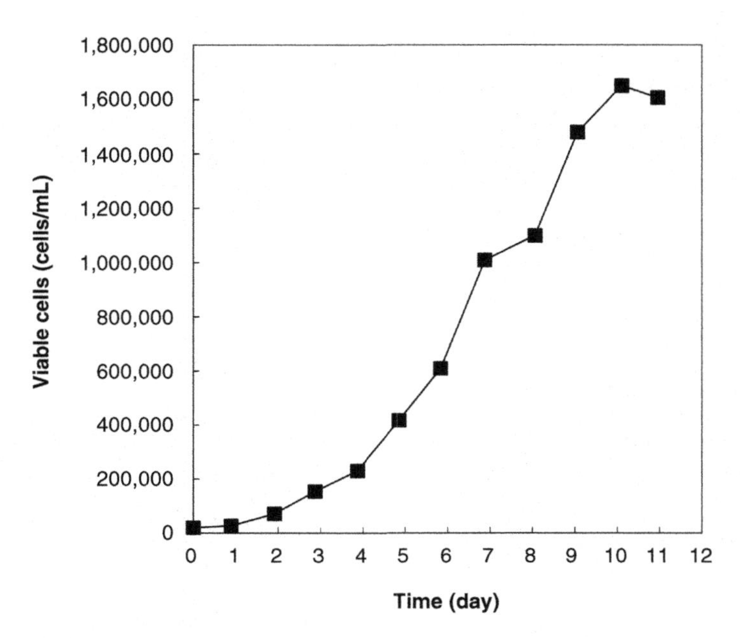

■ **FIGURE 6.6** Growth of bone marrow MSCs in microcarrier-based, ambr-15 bioreactor systems.

bioreactor (Sartorius Stedim Biotech GmbH, Goettingen, Germany), for cell therapy manufacturing process development and optimization should be useful. The ambr15 bioreactor system has been proven to support microcarrier-based MSC growth (Fig. 6.6) and is currently being used for the high-throughput process optimization of MSC culture under controlled and stirred environment.

It should also be noted that optimal conditions for cell attachment phase (typically initial 24 h) and growth phase may be different. Optimizing the initial culture conditions to maximize cell attachment and attachment homogeneity to microcarriers is an important requirement for subsequent growth of cells, avoiding a prolonged lag phase. It has been well documented that cell attachment to the microcarrier beads is influenced by medium pH, medium components (eg, serum content), stirring protocol (eg, intermittent stirring regime), initial culture volume as well as the type and property of microcarriers [14]. In addition to the need for high cell attachment efficiency (ie, ratio of the number of cells attached to microcarriers to the number of cells inoculated), achieving a uniform distribution of cells on microcarriers during the inoculation phase should be a requirement to maintain a constant growth of cells. Assuming that the probability of cell attachment to microcarriers is described by a Poisson distribution [20], it is important to identify an optimal cell-to-bead ratio to minimize the number of empty microcarrier beads after the initial cell attachment phase. Those parameters that are particularly important for the initial cell attachment phase are marked in Table 6.5.

Finally, certain factors are relevant to high cell density culture. For example, the medium pH is typically maintained in cell culture by a buffer system of bicarbonate (provided as a medium component) and dissolved CO_2 (diffused from enriched CO_2 gases) [21]. At high cell densities, however, the cultures become acidic due to the release of acidic metabolites (eg, lactate) and CO_2 from cells into culture medium. As a result, the buffering capacity may become insufficient to maintain the desired pH and thus one may need to add base (ie, a concentrated base solution to be added at low volume). In addition, for high cell density cultures, surface aeration would be inadequate to provide cells with a sufficient amount of oxygen, requiring a direct aeration by sparging. However, the sparging results in the generation of a significant amount of gas bubbles, and the collapse of gas bubbles at or near the culture surface are associated with cell lysis and also protein denaturation. In particular, sparging through culture media containing high protein content (eg, serum-supplemented media) causes excessing foaming. Therefore, it may be necessary to add an antifoaming agent as necessary to avoid or at least minimize the foaming.

6.9 FEEDING CONTROL

Under cell culture conditions in which all the nutritional and regulatory requirements for the maintenance and growth of cells are provided, cell growth becomes limited typically by two major reasons—the depletion of key nutrients, particularly those consumed by cells rapidly to generate energy and biosynthesis substrates, and/or the accumulation of toxic metabolic by-products released by cells. A carbohydrate source is essential for cell growth in culture as a source of energy and a pool of intracellular intermediates that are used in biosynthesis. Glucose is the most commonly provided carbohydrate and is normally included in culture medium at a concentration of 5–25 mM. Lactate is an end-product of cellular metabolism that is released into culture medium as a result of anaerobic glycolysis and largely derived from glucose. The production of lactate often exceeds the pH buffering capacity of culture medium, which may cause inhibitory effects on cell growth.

Amino acids are essential elements of media as building blocks for protein synthesis. In particular, glutamine also plays major roles in many metabolic pathways, that is, acting as a precursor for tricarboxylic acid cycle intermediates. Thus, glutamine is normally included in culture medium at a higher concentration than other amino acids. The metabolic breakdown of glutamine produces ammonia. Ammonia is also released by spontaneous decomposition of glutamine, which is labile in cell culture media. It is well known that ammonia causes inhibitory effects on cell growth.

As cell growth is often limited by a premature exhaustion of key nutrients or growth factors, cell culture media should first be carefully selected for the cell type of interest and are often further optimized to avoid the depletion of key nutrients and growth factors at an early time point in culture. In addition, a well-designed medium feeding regime should be developed to maintain the concentration of each of those key medium components within a range that supports an optimal cell growth during the entire culture period. The design of medium formulation and feeding regime should also involve strategies to maintain a low concentration of toxic metabolic by-products during the culture period, as the accumulation of such metabolites above certain levels will negatively affect the growth and desired properties of cells.

6.9.1 Batch Versus Fed-Batch Versus Continuous Perfusion

An important benefit of using microcarrier-based bioreactors for adherent cell culture is that cell culture processes can readily be operated at different modes of medium feeding as needed, that is, batch, fed-batch, or continuous perfusion. In a batch mode, cells are inoculated and allowed to grow without medium feeding or removal until a desired cell density is reached. Static planar culture systems for MSC manufacturing are often designed to be operated in a batch mode. This is mainly because such planar systems involve the use of a large number of vessels and handling each vessel individually for medium feeding, which is labor intensive and increases risk of contamination. In addition, the limited growth surface per unit culture volume in the planar systems would not support high-density cultures, thereby requiring cell harvest after only several days prior to experiencing an exhaustion of nutrients and a significant accumulation of toxic by-products. Therefore, no medium feeding, or only a very simple medium feeding if needed, is normally required for the culture of MSCs in planar culture systems.

Fed-batch culture involves controlled feeding of medium or selected medium components during the course of culture, either by partial medium changes or by the addition of specific nutrients at predetermined intervals or based on monitoring of growth-limiting factors. Therefore, it is necessary to first identify the growth-limiting factors that are either depleted nutrients or accumulated toxic by-products through cell growth characterization studies. Once the growth-limiting factors are identified, the correlation of their concentration changes with cell growth profiles should be investigated to discover the range of the growth-limiting factor concentration for maintaining desired cell growth. Then, the concentration of the growth-limiting factors should be monitored to design an optimal medium feeding strategy.

To enable medium feeding according to the monitoring of growth-limiting factors, the level of such factors should be either monitored directly or estimated through the measurement of a "reference" component that can be readily and accurately measured (eg, glucose, lactate, etc), based on the availability of relevant analytical techniques [21,22]. Typically, the concentration of glucose, glutamine, lactate, and ammonia can be measured by taking a small sample of culture medium and quickly analyzing it using an automated metabolite analyzer (eg, BioProfile FLEX Analyzer). In particular, either glucose or lactate is often used as a reference component to determine the time points for medium feeding.

Fed-batch modes can be readily implemented for microcarrier culture without the need for special devices to separate cells (attached to the microcarriers) from medium, as a temporary interruption of impeller stirring will lead to the rapid sedimentation of microcarriers, which allows for medium change. For this reason, fed-batch modes have been widely used for manual medium change for microcarrier-based MSC culture, particularly during the phase of process development using spinner flasks or small and pilot scale bioreactors. Fig. 6.7 shows microcarrier-based growth profiles of MSCs under a fed-batch mode in comparison with a batch mode. Cells grew over 9 days in the fed-batch mode achieve greater than 1.8×10^6 cells/mL, while cell growth stopped on day 7 in the batch mode reaches less than 0.6×10^6 cells/mL.

An alternative method for medium replenishment is a perfusion mode, in which fresh medium is pumped in continuously at a given perfusion rate (either predetermined or varied according to a reference component) using a peristaltic pump, and the spent medium is pumped out at the same rate, while the cells are retained in the bioreactor. The implementation of continuous perfusion processes is becoming increasingly common in the biotech and pharmaceutical industry for large-scale manufacturing of therapeutics, as the continuous supply of nutrients and other required factors and removal of toxic by-products may ensure high productivity and quality of products. This is considered particularly important for MSCs and other stem

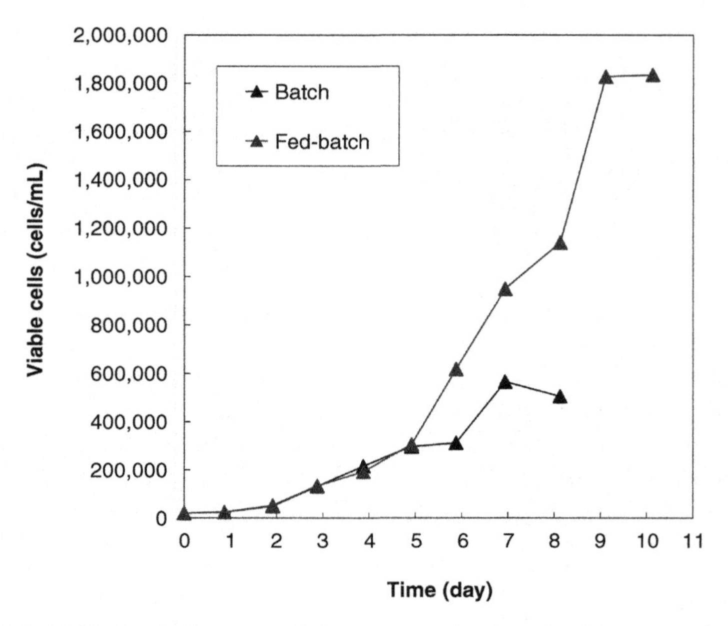

FIGURE 6.7 Growth of bone marrow MSCs in microcarrier-based stirred-tank bioreactors under batch versus fed-batch modes.

cells, since maintaining the concentration of such nutrients/growth factors and metabolic products at optimal levels may be crucial in regulation of cell growth, differentiation and other critical attributes. Cell retention devices are required for perfusion processes to separate cells from the medium, which is continuously pumped out from the culture. Cells can be easily separated from the medium in microcarrier-based suspension culture by sedimentation employing a simple settling device, such as a column separator [23]. It has been demonstrated that MSCs tend to form aggregates as they grow in microcarrier culture, which should lead to high sedimentation rates in the settling device, making the separation of cells grown on microcarriers from the effluent medium readily applicable.

Once the cell retention is ensured, medium perfusion rates should be optimally determined to maintain the nutrient and metabolite levels at desired ranges, while utilizing the medium as little as possible to minimize medium cost. Using a different type of bioreactor (ie, hollow-fiber bioreactor system), Hanley and colleagues implemented a regime of medium perfusion rates for the culture of MSC in which the perfusion rates were adjusted according to the concentration of lactate [24]. Similar medium perfusion strategies can be designed for microcarrier-based MSC culture, based on online or offline measurement of the concentration of a key growth-limiting factor or a reference component.

6.10 CONCLUSION

In summary, suspension-based bioreactor culture for the expansion of therapeutic hMSC has many advantages—the ability to produce much higher and needed cell numbers thanks to a higher surface area to volume ratio, greater scalability, increased automation, and increased process control. For hMSC, a bioreactor process development program should focus on selecting the following components—microcarrier, culture media, controller, and single-use vessel. The critical quality attributes of the cells should be determined upfront so that these can be used as a quantitative scale to assess the various platform components. Finally, one should strive to continuously improve the process components, especially the monitoring capabilities of the bioreactor system.

Per the expected dose and patient population, an analysis of the commercialization objectives and implications on the needed manufacturing platform is highly recommended. Generally speaking, the earlier in process development that suspension culture is adopted for cell expansion, the better. This is due to the fact that, especially for hMSC, change to the process may result in change to cell biology, putting into question the ability to change the process in late stages of development and resulting, at the least, in the need for complex comparability studies. Therefore, use of a commercial platform-similar process early-on is the best approach.

REFERENCES

[1] Kirouac DC, Zandstra PW. The systematic production of cells for cell therapies. Cell Stem Cell 2008;3(4):369–81.

[2] Madrigal M, Rao KS, Riordan NH. A review of therapeutic effects of mesenchymal stem cell secretions and induction of secretory modification by different culture methods. J Transl Med 2014;12:260.

[3] GE Healthcare. Microcarrier cell culture: principles and methods. 2005. http://www.gelifesciences.com/file_source/GELS/Service%20and%20Support/Documents%20and%20Downloads/Handbooks/pdfs/Microcarrier%20Cell%20Culture.pdf

[4] YSI. YSI 2900 Series Biochemistry Analyzers Operations and Maintenance Manual 2015. Available at: https://www.ysilifesciences.com/File%20Library/Documents/Manuals/YSI-2900-Series-Manual.pdf.

[5] Nova biomedical. NOVA BioProfile FLEX Instruction Manual. http://www.novabiomedical.com

[6] Roche. Application of the Roche cedex bio analyzer in the culture of pancreatic islets and adult human mesenchymal stem cells. 2015. http://custombiotech.roche.com/content/dam/internet/dia/custombiotech/custombiotech_com/en_GB/pdf/Application%20note%20Cedex%20Bio%20in%20Human%20Pancreatic%20Islets%2001_15.pdf

[7] Precision Sensing. Online pH and DO Measurements in Microcarrier-Based hMSC Cultivations. http://www.presens.de/uploads/tx_presensapplicationnotes/140929_APP_hMSC_Monitoring_in_Spinner_Flasks_w.pdf

[8] Fehrenbach R, Comberbach M, Petre JO. On-line biomass monitoring by capacitance measurement. J Biotechnol 1992;23(3):303–14.

[9] Pattasseril J, Varadaraju H, Lock L, Rowley JA. Downstream technology landscape for large-scale therapeutic cell processing. BioProcess Int 2013;11(3):38–47.

[10] Nienow AW, Rafiq QA, Coopman K, Hewitt CJ. A potentially scalable method for the harvesting of hMSCs from microcarriers. Biochem Eng J 2014;85:79–88.

[11] Rafiq QA, Brosnan KM, Coopman K, Nienow AW, Hewitt CJ. Culture of human mesenchymal stem cells on microcarriers in a 5 l stirred-tank bioreactor. Biotechnol Let 2013;35(8):1233–45.

[12] Jung S, Sen A, Rosenberg L, Behie LA. Human mesenchymal stem cell culture: rapid and efficient isolation and expansion in a defined serum-free medium. J Tissue Eng Regen Med 2012;6(5):391–403.

[13] Eibes G, dos Santos F, Andrade PZ, Boura JS, Abecasis MM, da Silva CL, et al. Maximizing the ex vivo expansion of human mesenchymal stem cells using a microcarrier-based stirred culture system. J Biotechnol 2010;146(4):194–7.

[14] Jung S, Panchalingam KM, Wuerth RD, Rosenberg L, Behie LA. Large-scale production of human mesenchymal stem cells for clinical applications. Biotechnol Appl Biochem 2012;59(2):106–20.

[15] Santos F, Andrade PZ, Abecasis MM, Gimble JM, Chase LG, Campbell AM, et al. Toward a clinical-grade expansion of mesenchymal stem cells from human sources: a microcarrier-based culture system under xeno-free conditions. Tissue Eng Part C Methods 2011;17(12):1201–10.

[16] Sart S, Schneider YJ, Agathos SN. Ear mesenchymal stem cells: an efficient adult multipotent cell population fit for rapid and scalable expansion. J Biotechnol 2009;139(4):291–9.

[17] Sart S, Schneider YJ, Agathos SN. Influence of culture parameters on ear mesenchymal stem cells expanded on microcarriers. J Biotechnol 2010;150(1):149–60.

[18] Papoutsakis ET. Fluid-mechanical damage of animal cells in bioreactors. Trends Biotechnol 1991;9(12):427–37.

[19] Park Y, Subramanian K, Verfaillie CM, Hu WS. Expansion and hepatic differentiation of rat multipotent adult progenitor cells in microcarrier suspension culture. J Biotechnol 2010;150(1):131–9.

[20] Frauenschuh S, Reichmann E, Ibold Y, Goetz PM, Sittinger M, Ringe J. A microcarrier-based cultivation system for expansion of primary mesenchymal stem cells. Biotechnol Prog 2007;23(1):187–93.

[21] Butler M. Animal cell culture and technology: the basics. 1st ed. New York: BIOS Scientific Publishers, Oxford; 1996.

[22] Hu WS. Cell culture bioprocess engineering. Minnesota: University of Minnesota; 2012.

[23] Butler M, Imamura T, Thomas J, Thilly WG. High yields from microcarrier cultures by medium perfusion. J Cell Sci 1983;61:351–63.

[24] Hanley PJ, Mei Z, Durett AG, Cabreira-Harrison Mda G, Klis M, Li W, et al. Efficient manufacturing of therapeutic mesenchymal stromal cells with the use of the Quantum Cell Expansion System. Cytotherapy 2014;16(8):1048–58.

Chapter 7

GMP Requirements

A.P. Gee

Professor of Medicine and Pediatrics, Center for Cell and Gene Therapy, Baylor College of Medicine, Houston, TX, United States

CHAPTER OUTLINE

7.1 INTRODUCTION

Mesenchymal stromal cells (MSCs) are under intensive study for a wide variety of clinical indications that involve their immunosuppressive properties and their activities in regenerative medicine [1]. This has been facilitated by the fact that MSCs are relatively easy to grow from a variety of starting sources (marrow, adipose tissue, placenta, umbilical cord blood, Wharton's jelly, etc. [2]), although it is still not clear whether MSCs produced from different tissues and culture methods are essentially identical [3]. In all cases, however, production of the cells requires ex vivo culture expansion and clinical

Mesenchymal Stromal Cells. http://dx.doi.org/10.1016/B978-0-12-802826-1.00007-6

use of the cells in organs and tissues where they would not be expected to be exerting their normal function(s)—nonhomologous use. Under FDA regulations this clearly places a clinical trial using these cells into one that requires an Investigational New Drug (IND) approval [4]. An IND application must be filed by the sponsor of the study and provide comprehensive information that includes preclinical data on the therapeutic concept, safety and toxicity information, clinical trial design, and preparation and testing of the cell product. Under IND regulations this product must be manufactured under current Good Manufacturing Practice (cGMP) regulations.

The cGMP laws were implemented following a number of catastrophic incidents in which patients were severely harmed or died after receiving medicinal products [5]. The intent of the regulations was that drugs must be manufactured using a controlled, reproducible, auditable procedure that resulted in a safe and efficacious product. cGMP addresses all aspects of manufacturing of the drug, including management of raw materials, equipment, facilities, staff, procedures to be used, packaging, labeling, storage and distribution, quality programs, etc. (Fig. 7.1). The regulations are published annually in Volume 21 of the Code of Federal Regulations (CFR) in different Parts that address specific types of drugs, for example, blood and blood components, medicated feeds, finished pharmaceuticals, etc. Commercial manufacturers are very familiar with these laws, but they are more of a challenge to academic investigators, who are generally not used to thinking of themselves as drug manufacturers. In the following sections, the basics of cGMP as they relate to manufacturing of MSCs for a clinical trial are reviewed.

7.2 THE ELEMENTS OF cGMP

The major components of cGMP as they relate to cellular products can be found in Title 21 CFR Parts 210 and 211 "cGMP in Manufacturing, Processing, Packing, or Holding of Drugs and Finished Pharmaceuticals" [6]. Parts 600–680 of 21CFR relate to cGMP for Blood and Blood Components and, at first glance, may appear to be more relevant; however, cellular therapy products are classified as drugs for regulatory purposes.

For the uninitiated a first reading of this material may be a challenge, since the regulations use a distinct form of English that is constructed to convey a very specific message. Once you crack this code you will see the same phrases and themes repeatedly and their message will become clear. This process is helped by the list of definitions provided at the start of each part, which precisely explains the meaning of terms. For the academic manufacturer a lot of the information in Parts 210 and 211 covers areas that are specific to

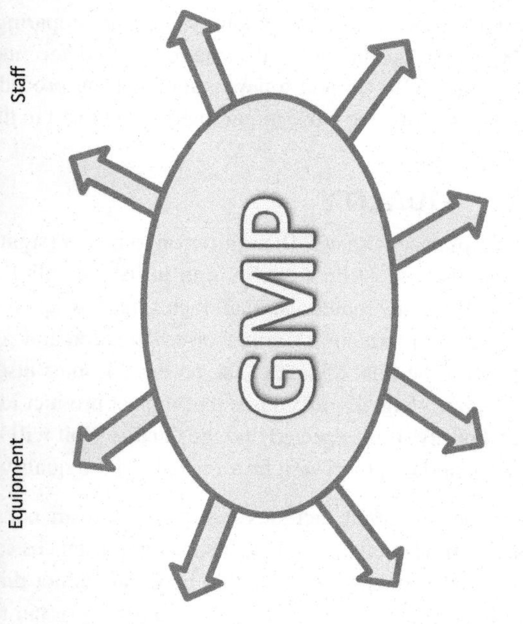

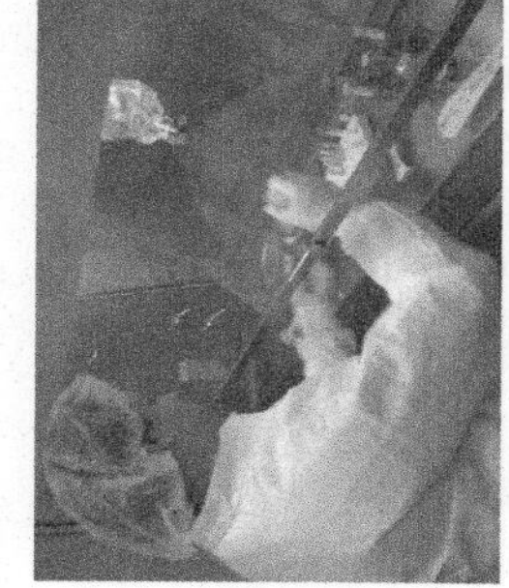

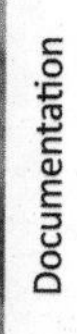

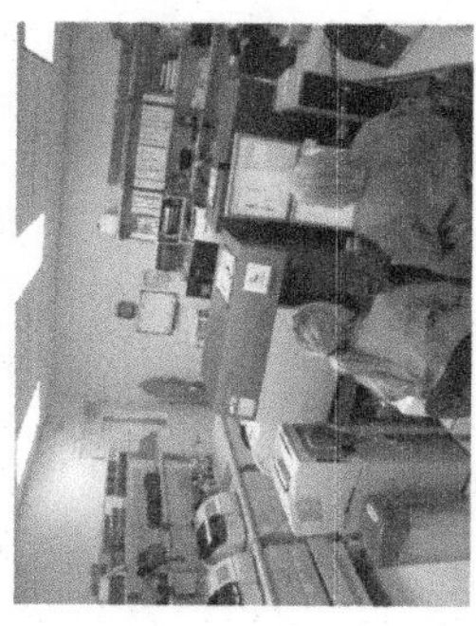

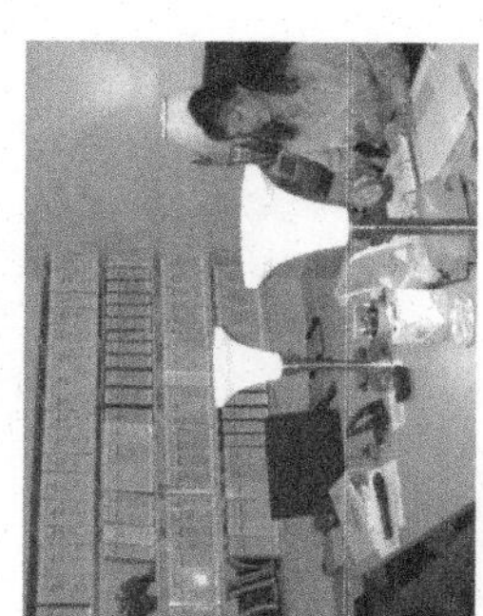

■ **FIGURE 7.1 Main components of cGMP manufacturing.**

small molecule drugs manufactured in large batches and intended for sale to physicians and/or the public. The FDA recognizes that academic physicians are usually performing early phase (Phase 1/Early Phase 2) clinical trials, and manufacturing activities will be on a much smaller scale. For this reason they published a guidance "Guidance for Industry cGMP for Phase 1 Investigational Drugs" in 2008 [7]. This explains[1] the elements of cGMP that the FDA expects to be in place for these types of trials. As these studies proceed toward an application for licensure of the drug, the expectation for complete cGMP compliance increases. Additional information is available in a later draft guidance "Considerations for the Design of Early-Phase Clinical Trials of Cellular and Gene Therapy Products" published in 2015, which addresses some of the problems specific to preparing these types of products [8]. These documents enable the reader to determine what elements need to be in place to support a new manufacturing procedure. The following sections describe the basic requirements contained in the elements (Fig. 7.1).

7.3 QUALITY

Compliance with cGMP requires ongoing oversight of quality. For this purpose there must be a quality unit to oversee all activities, from control of incoming raw materials, manufacturing, testing, staff, equipment, labeling, review of production records, error detection and correction, and authorization for the final release of the product. It must also have the responsibility for approving all procedures that impact product identity, strength, quality, and purity. It is expected that the Quality Unit will have an active audit program and that there will be a formal written quality plan.

In small academic facilities sometimes there are not sufficient staff to appoint a full-time quality unit. Under such circumstances an individual may provide quality oversight provided that he or she is not directly involved with any step of the manufacturing process under his or her review. The Quality Unit generally is responsible for compliance with cGMP regulations and with the standards from any professional organization from which the facility seeks accreditation, for example, Foundation for the Accreditation of Cellular Therapy (FACT) and the College of American Pathologists (CAP).

7.4 STAFF QUALIFICATIONS

The expectation is that staff involved with all aspects of manufacturing will have adequate education, training, and experience to perform their assigned functions. The training must include ongoing education in cGMP

[1]The official title actually uses a capital C.

regulations performed by qualified trainers. The staff must also be of sufficient number to perform and supervise the activities conducted by the manufacturer. In reality, the expectation is that each position and its associated responsibilities should be formally described, including the necessary qualifications and training. The staff member must have documented training on all procedures to be performed, including relevant health and safety training and initial orientation. In addition, there must be training on aseptic technique and appropriate gowning and health requirements to protect the product from contamination. Formal training on cGMP regulations is also required. There must be an ongoing system to evaluate staff proficiency and competence.

7.5 FACILITIES

Many new investigators equate cGMP regulations with using a clean room environment for manufacturing. This is not mandated by the FDA, although most new academic cellular therapy facilities have chosen to manufacture products in a clean room environment [9]. The regulations simply state that the facility must be of adequate size, construction, and location to permit cleaning, maintenance, and operations, and there be adequate systems for plumbing, lighting, sewage and refuse, and washing and toilet facilities. The design should prevent contamination and be of adequate space to prevent mix-ups of materials, reagents, products, documentation, etc. This is achieved by designating separate areas for different activities, including a holding area for items and materials that are under quarantine, before being released by the Quality Unit for use. Aseptic manufacturing areas must have hard, smooth, cleanable surfaces to facilitate cleaning.

If a clean room is to be used the air must be supplied through HEPA filters under positive pressure and there must be documentation to demonstrate that the clean room classification is being maintained. There must be a program in place for cleaning and maintaining equipment and for disinfection of rooms. A system for monitoring environmental conditions is also required.

As was stated earlier, the cGMP regulations were initially implemented to assure the safety and purity of medicines rather than cellular therapy products. Small-molecule drugs are often manufactured in a large clean room using open procedures. Under these circumstances environmental quality is critical to avoid airborne contamination of the product. In contrast, cellular therapy products are usually manufactured within an International Standards Organization (ISO) biological safety cabinet (BSC), and the cells are usually in closed containers while not inside the cabinet. The use of a classified environment outside the cabinet provides an additional level of assurance, but probably does not contribute to protection of the cells. If the BSC is placed in a classified environment the level of classification will depend on the regulatory authority. In the United States an ISO5 BSC may be placed within an ISO7 clean room. In countries belonging to the European Union this is not permitted, and the cabinet must be within an ISO6 room. When manufacturing for early phase clinical trials the environment must be described in the IND application and the regulatory authority will determine its suitability. The demand for a large number of therapeutic MSCs has prompted companies to develop bioreactors for their generation. These are functionally closed systems in which the cells are not exposed to the environment after the culture components are seeded into the reactor. Sterile connect devices and tubing welders can be used to maintain aseptic handling. It remains to be determined whether regulatory authorities will permit the use of such systems in unclassified space. Even if they do, downstream processing of the cells for testing and cryopreservation will usually require a BSC, unless suitable closed systems become available. Some manufacturers have chosen to use isolators in which the major manufacturing equipment is enclosed within a stand-alone classified environment. In effect this is a "clean room in a box" and multiple isolators may be placed in a single production area. An issue is the cost and the necessity to duplicate equipment in each isolator rather than sharing equipment (with appropriate changeover procedures) that is possible with conventional clean rooms.

If classified manufacturing space is chosen it must be supported by data to show, on an ongoing basis, that the classification is met. This could be achieved by instituting a monitoring program that measures both airborne viable and nonviable particles and establishes alert and alarm limits. The former is indicative of a developing problem that must be addressed immediately. The latter defines the conditions under which manufacturing activities must be discontinued until the problem is resolved and a corrective action implemented and its efficacy verified. Information on the design and operation of the Heating, Ventilation and Air Conditioning (HVAC) system should be available together with records of its maintenance and calibration.

Whatever manufacturing environment is selected, the emphasis will be on documentation that it is appropriately cleaned and managed. Room disinfection procedures should be based upon removal of microbes regularly detected in the room, and should include rotation of disinfectants to avoid build-up of resistance. The cleaning procedure must be formalized in a Standard Operating Procedure (SOP) that describes the preparation of disinfectants, frequency of and methods for cleaning, removal of residual cleaning agents, and the plan to monitor cleaning efficacy.

The same is true for cleaning of equipment, which can be based on recommendations from the manufacturer. There must be documentation of each cleaning procedure and how it relates to use of the equipment for manufacturing procedures. A system is required to be able to track the identity of equipment used to manufacture a specific product and, conversely, all the products manufactured by a specific piece of equipment. There must be a program for equipment calibration, maintenance and repair, and documentation that the staff are trained on the equipment that they use. Each piece of equipment should undergo a qualification procedure [10]. This documents how the equipment was selected, for example, what features and attributes must it have (design qualification), how it was received and installed (installation qualification), a check on its function (operational qualification), and whether it is able to achieve the required endpoints in the specific application in which it will be used (performance qualification). Once the equipment has undergone this process it can be released for manufacturing use by the Quality Unit.

Within the manufacturing area a primary concern must be the potential for cross-contamination between products [11]. This can be addressed by implementing changeover procedures. Manufacturing activities generally fall into two types—ongoing and campaign. The former is most frequently seen in facilities where multiple products must be handled in the same space, for example, autologous MSC are being prepared for several recipients. The most effective solution is to allocate each product to a different manufacturing suite; however, facility capacity may not allow this. Under these circumstances an alternative strategy would be to allocate each product to a different incubator within the room. If this approach is chosen then there must be a procedure in place to prevent cross-contamination between these products. This would address issues such as container labeling, sequential handling of the products with line clearance between each. Line clearance would include items such as removing the first product from the manufacturing area, putting away the associated documentation, cleaning equipment and working areas, before starting to work on the second product.

Campaign manufacturing can be used when a single product will be assigned to a manufacturing space and will remain there until the final product has been prepared. This allows precleaning of the manufacturing area and equipment, stocking with dedicated reagents and materials, and release for manufacturing by the Quality Unit after satisfactory environmental monitoring has been performed. In both cases monitoring will continue during product preparation.

7.6 MATERIALS MANAGEMENT

Whenever possible clinical grade materials and reagents must be used for manufacturing cellular therapy products. Most investigators develop new products using research grade reagents and then find that there is no pharmaceutical grade equivalent. For this reason, it is important to address reagent selection as early as possible in the development process. It is also important not to use reagents that would be incompatible with clinical use, unless it can be convincingly demonstrated that these have been completely removed from the final product. Antibiotics should also be avoided to prevent allergic reactions in sensitive recipients, and to prevent masking of low level product contamination.

Many cell therapy products require the use of reagents, such as cytokines, that have not yet been manufactured to pharmaceutical standards. The position should be to use the purest available reagent with the most detailed testing on the certificate of analysis from the manufacturer. This information should be provided in the Chemistry Manufacturing and Control (CMC) section of the IND application [12] and the regulatory agency will determine the suitability for manufacturing. It may require additional testing or information from the supplier. Copies of certificates of analysis (CofA) for all materials in contact with the cells should be provided in the CMC even if they are of pharmaceutical grade.

Once manufacturing starts, CofA should be kept on file for each lot of reagent used.

If the cells are to be prepared using a device such as a bioreactor that has not received regulatory approval for that indication, then the CMC must contain information on the design and operation of the reactor. Most often this can be obtained from the device manufacturer either directly, or, if some information is proprietary, via a Master File that the manufacturer has provided to the FDA. The agency can then examine the information in this file directly without revealing its contents to the end user of the equipment.

There must be a system for handling incoming reagents and materials and for quarantining those which fail to meet specifications set by the facility, or which require additional testing before use in manufacturing. Quarantined items must be clearly identifiable and must be segregated for released supplies to avoid mix-ups. Materials used for manufacturing each product lot must be recorded, as must the identity of all products manufactured using each lot of material. There must also be a system to ensure that older materials are used first (first-in first-out management) and that nothing is used past its expiration date. Expiration dates may be extended for a specific application if there is documentation that it remains fit for purpose upon extended stability testing.

In the case of MSCs, most early studies were performed using cells that had been cultured in media containing fetal bovine serum. To avoid use of xenogeneic protein other investigators used autologous serum, AB serum, or human platelet lysate (HPL), with varying degrees of success [13]. HPL is usually prepared from expired lots of platelets collected from donors screened for infectious diseases and collected at the local blood bank. The platelets are concentrated by centrifugation and then subjected to several freeze–thaw cycles. The resulting lysate is stored at −80°C and used to supplement the culture medium. With some exceptions [14], lot-to-lot variability is generally low and many researchers report improved results compared to using serum preparations. This is, however, a poorly characterized supplement, and the FDA has consistently encouraged use of defined culture medium for preparation of therapeutic products. As a result, a number of serum-free [15] and xeno-free media [16] are now available and have been shown to support cell growth. If the cells are to be cultured and implanted on a scaffold or matrix, the FDA has included recommendations on preclinical evaluation in a Guidance document [17], it has also provided specific recommendations on therapeutic cells derived from adipose tissue [18].

7.7 MANUFACTURING ACTIVITIES

The manufacturing procedure should be designed to minimize the risk of product contamination and cross-contamination. Wherever possible closed systems should be used. MSCs were originally cultured in T flasks (Fig. 7.2a) from mononuclear fractions isolated on Ficoll-Hypaque from fresh bone marrow aspirates. The nonadherent cells were typically removed 24–48 hours after initiation of the cultures, and the adherent cells were split at a 1:4 ratio when the monolayer was nearly confluent. Passaging was accomplished by rinsing the monolayer with phosphate-buffered saline to reduce residual protein, and then exposing it to Trypsin until the cells detached into a uniform cell suspension. Medium containing serum was then added to inactivate the enzyme, and the cells were split into four flasks and diluted to the appropriate concentration with fresh medium. To obtain a clinically relevant dose, this procedure had to be repeated many times, resulting sometimes in hundreds of large T flasks of cells that then had to be harvested and pooled (Fig. 7.2c). Such a procedure involves thousands of open operations each of which could result in product contamination [19]. An alternative method is to use multilayer flasks. Each of these contains several layers of

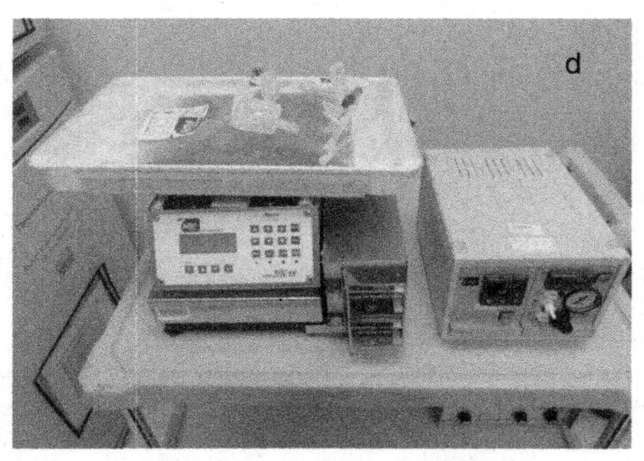

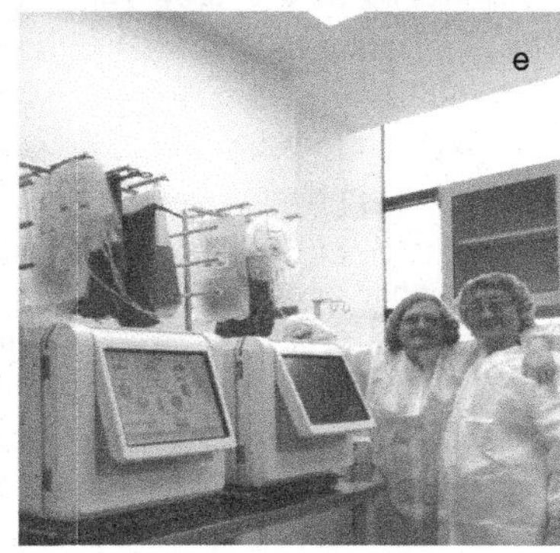

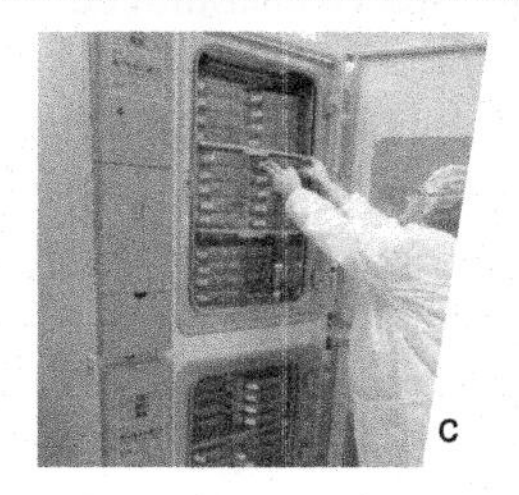

■ **FIGURE 7.2 Developments in MSC expansion.** (a) Traditional culture method using single or multilayer T flasks. (b) Manufacturing using 10-layer cell factories. (c) Numbers of T flasks required to provide a clinical MSC dose. (d) The Xuri (Wave) bioreactor from GE Healthcare. MSCs are grown on microcarriers suspended in the culture medium. (e) The quantum bioreactor from Terumo. MSC are grown on hollow fibers perfused with culture medium.

plastic surface on which the cells can grow, thereby dramatically increasing the numbers of cells obtainable from each flask, while simultaneously decreasing the number of open procedures required to grow them. These culture vessels range in size from that of a normal T flask to large cell factories containing 10 or more culture levels (Fig. 7.2b). The larger vessels tend to be a little more difficult to handle because of their physical size [19]. There are a number of adapters and closures that can be used to close up these culture systems and further reduce the risk of contamination.

A number of bioreactors have now been manufactured that can be used to grow adherent cells [20]. The Wave bioreactor (now the Xuri Bioreactor from GE Healthcare) was among the first. It was developed to culture nonadherent cells, but has since been modified so that the medium within the culture bag contains microcarriers to which adherent cells will attach (Fig. 7.2d). Subsequently, Terumo developed the Quantum hollow-fiber reactor [21] (Fig. 7.2e). This is an automated device in which the cells attach to fibronectin-coated hollow fibers, which are then perfused with the culture medium. The flow rates are adjusted depending on the rate of cell growth as determined by the content of lactate detected in the medium. Cells are harvested from the fibers by exposure to enzyme—usually a recombinant form of trypsin. They are then passaged into new bioreactor cartridges. Similar systems have been developed by investigators and by biotech companies [22–24]. At this time, none of these has received approval from the FDA for MSC expansion and the investigators will be required to provide information to the agency on the construction and use of the specific bioreactor. This will be included in the CMC section of the IND application. This information must address factors such as potential leachables from the materials used to construct the reactor, reliability of performance etc. Wherever possible the surfaces in contact with the cells should be disposable to avoid the need for the generation of extensive cleaning and sterilization.

The up-front preparation of the mononuclear cells from marrow can be performed in a functionally closed device using density gradient centrifugation (eg, Sepax from Biosafe), by nondensity gradient centrifugation [25], by enriching the MSCs by passage through a filter device (Kaneka Bone Marrow MSC Separation Device) [19], or completely avoided by seeding the bioreactor with unfractionated marrow (Quantum from Terumo) [21].

Whatever culture method is chosen, validation data should be presented in the CMC. The FDA has prepared a guidance document on process validation that provides useful information on study design [26]. In the validation document the characteristics of the final product required must be specified before starting the study. In the case of MSCs these characteristics will generally consist of the specified release testing requirements and the numbers of cells required. The study documents will indicate these as the validation requirements and describe in detail how the study is to be performed, the data that are to be collected, the statistical analysis to be performed etc. When the design is approved by the Quality Unit, the study may start. The resulting data will be returned to the Quality Unit for evaluation and they will determine whether the procedure has met the validation requirements. Usually before the validation starts, a draft SOP and accompanying worksheets will have been developed based on the translational experiments. These should be used to document performance of the validation experiments and they will form the basis of the batch record that will be used to record future manufacturing runs.

7.8 RELEASE TESTING

Cell therapy products intended for clinical use must be tested for purity, safety, and identity. By the start of Phase 3 clinical trials testing must also include a validated assay for potency that reflects the product's intended clinical utility. It is advisable, however, to include some type of potency assay in Phase 1 studies.

The following tests are normally required:

Sterility: Bacterial and fungal sterility testing is an absolute requirement. The traditional testing method is that described in 21 CFR Part 610.12. The results are obtained after a 14 day incubation period, this makes it unsuitable for use with fresh cell products. These may frequently be released using tests performed earlier in the manufacturing process, together with a Gram stain on the final product. However, a 14-day testing must still be performed on the final product, and there must be a procedure in place to deal with positive results obtained after the cells have been administered. This will generally involve immediate notification of the recipient's physician, identification of the organism, and testing for antibiotic sensitivity. Sensitivity results are immediately reported to the physician. An investigation of the source of the contaminant should also be performed by the Quality Unit. Rapid culture methods (eg, Bactec and BacTAlert may be used with permission of the FDA).

Mycoplasma: Mycoplasma is usually tested using a culture-based method that takes several weeks to perform [27]. For this reason the FDA is willing to consider use of rapid methods, such as PCR [28] and biochemical detection methods, for example, MycoAlert from Lonza [29]. If these are chosen the agency may require validation data to show equivalent specificity and sensitivity to the approved methods. These are the types of issues that should be covered in the pre-IND meeting.

Adventitious Agent Testing: Adventitious agents of most concern to the FDA are viruses. These may be tested using in vivo and in vitro methods. These tests have become of particular concern as the use of MSCs has changed from small numbers of autologous cells, to preparation of large allogeneic cell banks that could treat many patients. The latter pose a much larger potential risk than the former. The question then becomes how large can an allogeneic study be to require full master cell bank (MCB) testing for viruses? A Phase 1 study of 10 patients, for example, would not be feasible if the cells had to undergo full MCB viral testing, because of the time required and the costs involved. Although a definitive recipient number that marks transition to the requirement for full adventitious agent testing has not been yet established, it appears to lie somewhere between 20 and 35 patients. Below this number in vitro adventitious viral testing (a single test for multiple viruses) is sufficient. Above it, both full in vitro (multivirus and individual virus testing) and in vivo viral testing may be required. In all cases the donor of the starting tissue must have undergone formal donor eligibility assessment as described in 21 CFR Part 1271 Subpart C [30,31]. This includes testing for specified infectious diseases and health screening/history. This is designed to identify behaviors that may potentially increase the risk of infectious diseases, for example, travel to certain areas. The use of ineligible donors for preparation of cells to be used under an IND is only permitted under special FDA approval and where there is urgent medical need. This would not be the case for most MSC products.

Identity: Identity testing is used to confirm that the final product preparation consists of the right cells and to provide information on the identity of any contaminating cell populations. The most widely used test for this purpose is immunophenotyping using panels of product-specific (associated) monoclonal antibodies, supplemented by antibodies against likely contaminants. Many investigators and professional groups have developed antibody panels to characterize MSCs and the majority agree on the major panel components [32,33]. The lack of absolute specificity of these markers and the likelihood that they will change as more is known about MSCs has been recognized by the FDA [34]. For MSC preparations the purity is usually >90% based on these markers.

Purity: The final product must be free of contaminants. These may consist of reagents used during manufacturing, and this is a prime reason for not using media-containing antibiotics. The most widely tested contaminant is endotoxin. The upper limit in the product is 5 EU/kg/dose. Endotoxin testing is notoriously tedious and somewhat unreliable; however, this has been dramatically changed by the availability of a rapid automated system (Endosafe from Charles River) [35]. This provides results within an hour and has been approved by the FDA. The method is based upon the Limulus amebocyte system that is recommended in the CFR.

Potency: Although not formally required until the start of a Phase 3 clinical trial, it is advisable to include a potency assay in the release test panel in the earlier stages. Given the uncertainty about how MSCs may exert their functions in different indications, it is difficult to select an assay that will measure their mode of action(s).

In general, two broad categories have been chosen. The first is the immunosuppressive effect of MSCs. In this assay, MSCs are added to a mixed lymphocyte reaction in which T cells of different HLA types react against each other or to a single T cell population that has been stimulated to proliferate [36]. The proliferative response of T cells can then be modulated and measured by adding increasing numbers of MSCs to the cultures. Another characteristic of MSC is their potential to differentiate along three lineages (adipogenic, chondrogenic, and osteogenic). This can be demonstrated using in vitro assays in which growth conditions are controlled to allow the cells to mature along each lineage. This latter assay is more appropriate if MSCs are used, for example, for repair of cartilage or bone. The duration of both of these assays makes them unsuitable for release of fresh cells, although they can be used to obtain potency data retrospectively. They are more suitable for characterizing cryopreserved cell banks. As more is known about MSC it is likely that multiple assays may be required to assess their functional activity [37].

Additional Release Tests: Cell viability should be included in release testing. Although Trypan blue exclusion is widely used, it is generally less accurate and more subjective than 7-Aminoactinomycin D (7-AAD) staining quantitated on a flow cytometer. This is particularly true when the cells are thawed and tested. The FDA generally sets the minimum viability at 70%.

Cell dose must be specified in the protocol and this information will be required when releasing the cells for patient administration.

7.9 TESTING RESOURCES

Many academic manufacturing facilities are located in medical schools that have an affiliated hospital. Depending on the nature of the test, and approval of its use by the FDA, some assays may be performed by the hospital clinical laboratories, for example, sterility testing, Gram stains etc. The manufacturer should ensure that copies of testing protocols are on file and that there is a system for tracking test samples and results. Some regularly used assays are now available as test kits from commercial sources, for example, endotoxin and mycoplasma assays. Their use for release testing should be cleared with the FDA, and the tests should not be performed by staff with a direct interest in clinical use of the product.

The use of commercial testing companies is often complicated by turnaround times; however, they are a good choice for more complex tests that are sufficiently infrequent that it would be inefficient and cost prohibitive to bring them in house and maintain staff competency and proficiency in their use. In contrast, some Phase 1 potency tests are unique to a particular product and manufacturer and these are normally set up in house while their long-term value is determined.

Stability: Product stability testing should be performed to provide information on the functionality and lifespan of the cells under two different conditions. The first is for cells maintained in cryostorage. A program should be developed to thaw cells after different times in storage and perform testing that reflects their viability and potency over time. This information may not be available at the start of the Phase 1 trial; however, a plan should be in place to collect the data. The second stability study is performed on fresh and/or thawed cells released for patient administration. It is important for the physician to have information on how long these cells may be held prior to receipt by the patient. This information should be available before the start of the clinical trial, and the holding conditions must be clearly specified, e.g., cell concentration, excipient, temperature, container etc. There is conflicting evidence that cryopreservation affects the functional activity of MSC [38,39].

7.10 THE CHEMISTRY, MANUFACTURING, AND CONTROL SECTION OF THE IND

The CMC section of the IND is intended to provide the FDA with information on all aspects of the cellular therapy product (Fig. 7.3). This includes reagents, materials, manufacturing and testing, labeling, storage, distribution, etc. It provides assurance that the product can be made safely and reproducibly by a controlled and auditable process. The preparation of this document often caused confusion as to what the specific requirements were. This was resolved in 2003 by publication of a Draft Guidance "Instructions and Template for Chemistry, Manufacturing, and Control (CMC) Reviewers of Human Somatic Cell Therapy Investigational New Drug Applications (INDs)" [12]. This document was intended to provide FDA reviewers with a guide to what information should be included in the CMC section of a cell therapy IND. As such, it can be used by the manufacturer as a template when providing this information. It gives step-by-step instructions as to the types of information to be provided, the format, and even provides templates for the submission of certain data. Adherence to these instructions will facilitate approval of the IND. As a generic document for cellular therapy products, some sections will not apply to MSCs; however, particular attention should be paid to the sections on cell banking

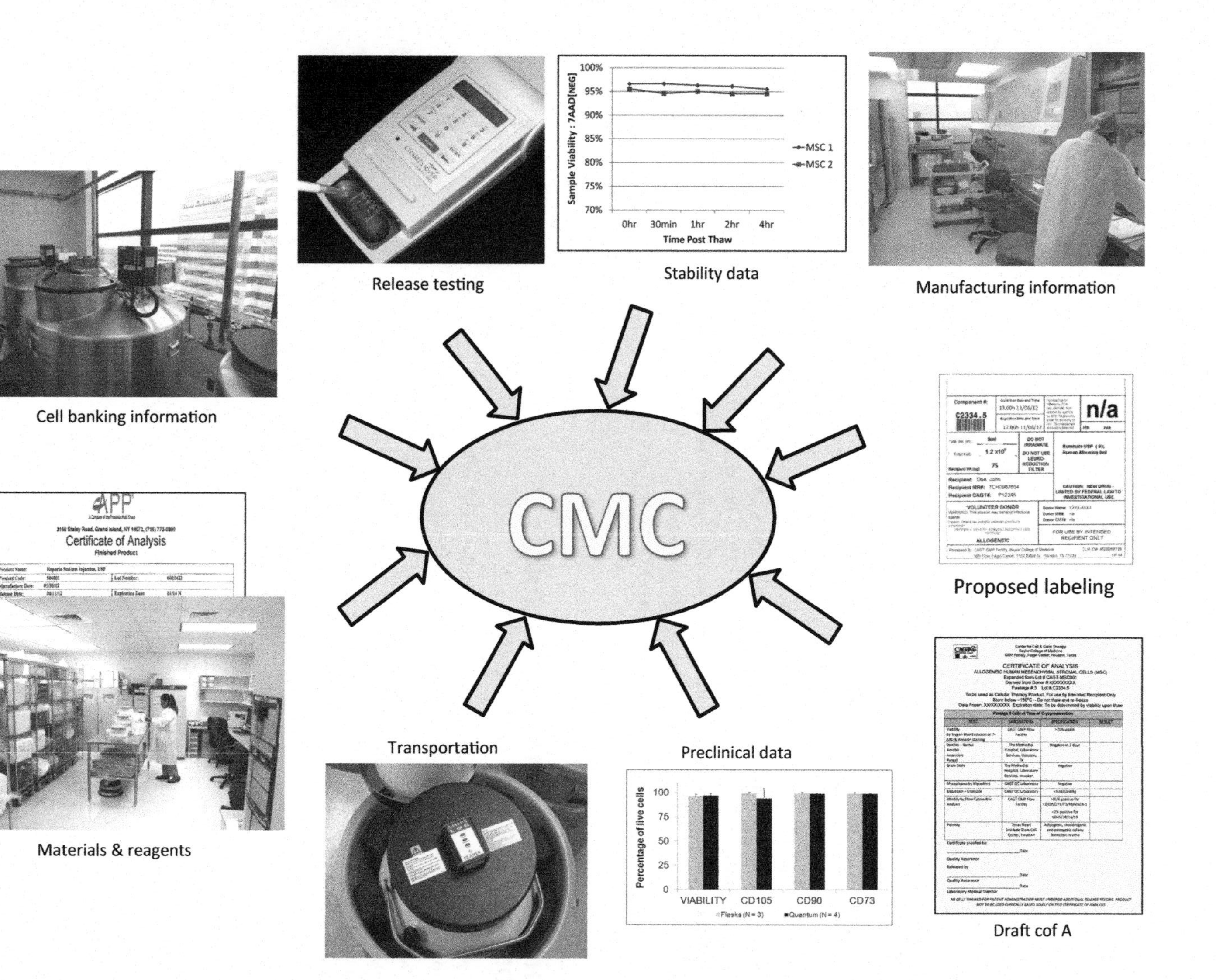

FIGURE 7.3 Main components of the chemistry manufacturing and control (CMC) section of an IND application.

and release testing. The manufacturer should expect to describe the system in use to track the cell donor, intended recipient and product throughout collection, manufacturing, storage, release, distribution, and follow-up. Information on this system should also be provided to the end-user to allow additional tracing and tracking. Copies of the proposed certificate of analysis that documents the release testing results, and of the final and intermediate product labels should be submitted in the CMC. Labeling should adhere to FDA requirements, which require specific language, for example, "For Autologous Use Only" and "New Drug for Investigational Use Only" etc.

The FDA will also require data to show that the procedure used to administer the cells, for example, a catheter, will not adversely affect the cells [17]. This will require information on the viability, identity, recovery, and functionality of cells passed through the catheter. If the final product is to be shipped or transported for administration [40], data on preservation of cell integrity during shipping must be provided.

Prior to submission of the IND application to the FDA the sponsor has the opportunity to hold a number of meetings with the agency [41]. These "pre-pre" and "pre-IND" conference calls are encouraged by the FDA and provide the sponsor with the change to discuss problem areas, use of particular approaches, etc. Additional guidance is available from the FDA on preparation of the electronic IND submissions [42]. The FDA must respond to the IND application within 30 days of receipt. No response is indicative of approval without modification, but this should be confirmed in writing. They have the option to put the application on hold pending response from the applicant to specific issues [43]. Alternatively, there may be questions that require a response from the applicant but do not prevent the clinical study from starting (no-hold issues). Once the study is started, the manufacturer is expected to keep detailed documentation on all aspects of product manufacturing and distribution. They must also be kept informed of any adverse reaction in the recipient to product use, and any complaints from the customer (physician etc.). An adverse reaction to the product will, depending on its severity, have to be reported to the FDA within a defined time. The manufacturer will need to conduct an investigation as to whether it is potentially related to manufacturing, determine what steps must be taken (eg, putting the remaining cells into immediate quarantine etc.), what corrective actions are required, their implementation and a follow-up as to their efficacy. The reporting responsibility is with the IND sponsor, so there must be open and ongoing communication between the two throughout the clinical trial.

7.11 DOCUMENTATION

Compliance with cGMP is primarily available to the FDA through the manufacturer's documentation system. This must address all aspects of operations so that an audit can be performed that follows all activities related to manufacturing of the product. Similarly, any component of manufacturing, for example, equipment management must be fully auditable. Design of an appropriate documentation system is laborious and, once designed, staff must be trained in its use and maintenance. There are formal procedures that must be followed for completion of documents; these include elements such as identification of the staff member performing each significant step of a procedure. All corrections must be signed and dated and made so as not to obscure the original information. The FDA recommends that the worksheets or batch records correspond closely with the major steps described in the SOP, any changes must be documented by means of a variance or deviation, unless specifically permitted in the SOP. A variance may be prospective, in which case it has been preapproved by the Quality Unit; or retrospective, in which case it was detected after it occurred and an explanation of what happened and its impact is formally assessed in collaboration with the Quality Unit. Retrospective variances should include a plan to prevent further occurrences and an assessment of the efficacy of that plan.

7.12 AUDITS

Audits are an essential component of operations. They provide additional assurance that operations are within specifications. Routine internal audits should be performed by the Quality Unit to evaluate major components of facility practices, for example, compliance with manufacturing SOPs, equipment management, etc. For cause audits may be performed when there is some reason to suspect that there is a developing problem, or as the result of a significant variance. An active and thorough audit program provides evidence to the FDA that the quality program is effective and efficient.

Importantly, the FDA has the right to audit a manufacturer at any time. These may be random, or the result of an issue that has attracted the Agency's concern. The manufacturer should have a formal procedure to handle FDA audits, so that they are managed efficiently, the staff know what to expect and that the appropriate records are kept. The plan must be to be open and honest with the auditors. Staff should continue their work and respond to questions accurately. If they do not know how to respond they should indicate this, but also inform the auditor that they intend to obtain the requested information from their supervisors. Copies of documentation should be provided when requested and a record of which documents were provided must be kept.

The audit will start with the FDA staff identifying themselves and stating a reason for their visit. At the conclusion of the audit they will hold a closing meeting during which they inform the manufacturer of their findings. If there are significant findings they are presented as a Form 483 "Notice of Inspectional Observations" [44], which lists the issues that require action by the manufacturer. A time period is set by which a response is required. In the formal response the manufacturer must address each finding specifically and include a plan on how it will be resolved and within what timeframe. The plan should include identification of the root cause of the issue. If the plan has already been completed, evidence of its efficacy should also be included. If the response is not satisfactory the FDA can escalate its actions up to closure of the facility and arrest of the manufacturer.

7.13 POST-CMC MANUFACTURING CHANGES

The FDA recognizes that there may be changes to the manufacturing procedure after the IND has been approved. It is up to the manufacturer and the IND sponsor to determine the impact of the change on the product safety, purity, potency, and potential efficacy. Minor changes with no potential impact may be made without formal FDA approval, but they must be described in the IND Annual Report. If a change has likely impact it should be reported to the FDA and they may allow manufacturing to continue without requiring additional information. Alternatively, they may suspend manufacturing until formal data is submitted to demonstrate the equivalence of products manufactured using the new and old procedures. To avoid confusion the manufacturer is advised to seek Agency input on any proposed change to the manufacturing method.

7.14 SUMMARY

Academic investigators often find the transition to cGMP manufacturing somewhat traumatic. As soon as research indicates that a cell product may be of therapeutic interest, contact should be made with a cGMP Facility to provide input on how the transition should occur. This will assist the researcher in selection of materials, methods, and equipment that will facilitate translation to clinical manufacturing. Cellular therapies are still relatively new and there was some confusion as to how they were regulated. This situation is now much clearer, and the FDA has made a tremendous effort to develop and publicize their risk-based approach to regulation of cellular and gene therapy products. This is complemented by their advice through various types of IND meetings to provide input to investigators proposing to file a new application. Assistance is also available through a number of draft and finalized guidance documents and the Agency's participation at meetings of professional societies. In spite of this, there is often considerable trepidation about the regulatory requirements and the ability to support clinical trials without building a new manufacturing facility. This is best overcome by taking the time to read the regulations and guidance documents, and by speaking with FDA staff to determine their expectations from the manufacturer.

ACKNOWLEDGMENTS

The author thanks Zhuyong Mei, Mariola Klis and the staff of the GMP Facility at the Baylor College of Medicine Center for Cell and Gene Therapy for all of their assistance. Additional thanks to Sara Richman and April Durett for their help with the manuscript, and for partial support to grant # RP130256 from the Cancer Prevention and Research Institute of Texas.

REFERENCES

[1] Glenn JD, Whartenby KA. Mesenchymal stem cells: emerging mechanisms of immunomodulation and therapy. World J Stem Cells 2014;6(5):526–39.

[2] Hass R, et al. Different populations and sources of human mesenchymal stem cells (MSC): a comparison of adult and neonatal tissue-derived MSC. Cell Commun Signal 2011;9:12.

[3] Phinney DG. Functional heterogeneity of mesenchymal stem cells: implications for cell therapy. J Cell Biochem 2012;113(9):2806–12.

[4] Are my HCT'Ps regulated solely under section 361 of the PHS Act and the regulations in this party, and if so what must I do? Code of Federal Regulations, 2015; Title 21, Part 1271.10.

[5] Immel BK. A brief history of the GMPs for pharmaceuticals. Pharm Technol 2001;48–52. http://gmpnews.ru/wp-content/uploads/2010/05/History-gmp.pdf.

[6] cGMP in Manufacturing, Processing, Packing, or Holding of Drugs and Finished Pharmaceuticals. Code of Federal Regulations, 2015, Title 21, Parts 210 and 211.

[7] U.S. Department of Health and Human Services Food and Drug Administration Center for Drug Evaluation and Research (CDER) Center for Biologics Evaluation and Research (CBER) Office of Regulatory Affairs (ORA). Guidance for industry CGMP for phase 1 investigational drugs. Available at: http://www.fda.gov/downloads/drugs/guidancecomplianceregulatoryinformation/guidances/ucm070273.pdf\; 2008.

[8] U.S. Department of Health and Human Services Food and Drug Administration Center for Biologics Evaluation and Research. Considerations for the design of early-phase clinical trials of cellular and gene therapy products. Available at: http://www.fda.gov/BiologicsBloodVaccines/GuidanceComplianceRegulatoryInformation/Guidances/default.htm; 2015.

[9] Gee AP, editor. Cell therapy: cGMP facilities and manufacturing. New York: Springer; 2009.

[10] Griffin DL. Facility equipment. In: Gee AP, editor. Cell therapy: cGMP facilities and manufacturing. New York: Springer; 2009. p. 171–5.

[11] U.S. Department of Health and Human Services Food and Drug Administration Center for Biologics Evaluation and Research. Current Good Tissue Practice (CGTP) and additional requirements for manufacturers of human cells, tissues, and cellular and tissue-based products (HCT/Ps). Draft guidance. Available at: http://google2.fda.gov/search?q=cache:PqLovMDvqnkJ:www.fda.gov/downloads/biologicsbloodvaccines/guidancecomplianceregulatoryinformation/guidances/tissue/ucm091408.pdf+cross+contamination&proxystylesheet=FDAgov&output=xml_no_dtd&site=FDAgov-Section-VBB&client=FDAgov&ie=UTF-8&access=p&oe=UTF-8; 2009.

[12] U.S. Department of Health and Human Services Food and Drug Administration Center for Biologics Evaluation and Research. Content and review of chemistry, manufacturing, and control (CMC) information for human somatic cell therapy investigational new drug applications (INDs). Guidance for FDA reviewers and sponsors. Available at: http://www.fda.gov/downloads/BiologicsBloodVaccines/GuidanceComplianceRegulatoryInformation/Guidances/Xenotransplantation/ucm092705.pdf; 2008.

[13] Ben Azouna N, et al. Phenotypical and functional characteristics of mesenchymal stem cells from bone marrow: comparison of culture using different media supplemented with human platelet lysate or fetal bovine serum. Stem Cell Res Ther 2012;3(1):6.

[14] Shih DT, Burnouf T. Preparation quality criteria, and properties of human blood platelet lysate supplements for ex vivo stem cell expansion. N Biotechnol 2015;32(1):199–211.

[15] Al-Saqi SH, et al. Defined serum-free media for in vitro expansion of adipose-derived mesenchymal stem cells. Cytotherapy 2014;16(7):915–26.

[16] Swamynathan P. Are serum-free and xeno-free culture conditions ideal for large scale clinical grade expansion of Wharton's jelly derived mesenchymal stem cells? A comparative study. Stem Cell Res Ther 2014;5(4):88.

[17] U.S. Department of Health and Human Services Food and Drug Administration Center for Biologics Evaluation and Research. Preclinical assessment of investigational cellular and gene therapy products. Available at: http://www.fda.gov/downloads/BiologicsBloodVaccines/GuidanceComplianceRegulatoryInformation/Guidances/CellularandGeneTherapy/UCM376521.pdf; 2013.

[18] U.S. Department of Health and Human Services Food and Drug Administration Center for Biologics Evaluation and Research (CBER) Center for Devices and Radiological Health (CDRH) Office of Combination Products in the Office of the Commissioner (OCP). Human cells, tissues, and cellular and tissue-based products (HCT/Ps) from adipose tissue: regulatory considerations. Draft guidance. Available at: http://www.fda.gov/downloads/BiologicsBloodVaccines/GuidanceCompliance-RegulatoryInformation/Guidances/Tissue/UCM427811.pdf; 2014.

[19] Hanley PJ, et al. Manufacturing mesenchymal stromal cells for phase I clinical trials. Cytotherapy 2013;15(4):416–22.

[20] Godara P. Mini-review: design of bioreactors for mesenchymal stem cell tissue engineering. J Chem Technol Biotechnol 2008;83:408–20.

[21] Hanley PJ, et al. Efficient manufacturing of therapeutic mesenchymal stromal cells with the use of the Quantum Cell Expansion System. Cytotherapy 2014;16(8): 1048–58.

[22] Carmelo JG, et al. Scalable ex vivo expansion of human mesenchymal stem/stromal cells in microcarrier-based stirred culture systems. Methods Mol Biol 2015;1283:147–59.

[23] Nold P, et al. Good manufacturing practice-compliant animal-free expansion of human bone marrow derived mesenchymal stroma cells in a closed hollow-fiber-based bioreactor. Biochem Biophys Res Commun 2013;430(1):325–30.

[24] Neumann A, et al. Characterization and Application of a Disposable Rotating Bed Bioreactor for Mesenchymal Stem Cell Expansion. Bioengineering 2014; 1:231–45.

[25] Dal Pozzo S, et al. High recovery of mesenchymal progenitor cells with non-density gradient separation of human bone marrow. Cytotherapy 2010;12(5):579–86.

[26] U.S. Department of Health and Human Services Food and Drug Administration Center for Drug Evaluation and Research (CDER) Center for Biologics Evaluation and Research (CBER) Center for Veterinary Medicine (CVM). Process validation: general principles and practices. Guidance for industry. Available at: http://www.fda.gov/downloads/Drugs/Guidances/UCM070336.pdf; 2011.

[27] Chandler DKF, et al. Historical overview of mycoplasma testing for production of biologics. Am Pharm Rev 2011;14(4):37370.

[28] Uphoff CC, Drexler HG. Detecting mycoplasma contamination in cell cultures by polymerase chain reaction. Methods Mol Biol 2011;731:93–103.

[29] Mariotti E, et al. Rapid detection of mycoplasma in continuous cell lines using a selective biochemical test. Leuk Res 2008;32(2):323–6.

[30] U.S. Department of Health and Human Services Food and Drug Administration Center for Biologics Evaluation and Research. Eligibility determination for donors of human cells, tissues, and cellular and tissue-based products (HCT/Ps). Guidance for industry. Available at: http://www.fda.gov/downloads/BiologicsBloodVaccines/GuidanceComplianceRegulatoryInformation/Guidances/Tissue/ucm091345.pdf; 2007.

[31] US Food and Drug Administration. Donor eligibility final rule and guidance questions and answers. Available at: http://www.fda.gov/BiologicsBloodVaccines/TissueTissueProducts/QuestionsaboutTissues/ucm102842.htm; 2015.

[32] Krampera M, et al. Immunological characterization of multipotent mesenchymal stromal cells—The International Society for Cellular Therapy (ISCT) working proposal. Cytotherapy 2013;15(9):1054–61.

[33] Dominici M, et al. Minimal criteria for defining multipotent mesenchymal stromal cells. The International Society for Cellular Therapy position statement. Cytotherapy 2006;8(4):315–7.

[34] Mendicino M, et al. MSC-based product characterization for clinical trials: an FDA perspective. Cell Stem Cell 2014;14(2):141–5.

[35] Gee AP, et al. A multicenter comparison study between the Endosafe PTS rapid-release testing system and traditional methods for detecting endotoxin in cell-therapy products. Cytotherapy 2008;10(4):427–35.

[36] Bloom DD, et al. A reproducible immunopotency assay to measure mesenchymal stromal cell-mediated T-cell suppression. Cytotherapy 2015;17(2):140–51.

[37] Samsonraj RM, et al. Establishing criteria for human mesenchymal stem cell potency. Stem Cells 2015;33(6):1878–91.

[38] Luetzkendorf J, et al. Cryopreservation does not alter main characteristics of Good Manufacturing Process-grade human multipotent mesenchymal stromal cells including immunomodulating potential and lack of malignant transformation. Cytotherapy 2015;17(2):186–98.

[39] Pollock K, et al. Clinical mesenchymal stromal cell products undergo functional changes in response to freezing. Cytotherapy 2015;17(1):38–45.

[40] Veronesi E, et al. cGMP-compliant transportation conditions for a prompt therapeutic use of marrow mesenchymal stromal/stem cells. Methods Mol Biol 2015;1283: 109–22.

[41] U.S. Department of Health and Human Services Food and Drug Administration Center for Drug Evaluation and Research (CDER) Center for Biologics Evaluation and Research (CBER). Formal meetings between the FDA and sponsors or applicants. Guidance for industry. Available at: http://www.fda.gov/downloads/Drugs/Guidances/ucm153222.pdf; 2009.

[42] U.S. Department of Health and Human Services Food and Drug Administration Center for Biologics Evaluation and Research (CBER). Providing regulatory submissions to CBER in electronic format—Investigational new drug applications (INDs). Guidance for industry. Available at: http://www.fda.gov/downloads/biologicsblood-vaccines/guidancecomplianceregulatoryinformation/guidances/general/ucm150028.pdf; 2002.

[43] U.S. Department of Health and Human Services Food and Drug Administration Center for Drug Evaluation and Research (CDER) Center for Biologics Evaluation and Research (CBER). Submitting and reviewing complete responses to clinical holds. Available at: http://www.fda.gov/downloads/RegulatoryInformation/Guidances/ucm127538.pdf; 2000.

[44] US Food and Drug Administration. FDA Form 483 frequently asked questions. Available at: http://www.fda.gov/ICECI/Inspections/ucm256377.htm; 2015.

Chapter 8

Mesenchymal Stromal Cells and the Approach to Clinical Trial Design: Lessons Learned From Graft Versus Host Disease

N. Dunavin, A.J. Barrett and M. Battiwalla
Hematology Branch, National Heart, Lung, and Blood Institute, National Institutes of Health, Bethesda, MD, United States

CHAPTER OUTLINE

Mesenchymal Stromal Cells. http://dx.doi.org/10.1016/B978-0-12-802826-1.00008-8

8.1 INTRODUCTION

Mesenchymal stromal cells (MSCs) are multipotent cells that differentiate into cells of mesenchymal origin (adipocytes, chondrocytes, and osteoblasts), support the growth and differentiation of adjacent cell populations, and regulate the immune response. Because of demonstrable tissue regenerative and immune modulatory functions, nonimmunogenicity, and a history of clinical safety, MSCs are currently being administered to human subjects in hundreds of clinical studies worldwide for diverse indications. MSCs are under development as a new treatment for regeneration of damaged bone, cartilage, and tendons; for cardiovascular diseases including myocardial infarction and peripheral vascular diseases; for pulmonary diseases including chronic obstructive lung disease and acute respiratory distress syndrome; for neurologic diseases including amyotrophic lateral sclerosis and spinal cord injury; and, for immune-mediated disorders including graft versus host disease (GVHD), multiple sclerosis, and inflammatory bowel disease.

There are enormous advantages to MSCs that have made them the leading cellular therapy candidate closest to regulatory approval for a number of indications. These advantages include their ease of manufacture, an impressive track record of safety, nonimmunogenicity allowing the generation of an off-the-shelf allogeneic universal third-party product, and several advanced clinical trials. Nevertheless, the road to clinical approvals is proving to be challenging and uncertain. Since regulatory approval of cellular products including MSCs is indication-specific, there will be different considerations for each indication including the choice of preclinical models, manufacture technique, release criteria, potency assays, and in clinical trial design based upon differences in risk versus benefit and response endpoints. Furthermore, there will be additional challenges postregulatory approval in terms of reimbursement, manufacturing, cryopreservation, shipping and handling, and in administration outside the clinical trial setting. This chapter will focus only on the translational aspects of MSC implementation in the field of acute GVHD, an indication closest to broad clinical approval. We will not discuss the challenges after regulatory approval. For other indications, the reader is referred to excellent reviews [1,2].

8.2 MSCs AND GVHD

8.2.1 Diagnosis and Immunobiology of Acute GVHD

Acute GVHD is a complication of allogeneic hematopoietic stem cell transplant that begins when donor-derived T lymphocytes gain specificity for disparate major and minor histocompatibility antigens in host tissues. When primed T cells encounter exposed antigens in damaged tissue, they differentiate into effector T cells, proliferate, and further damage host tissue through direct cytolysis or by releasing inflammatory molecules that recruit additional effector cells of the immune system. These initial steps initiate a cascade of tissue damage and inflammation that can result in cytokine storm. Thenafter, a spectrum of injuries follow that can affect nearly every organ system. Classic manifestations involve skin, gut, and liver. Skin involvement is usually the first sign of acute GVHD that varies in extent from a rash affecting extensor surfaces, face, and hands to a generalized erythroderma affecting the entire body. Acute GVHD of the gastrointestinal system can affect the entire digestive tract leading to diarrhea, blood loss, vomiting, nutritional failure, weight loss, and secondary infection. Hepatic GVHD primarily affects the bile canaliculi with pericanalicular lymphocyte infiltration and damage to biliary endothelium. Tissue damage is not limited to these sites only; endothelium, lung, central nervous system, thymus, and bone marrow can all be targets of alloreactivity [3,4].

8.2.2 Current Preemptive Strategies

The strategies used for prophylaxis depend on the intensity of conditioning regimen, stem cell source, and degree of HLA mismatch. The initial approach to minimizing acute GVHD involves selection of the best matched donor, avoidance of female-to-male donor combinations, using bone marrow rather than peripheral blood as a stem cell source, and reducing conditioning intensity when appropriate to the clinical situation. Prophylaxis consists of one or several immunosuppressive medications. In myeloablative conditioning regimens GVHD prophylaxis can be achieved with a short course of intravenous methotrexate (MTX) and several months of oral cyclosporine or tacrolimus. Replacement of MTX with mycophenolate mofetil can achieve similar outcomes with less mucositis and hepatotoxicity; this is a common strategy used in reduced-intensity transplants. Potent anti-T-cell therapy with antithymocyte globulin or alemtuzumab (Campath) can improve the rates of acute and chronic GVHD in higher risk unrelated donor transplants at the expense of more risk for infection. In some instances, standard calcineurin inhibitor-based GVHD prophylaxis has been refined through the use of the newer pharmacologic agent sirolimus [5]. There are wide variations in drug schedule and dosing based on patient characteristics and institutional standard clinical practice.

The continued growth of registries such as the Be the Match Registry of the National Marrow Donor Program has permitted better unrelated donor selection, such that GVHD outcomes with a well-matched unrelated donor approximate those in matched related donor transplantation [6]. Haploidentical transplants are rapidly improving such that outcomes may approximate those seen with fully matched donors [7]. Consequently, a suitable donor can be found for most patients who require an allogeneic transplant regardless of their age or ethnic background. However, despite these advances, the authors estimate that acute GVHD of varying severity can be expected in 20–50% of transplant recipients based on the clinical scenario. Thus, GVHD remains a challenging problem for transplant physicians due to the lack of effective treatment options once it is established.

8.2.3 Current Therapeutic Strategies

The standard initial therapy for acute GVHD is high-dose intravenous corticosteroids [8]. However, only half of patients with acute GVHD have a lasting response to corticosteroids alone, and survival of those who do not respond is poor [9]. Several agents have been administered along with upfront steroids in clinical trials to improve response rates, most notably pentostatin, mycophenolate mofetil, etanercept, and denileukin diftitox [10]. Unfortunately, no agent or combination of agents has been proven to be more effective than steroids alone. The response rate to single-agent corticosteroid therapy, when analyzed retrospectively in a large multicenter cohort, is approximately 50% [11]. In a randomized multicenter prospective study comparing mycophenolate mofetil plus steroids and steroids alone (BMT-CTN 0802), the contemporary response rate to upfront steroids was 50.4% at day 56 in the steroids alone arm and the overall transplant-related mortality was 21.5% at 12 months [12].

Steroid-refractory disease is usually defined as the absence of improvement after 5 days or a progression of stage within 72 h of high-dose intravenous corticosteroids [13]. While improvements have been made in preempting GVHD, mortality is reported to be as high as 50–70% in patients who are refractory to steroids [14,15]. Thus, refractory acute GVHD represents one of the most challenging situations physicians will encounter posttransplant. The reader is referred to several excellent reviews on the immunobiology, diagnosis, prevention, and treatment of acute GVHD [16–20].

8.2.4 MSCs in GVHD

Cellular therapy with MSCs represents a novel approach to the prevention and treatment of acute GVHD. MSC infusions have been administered in numerous GVHD studies. Strategies for using MSCs vary widely and typically fall into the categories of (1) coinfusion of MSCs for GVHD prophylaxis at the time of stem cell infusion or at the time of engraftment, (2) infusion of MSCs at acute GVHD onset, and (3) treatment with MSCs at the time when GVHD is determined to be steroid-refractory or refractory to second line immunosuppressive therapies. As discussed previously, outcomes in refractory GVHD are dismal without effective therapy; thus altogether more than 500 patients in more than 30 cohorts worldwide have been given MSCs in this clinical scenario (Table 8.1) [21–46]. In the majority of these studies, responses were observed with no adverse reactions. Given the promising results of early phase studies for the treatment of steroid-refractory disease, Kebriaei et al. conducted the first prospective trial of third-party MSCs for the treatment of de novo acute GVHD using the premanufactured, universal donor formulation of bone marrow-derived MSCs marketed by Osiris Therapeutics, Inc. as Prochymal [47]. This industry sponsored, randomized, multicenter trial compared two different dose levels of MSCs (2 or 8×10^6 MSC/kg) combined with standard corticosteroid therapy. Subjects were infused with MSCs within 72 h of acute GVHD diagnosis and they received one additional dose 3 days later. Of 31 adult patients, 71% had a complete response (CR) at day 28 after therapy (primary endpoint), and the initial overall response rate was 94%. No evidence of a dose response relationship was seen. These promising results led to the use of Prochymal in a large industry-sponsored phase III trial for de novo acute GVHD (Protocol 265, Osiris Therapeutics, Inc). The primary endpoint was CR within 28 days of treatment administration, and treatment failures included patients without CR at 28 days and those requiring increased doses of corticosteroids or additional immunosuppressive therapy. This study failed to reach significance in the primary endpoint, and the results have not been published except in press release and in abstract form [27,48]. The overall response rate was 82% in the MSC-treated group versus 73% in the placebo group ($p = 0.12$). The higher than historically reported response to steroid therapy, the use of late passage MSC, the predominance of skin GVHD, and the choice of primary endpoint at day 28 are widely attributed to have been the cause for failure to meet the primary endpoint [49,50]. Despite this early setback, many groups have now devised strategies to improve MSC therapy in GVHD, focusing on modifying the cellular therapy or selecting the appropriate patient. Moreover, the continued development of MSC products in academic medical centers is justified by the presumed dissimilarity of MSC products between Prochymal versus the so-called "early passage" MSCs that were pioneered in Europe and are generated in most academic centers. MSCs proliferate rapidly in culture, but with prolonged expansion they may lose potency. Passage, or subculture, refers to the time when cells in culture are transferred to a new vessel at a lower seeding density to allow for further expansion. Early passage is generally considered to be at the fifth passage or less, whereas late passage MSCs are sometimes expanded well past the fifth passage. It was shown by Ren et al. that human MSCs expanded beyond fifth passage proliferated less efficiently and demonstrated a unique gene

Table 8.1 One Decade of Human Studies Using MSCs for Refractory Acute GVHD

Year Published	First Author, Country	Number of Patients	Source of MSC	GVHD Grade	MSC Passage	MSC Dose (Cells/kg), Number of Infusions	Response	Milestones
2004	Le Blanc et al., Sweden.	1	Haplo bone marrow	IV	Early	$1–2 \times 10^6$, 2 infusions	1 CR (100%)	First in human use of bone marrow-derived MSCs to treat acute GVHD.
2006	Ringden et al., Sweden	8	MRD, haplo, or third-party bone marrow	III–IV	Early	$0.7–9 \times 10^6$, 1–2 infusions	6 CR (75%)	
2007	Fang et al., China	6	Haplo or third-party adipose	III–IV	Early	1×106, 1 infusion	5 CR (83%)	First in human use of adipose-derived MSCs to treat acute GVHD.
2008	Le Blanc et al., Sweden, The Netherlands, Italy, Australia	55	MRD, haplo, or third-party bone marrow	II–IV	Early	$0.4–9\times10^6$, 1–5 infusions	30 CR (55%) 9 PR (16%)	
2008	Muller et al., Germany	2	Haplo bone marrow	III–IV	Early	$0.4–3\times10^6$, 1 infusion	1 CR (50%)	
2008	von Bonin et al., Germany	13	Third-party bone marrow	III–IV	Early	$0.6–1.1 \times 10^6$, 1–5 infusions	1 CR (8%) 1 PR (8%)	First in human use of platelet lysate expanded bone marrow-derived MSCs to treat acute GVHD.
2009	Martin et al. (abstract), United States, Canada, Australia	163	Prochymal third-party bone marrow	II–IV	Late	1×10^6, 8–12 infusions planned	ORR 82% BM-MSC v. 73% placebo ($p = 0.12$)	First study of Prochymal MSCs; performed no better than placebo in this large randomized study. Better responses in visceral GVHD. Results not published.
2010	Lim et al., China	1	Third-party bone marrow	III–IV	Early	2×10^6, 2 infusions	1 CR (100%)	
2010	Arima et al., Japan	3	MRD or haplo bone marrow	III	Early	0.5×10^6, 1 intraarterial infusion	2 PR (66%)	Intraarterial infusions not shown to be effective.
2010	Luccini et al., Italy	8	Third-party bone marrow	I–IV	Early	$0.7–3.7 \times10^6$, 1–5 infusions	4 CR (50%) 2 PR (25%)	
2011	Prasad et al., USA	12	Prochymal third-party bone marrow	III–IV	Late	$2–8\times10^6$, 2–21 infusions	7 CR (58%) 2 PR (17%)	First study of Prochymal in pediatric GVHD leading to approvals in Canada and New Zealand.

2011	Perez-Simon et al., Spain	10	MRD, haplo, or third-party bone marrow	II–IV	Early	1–2×10[6], 1–4 infusions	1 CR (10%) 6 PR (60%)	
2011	Wernicke et al., Germany	2	Third-party bone marrow	IV	Early	0.9–1.9 ×10[6], 1 infusion	2 CR (100%)	
2011	Wu et al., Taiwan	2	Third party, umbilical cord	IV	Early	3.3–4.1 ×10[6], 1 infusion	2 CR (100%)	First in human use of umbilical cord-derived MSCs for acute GVHD.
2011	Herrmann et al., Australia	12	MRD, haplo, or third-party bone marrow	I–IV	Early	1.7–2.3 ×10[6], 1–19 infusions	7 CR (58%) 4 PR (33%)	
2012	Dander et al., Italy	6	Third-party bone marrow	II–IV	Early	0.9–1.9 ×10[6], 2–3 infusions	1 CR (17%) 4 PR (66%)	
2013	Muroi et al., Japan	14	Third-party bone marrow	II–III	Late	2×10[6], 8–12 infusions	8 CR (57%) 5 PR (36%)	
2013	Resnick et al., Israel	50	MRD, haplo, or third-party bone marrow	II–IV	Early	0.3–4.3×10[6], 1–4 infusions	17 CR (34%)	
2013	Ball et al., Sweden, The Netherlands, Italy, Canada	37	Third-party bone marrow	III–IV	Early	0.9–3.0 ×10[6], 1–11 infusions	24 CR (65%) 8 PR (22%)	
2013	Introna et al., Italy	40	Third-party bone marrow	II–IV	Early	0.8–3.1 ×10[6], 2–11 infusions	CR 11 (27.5%) PR 16 (40%)	
2014	Sanchez-Guijo et al., Spain	25	Third-party bone marrow	II–IV	Early	0.7–1.3 ×10[6], 4 infusions	11 CR (44%) 6 PR (27%)	
2014	Yin and Battiwalla et al., United States	7	Third-party bone marrow	II–IV	Early	2×10[6], 3 infusions	5 CR (71%)	
2014	Kurtzberg et al., United States, Canada	75	Prochymal third-party bone marrow	II–IV	Late	2–8×10[6], 8–12 infusions	ORR 61.3% at day +28	
2014	Silla et al., Brazil	8	Third-party bone marrow	II–IV	Early	1–3×10[6], 1–3 infusions	CR 5 (62.5%)	
2015	Zhao et al., China	28	Third-party bone marrow	II–IV	Early	1×10[6], 2–8 infusions	CR 17 (61%) PR 4 (14%)	

CR, complete response; GVHD, graft versus host disease; haplo, haploidentical; MRD, matched related donor; MSC, mesenchymal stromal cell; ORR, overall response rate; PR, partial response expression profile [51]. Differentiation and senescence by continued culture expansion is one of the postulated differences between Prochymal and the MSCs made in most academic centers. Factors such as donor variance and cryopreservation may also play a role [52].

8.3 TRANSLATIONAL CHALLENGES OF MSCs FOR GVHD

The heterogeneity of MSCs remains an important challenge. The diversity of donor sources (autologous versus random), tissue origin (marrow, adipose tissue, or umbilical cord), and desired functionality (immune/inflammatory versus repair/regeneration) has led to abandoned efforts to generate a universal product. The multiple potential therapeutic applications of MSCs have conflicted with the orderly development of MSCs. For instance, it may be preferable to use an autologous product for tissue repair/regenerative purposes, but an allogeneic (random donor) product may be acceptable in a transplant recipient with GVHD. In this complex area of translation, areas of intersecting interest impact scientists, cell manufacturers, clinical investigators, and regulatory agencies (Fig. 8.1).

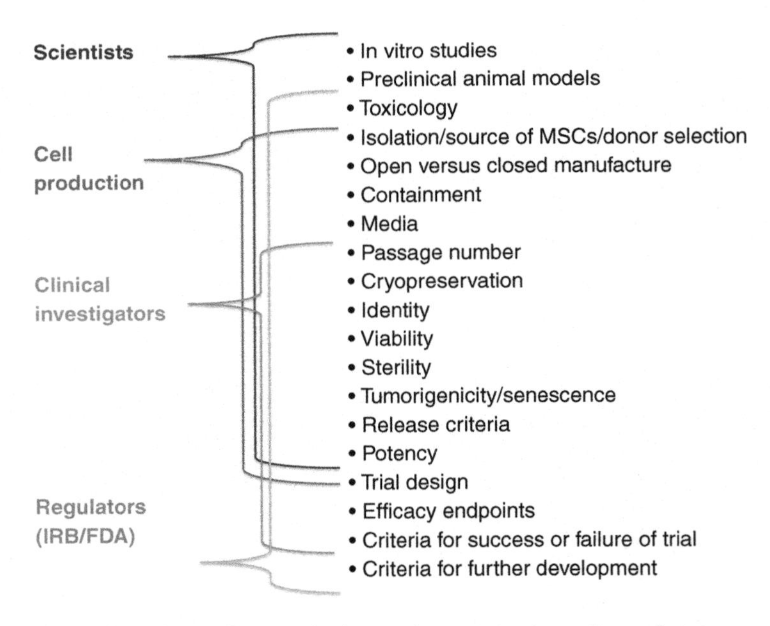

■ FIGURE 8.1 Areas of intersecting interest impact scientists, cell manufacturers, clinical investigators, and regulatory agencies.

8.3.1 Relevance of Preclinical Studies of MSCs in GVHD

The potency of third-party MSCs in reducing T-lymphocyte responses in an HLA-independent fashion has been well documented in vitro [53–56]. Human trials of MSCs for GVHD have advanced either concurrently or sometimes ahead of the development of corresponding animal models. In this setting, the role of animal models has been mostly to understand dose–schedule relationship, cell tracking, and understanding of biology. While bearing in mind the important differences between human and murine MSCs [57] and the caveat regarding murine models for GVHD in general [58], they are still very useful. For instance, examining the role, if any, of MSCs in promoting tumorigencity may only be possible in animal models. Another example of the utility of animal models was the establishment of a requirement for IFN-gamma licensing of MSCs in vivo [59]. Preclinical models have recently challenged the nonimmunogenic nature of third-party MSCs, an established tenet for "off-the-shelf" administration [60,61]. This could suggest that nonimmunosuppressed hosts may develop resistance to repeated dosing [62–64]. Moving forward, preclinical models will serve to identify the most important mechanisms of action, identify the critical regulators that may eventually result in a more targeted approach, help understand cell trafficking and longevity, and compare potency of different sources of MSCs of relevance to human use.

8.3.2 Source of MSCs and Donor Selection in GVHD

Primary human MSCs are known to have multiple subsets, which differ in function and frequency [65–67]. The source of MSCs adds a further dimension of complexity to this heterogeneity of primary human MSCs. Currently human MSCs are typically manufactured from three sources: bone marrow, umbilical cord, and adipose tissue. Most studies for GVHD have utilized marrow and rarely umbilical cord (Table 8.1) as there is greatest familiarity and track record from these sources in the setting of allogeneic transplant. Based on the degree of expansion and passage number, intrasubject variability is expected [51]. Autologous, match-related (including haploidentical) as well as unrelated (third-party) allogeneic donors have been utilized. Autologous products are typically modestly expanded for reinfusion into the same patient, whereas third-party products may undergo massive expansion to treat a number of subjects. These considerations have significant impact upon extent of differentiation, activity, potency, and safety. Most trials in the United States have focused on well-characterized third-party products. For example, the Osiris Prochymal MSCs used a well-characterized third-party product that had been extensively passaged on an industrial scale. The obvious advantages of such a strategy include infrequent testing for adventitious agents and a more homogenous cellular product with predictable biological characteristics. On the other hand, the multicenter European trials have utilized MSCs from several sources [25].

8.3.3 Fresh or Frozen?

Although viability may not be appreciably impacted by a freeze–thaw cycle with optimal conditions [68], this does not rule out a decline in potency. Galipeau et al. have postulated that preclinical models have mainly used fresh, growing MSCs and the deleterious impact of cryopreservation has not been rigorously tested [52]. Freshly thawed cryopreserved MSCs upregulate heat-shock proteins are refractory to interferon (IFN)-gamma-induced upregulation of indoleamine 2,3-dioxygenase, and are compromised in suppressing CD3/CD28-driven T-cell proliferation. Immune suppressor activity, IFN-gamma responsiveness, and induction of indoleamine-2,3-dioxygenase were fully restored following 24 h of MSC tissue culture postthaw [69]. It is possible that freshly thawed MSCs may undergo clearance in vivo prior to recovery of full functional capacity observed in 24 h in vitro. These findings of reduced potency by freeze–thawing have been confirmed by the Karolinska group [70]. Others have observed that high prefreeze senescence correlates to poor postthaw function [71,72], but neither report could substantiate altered potency. In conclusion, the impact of cryopreservation remains an area in dispute, but clearly fresh MSCs impose greater logistic challenges, increased costs, and additional regulatory concerns.

8.3.4 Potency and Release Criteria

Identity, viability, and sterility are the minimal release criteria in early-stage cellular therapy trials to which potency is added in late-stage trials. Sterility and viability (typically >70%) criteria are transparent, standardized, and easily achievable. Consensus markers to define identity have also been defined and are widely adopted [73] and are being extended [74] to general application. In contrast, robust, timely, and reproducible markers of MSC potency remain a challenge [75].

Potency is defined by the US 21 CFR 600.3(s) as "the specific ability or capacity of the product, as indicated by appropriate laboratory tests or by adequately controlled clinical data obtained through the administration of the product in the manner intended, to effect a given result." In other words, the potency assay for a cell therapeutic should measure a relevant biological property that predicts clinical benefit. The challenge arises from the diversity in cell source, heterogeneity inherent in cellular products, multiple mechanisms of action, dissimilar manufacture, differences in route/method of administration, and most importantly, the therapeutic indication. The US FDA has issued guidance to industry that mandates the development of meaningful potency tests to contribute to eventual postlicensing product release criteria [76]. Critical to the development of a potency marker is, therefore, correlation to efficacy, conclusive evidence of which has yet to be confirmed for MSCs in a phase III clinical trial.

The essential components of potency in the treatment of GVHD are the capacity of MSCs to immunomodulate and repair tissues. Flow cytometric markers of potency are attractive because the assays are robust, standardized, and yield immediate results. Wuchter et al. have proposed the use of static cell markers (CD73+, CD90+, CD105+, CD45−) as a potency assay from a German regulatory perspective within the EU [77]. But these excellent markers of identity do not indicate immunosuppressive or tissue regenerative potential, have not been validated, and may not pass muster with US regulations. The Prochymal product proposed using soluble tumor necrosis factor-receptor-1 as a potency marker [47], but theirs was a unique product and efficacy of the product was doubtful. Gene expression profiling offers comprehensive biological information but is too laborious for routine use [78]. In contrast, functional potency markers provide meaningful information with respect to extent of immunosuppression and, with effort, may be standardized [79,80]. These assays are laborious, measure the impact of MSCs on T-cell proliferation using standardized reagents, but remain susceptible to the biological variability in the source of T cells used for proliferation, may not be readily reproducible over time or at different centers, and need to be proven in the context of a clinical trial. In the end, choosing a test for potency depends entirely on the indication for treatment; in GVHD, there is still no reliable way to assess the immunosuppressive and tissue regenerative capacity of different MSC products.

8.3.5 Biodistribution, Fate, and Longevity

Essential questions regarding biodistribution, fate, and longevity of MSCs administered by various routes are important from a safety and regulatory standpoint. Animal models have conclusively shown that intravenously injected human MSCs are rapidly cleared from the circulation, and are sequestrated and ultimately destroyed in the lungs and liver within a few days [64,81–85]. This finding is not inconsistent with contact-independent (paracrine) methods of action at distant sites, but raises questions about the possible mechanisms of action, especially in GVHD. Paracrine pathways proposed include: direct release of antiinflammatory factors, interaction with the innate immune cells of the lung/liver to induce production of antiinflammatory factors, and extracellular vesicle release to catalyze production of antiinflammatory factors at the sites of inflammation.

In addition to species-specific biological differences [86], the biodistribution in animal models may be very different from humans. Differential aspects including the MSC cell size, MSC source, blood vessel lumen size, adhesive potential and porosity, MSC passage number, or the surface expression of adhesion molecules such as alpha 4 integrin, alpha 6 integrin, and fibronectin may be relevant [83]. Consequently, in vivo imaging in human subjects is going to be desirable, but not critical for regulatory approval. Methods for in vivo imaging of MSCs do exist but are cumbersome and technically challenging [87]. MRI-based detection of labeled MSCs in humans would require special MRI coils and would operate at the limits of sensitivity in the lungs.

8.3.6 Intellectual Property and Funding

Developmental funding in the biotech industry is critically dependent upon retention of intellectual property. However, efficacious MSCs may be easily manufactured with information in the public domain and this had resulted in enormous interest in academic centers with in-house cell manufacturing capability. Many of these efforts have floundered because academic funding rarely extends beyond phase II. While this translational "valley of death" applies to all therapeutics, it is particularly important for MSCs developed by academic centers because the absence of unique intellectual property (IP) in MSC manufacture and the threat of challenges to IP have disincentivized commercial funding.

8.3.7 Clinical Trial Parameters

The ease of manufacture, public technologic know-how, well-accepted safety profile, encouraging evidence of efficacy in the EU phase II and the Osiris Prochymal phase III studies, combined with the immense financial reward for an approved therapeutic for GVHD, have led to immense interest in development by academia and industry. However, it is important to recognize several pitfalls in clinical trial design in the arena of GVHD that could prove detrimental to developmental efforts. GVHD is a very difficult area in which to develop a new therapeutic. Despite enormous potential financial reward and increasing numbers of new antiinflammatory drugs, it is telling that there is not a single FDA- or EU-approved therapeutic for the indication of acute or chronic GVHD. Several sources of failure can be identified through careful analysis of past GVHD trials, notably:

- Absence of an appropriate control arm
- Broad patient eligibility including less severe forms of disease (ie, in GVHD, including skin versus ignoring skin)
- Inappropriate choice of primary endpoint and timing of endpoint assessment
- Inappropriate balancing by failing to stratify for patient characteristics.

8.3.8 **Control Arm**

In GVHD, response can often be greatly delayed by several weeks following initial intervention. Consequently, occurrence of response alone cannot appropriately provide a true measure of efficacy. In the phase III trials of Prochymal for steroid refractory GVHD, a 50% CR rate was observed in the placebo-control arm. Historical controls are not appropriate because there has been general improvement with supportive care alone over time. Efficacy trials without an appropriate control arm for GVHD are simply not interpretable. Similar considerations apply to the use of MSCs in other fields such as regenerative medicine, where clinical benefit is expected in some subjects with supportive care alone.

8.3.9 **Patient Eligibility**

Broader trial inclusion of all forms of GVHD in the case of a successful trial would lead to a licensing indication with potentially greater financial reward. However, the choice of including skin GVHD in the Prochymal phase III acute GVHD trial probably led to failure of the primary efficacy endpoint despite meeting secondary endpoints of improvement in the more dangerous liver and gastrointestinal GVHD over placebo. There was no statistical difference between Prochymal and placebo on the primary endpoints for either the steroid-refractory (35% vs. 30%, $n = 260$) or the first-line (45% vs. 46%, $n = 192$) GVHD trials. The primary endpoint for the steroid-refractory GVHD trial (durable complete response) for the per-protocol population approached statistical significance (40% vs. 28%, $p = 0.087$, $n = 179$). In patients with steroid-refractory liver GVHD, treatment with Prochymal significantly improved response (76% vs. 47%, $p = 0.026$, $n = 61$) and durable complete response (29% vs. 5%, $p = 0.046$). Prochymal significantly improved response rates in patients with steroid-refractory GI GvHD (88% vs. 64%, $p = 0.018$, $n = 71$) [27]. Therefore, by narrowing the indication and focusing on subjects who are most likely to benefit, the chances of successfully attaining the prespecified endpoint are greater. Trials in other indications should focus on the specific subgroup most likely to show an unambiguous and meaningful response that can only be attributed to the MSCs rather than the broadest eligibility to maximize profits.

8.3.10 Endpoint Selection

Consensus endpoints for trials in acute GVHD have been published [88]. Appropriate endpoints for GVHD response are day 28 or beyond, with earlier response endpoints being suspect. The latest BMT-CTN trial (BMT-CTN 0802) used a day 56 response endpoint. GVHD response alone is considered a valid primary endpoint for regulatory approval. However, because of competing risk with relapse, overall survival and nonrelapse mortality are also endpoints of equal importance. It is of no use if reduction of GVHD mortality is counterbalanced by increased relapse mortality. Consequently, trials need to be adequately powered to meet these endpoints of interest. Focusing on lethal forms of GVHD (severe liver or GI) is also more likely to produce a corresponding survival benefit. MSC trials in other indications would do well to focus on fully validated endpoints acceptable to both regulatory authorities as well as the broad spectrum of specialists in the field.

8.3.11 Stratification

Another consideration is appropriate stratification. Given a heterogeneous population of trial subjects, randomization is important for balancing unanticipated hazards. Sometimes, randomization fails as evidenced in the cyclosporine versus tacrolimus GVHD phase III study, when there was less GVHD in the tacrolimus arm but this did not translate into a survival benefit because of paucity of higher-risk subjects in the control cyclosporine arm [89]. In the current era, it is inconceivable of thinking about the impact of an intervention without stratification by biomarker levels, particularly ST2, the most determinant marker in predicting GVHD lethality [90]. Stratification is a useful clinical trial strategy to ensure homogeneity in comparator groups and should not be neglected in other indications.

8.4 OVERCOMING TRANSLATIONAL PITFALLS

While MSCs have beneficial effect in GVHD, MSCs are multifunctional cells and it is unknown which of their properties contributes most to their therapeutic benefits. By knowing the dominant mechanism by how MSCs exert immune suppression, investigators can determine the optimal timing, dose, and duration of MSC therapies. Moreover, given that many of the translational pitfalls listed above derive from the classification of MSCs as a

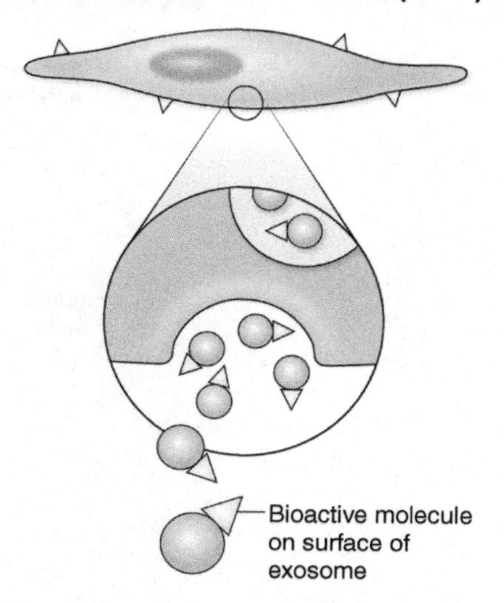

■ **FIGURE 8.2 MSC secrete exosomes that contain bioactive molecules.**

cellular therapy, one strategy is to explore further how to harness the mechanism of MSCs using a cell-free approach.

At least one such scenario does exist: MSCs are known to release virus-sized 40–100 nm bilipid membrane extracellular vesicles called "exosomes" (Fig. 8.2) [91]. Exosomes share membrane characteristics with their parent cell. Administration of exosomes intravenously recapitulates the beneficial effects of MSC treatment in murine models of injury and inflammation [92]. Recently, the first administration of MSC-derived exosomes for the treatment of refractory acute GVHD in a human subject was reported [93]. To explore this mechanism of action, Amarnath et al. introduced clinical-grade MSC products into a model of human-into-mouse xenogeneic GVHD (x-GVHD) mediated by human CD4+ Th1 cells [81]. It was found that MSC reversed established lethal x-GVHD through marked inhibition of Th1 cell effector function. Gene marking studies indicated MSC engraftment was limited to the lung. However in treated mice, circulating human CD73-expressing MSC exosomes were detected. CD73 hydrolyzes cyclic AMP to adenosine and phosphate [94]. In vitro CD73-expressing exosomes promoted the accumulation of adenosine, a potent suppressor of T cells. In the study by Amarnath, immune modulation mediated by MSC was fully abrogated by an adenosine A2A receptor antagonist. Therefore, one hypothesis is that the therapeutic efficacy of MSC in GVHD is mediated by circulating CD73+ exosomes (Fig. 8.3).

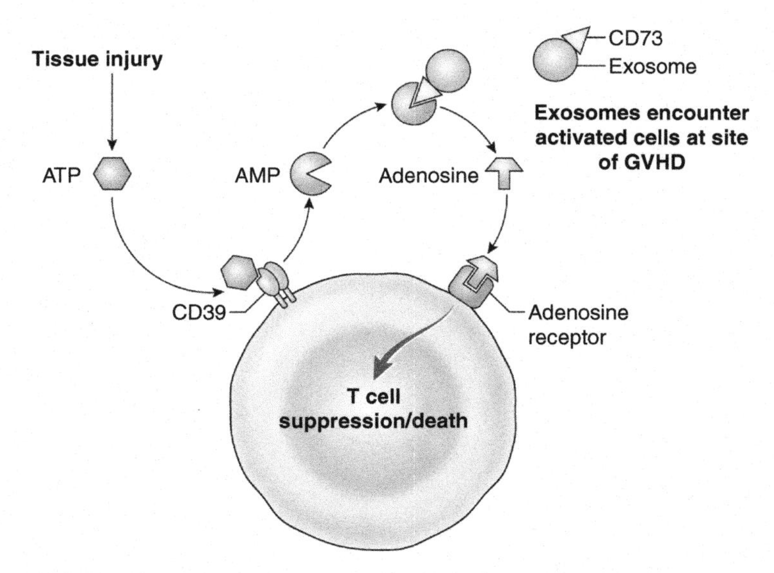

■ **FIGURE 8.3 Exosomes may exert immunosuppressive effect through adenosine signaling.**

Developing clinical-grade MSC-derived exosomes has some potential advantages over MSCs. Exosomes are non-self-replicating and thus have no malignant potential, and they are small enough to be sterilized by filtration. Thus, reproducibility, sterility, avoidance of classical alloimmune rejection, reduced concern for ectopic tissue formation, and greater potential for more streamlined regulatory approval and commercialized production are reasons to develop MSC-derived exosomes as a potential treatment for GVHD and diverse inflammatory conditions [95]. However, exosomes are also a heterogeneous product and many of the caveats that pertain to MSCs still apply. Reliable methods to characterize exosomes in terms of physical and biological characteristics have not been established. The development of widely accepted release criteria for clinical-grade exosome products is a priority for the field of cellular therapy. However, anticipating the future use of MSC-derived exosomes for the treatment of GVHD, one can hope that many of the pitfalls that were encountered with the clinical development of MSCs can be avoided by applying the lessons we learned in trial design.

8.5 CONCLUSIONS

MSCs are the foremost cellular therapy candidates for regulatory approval. MSCs are extremely heterogeneous and licensing will be influenced by source, manufacturing parameters, and clinical indication. GVHD is the indication closest to approval; however, the failure of clinical translation in GVHD after more than a decade of use in human subjects highlights the pitfalls in translating MSC treatments. Innovative approaches to regulatory management of new cellular therapeutics combined with enhanced understanding of the mechanism of action of MSCs promises to advance the field in the near future. New advances in biomarkers, a deeper understanding of the dominant mode of action in GVHD, and lessons learned from failed trials will allow us to focus on the most effective treatment strategies for the next generation of studies.

REFERENCES

[1] Thomsen GM, Gowing G, Svendsen S, Svendsen CN. The past, present and future of stem cell clinical trials for ALS. Exp Neurol 2014;262(Pt B):127–37.

[2] Griffin MD, Elliman SJ, Cahill E, English K, Ceredig R, Ritter T. Concise review: adult mesenchymal stromal cell therapy for inflammatory diseases: how well are we joining the dots? Stem cells 2013;31(10):2033–41.

[3] Lindemans CA, Hanash AM. The importance of bone marrow involvement in GVHD. Blood 2014;124(6):837–8.

[4] Saad AG, Alyea EP 3rd, Wen PY, Degirolami U, Kesari S. Graft-versus-host disease of the CNS after allogeneic bone marrow transplantation. J Clin Oncol 2009;27(30):e147–9.

[5] Ziakas PD, Zervou FN, Zacharioudakis IM, Mylonakis E. Graft-versus-host disease prophylaxis after transplantation: a network meta-analysis. PLoS One 2014;9(12):e114735.

[6] Flomenberg N, Baxter-Lowe LA, Confer D, Fernandez-Vina M, Filipovich A, Horowitz M, et al. Impact of HLA class I and class II high-resolution matching on outcomes of unrelated donor bone marrow transplantation: HLA-C mismatching is associated with a strong adverse effect on transplantation outcome. Blood 2004;104(7):1923–30.

[7] Wang Y, Liu QF, Xu LP, Liu KY, Zhang XH, Ma X, et al. Haploidentical vs identical-sibling transplant for AML in remission: a multicenter, prospective study. Blood 2015;125(25):3956–62.

[8] Martin PJ, Rizzo JD, Wingard JR, Ballen K, Curtin PT, Cutler C, et al. First- and Second-Line Systemic Treatment of Acute Graft-versus-Host Disease: Recommendations of the American Society of Blood and Marrow Transplantation. Biol Blood Marrow Transplant 2012;18(8):1150–63.

[9] Saliba RM, Couriel DR, Giralt S, Rondon G, Okoroji GJ, Rashid A, et al. Prognostic value of response after upfront therapy for acute GVHD. Bone Marrow Transplant 2012;47(1):125–31.

[10] Alousi AM, Weisdorf DJ, Logan BR, Bolanos-Meade J, Carter S, Difronzo N, et al. Etanercept, mycophenolate, denileukin, or pentostatin plus corticosteroids for acute graft-versus-host disease: a randomized phase 2 trial from the Blood and Marrow Transplant Clinical Trials Network. Blood 2009;114(3):511–7.

[11] MacMillan ML, Weisdorf DJ, Wagner JE, DeFor TE, Burns LJ, Ramsay NK, et al. Response of 443 patients to steroids as primary therapy for acute graft-versus-host disease: comparison of grading systems. Biol Blood Marrow Transplant 2002;8(7):387–94.

[12] Bolanos-Meade J, Logan BR, Alousi AM, Antin JH, Barowski K, Carter SL, et al. Phase III clinical trial steroids/mycophenolate mofetil vs steroids/placebo as therapy for acute graft-versus-host disease: BMT CTN 0802. Blood 2014;124(22):3221–7.

[13] Hoda D, Pidala J, Salgado-Vila N, Kim J, Perkins J, Bookout R, et al. Sirolimus for treatment of steroid-refractory acute graft-versus-host disease. Bone Marrow Transplant 2010;45(8):1347–51.

[14] Saliba RM, Couriel DR, Giralt S, Rondon G, Okoroji GJ, Rashid A, et al. Prognostic value of response after upfront therapy for acute GVHD. Bone Marrow Transplant 2012;47(1):125–31.

[15] Westin JR, Saliba RM, De Lima M, Alousi A, Hosing C, Qazilbash MH, et al. Steroid-Refractory Acute GVHD: Predictors and Outcomes. Adv Hematol 2011;2011:601953.

[16] Markey KA, MacDonald KP, Hill GR. The biology of graft-versus-host disease: experimental systems instructing clinical practice. Blood 2014;124(3):354–62.

[17] Holtan SG, Pasquini M, Weisdorf DJ. Acute graft-versus-host disease: a bench-to-bedside update. Blood 2014;124(3):363–73.

[18] Socie G, Ritz J. Current issues in chronic graft-versus-host disease. Blood 2014;124(3):374–84.

[19] Choi SW, Reddy P. Current and emerging strategies for the prevention of graft-versus-host disease. Nat Rev Clin Oncol 2014;11(9):536–47.

[20] Magenau J, Reddy P. Next generation treatment of acute graft-versus-host disease. Leukemia 2014;28(12):2283–91.

[21] Le Blanc K, Rasmusson I, Sundberg B, Gotherstrom C, Hassan M, Uzunel M, et al. Treatment of severe acute graft-versus-host disease with third party haploidentical mesenchymal stem cells. Lancet 2004;363(9419):1439–41.

[22] Ringden O, Uzunel M, Rasmusson I, Remberger M, Sundberg B, Lonnies H, et al. Mesenchymal stem cells for treatment of therapy-resistant graft-versus-host disease. Transplantation 2006;81(10):1390–7.

[23] Fang B, Song Y, Liao L, Zhang Y, Zhao RC. Favorable response to human adipose tissue-derived mesenchymal stem cells in steroid-refractory acute graft-versus-host disease. Transplant Proc 2007;39(10):3358–62.

[24] Fang B, Song Y, Zhao RC, Han Q, Lin Q. Using human adipose tissue-derived mesenchymal stem cells as salvage therapy for hepatic graft-versus-host disease resembling acute hepatitis. Transpl P 2007;39(5):1710–3.

[25] Le Blanc K, Frassoni F, Ball L, Locatelli F, Roelofs H, Lewis I, et al. Mesenchymal stem cells for treatment of steroid-resistant, severe, acute graft-versus-host disease: a phase II study. Lancet 2008;371(9624):1579–86.

[26] Muller I, Kordowich S, Holzwarth C, Isensee G, Lang P, Neunhoeffer F, et al. Application of multipotent mesenchymal stromal cells in pediatric patients following allogeneic stem cell transplantation. Blood Cell Mol Dis 2008;40(1):25–32.

[27] Osiris Therapeutics I. Osiris Therapeutics Announces Preliminary Results for Prochymal Phase III GvHD Trials 2009 [cited August 9, 2014]. Available at: http://files.shareholder.com/downloads/OSIR/3388040135x0x317779/7677da46-286a-47c4-865d-36c148119a1a/OSIR_News_2009_9_8_General.pdf.

[28] von Bonin M, Stolzel F, Goedecke A, Richter K, Wuschek N, Holig K, et al. Treatment of refractory acute GVHD with third-party MSC expanded in platelet lysate-containing medium. Bone Marrow Transplant 2009;43(3):245–51.

[29] Arima N, Nakamura F, Fukunaga A, Hirata H, Machida H, Kouno S, et al. Single intra-arterial injection of mesenchymal stromal cells for treatment of steroid-refractory acute graft-versus-host disease: a pilot study. Cytotherapy 2010;12(2):265–8.

[30] Lim JH, Lee MH, Yi HG, Kim CS, Kim JH, Song SU. Mesenchymal stromal cells for steroid-refractory acute graft-versus-host disease: a report of two cases. Int J Hematol 2010;92(1):204–7.

[31] Lucchini G, Introna M, Dander E, Rovelli A, Balduzzi A, Bonanomi S, et al. Platelet-lysate-Expanded Mesenchymal Stromal Cells as a Salvage Therapy for Severe Resistant Graft-versus-Host Disease in a Pediatric Population. Biol Blood Marrow Tr 2010;16(9):1293–301.

[32] Perez-Simon JA, Lopez-Villar O, Andreu EJ, Rifon J, Muntion S, Campelo MD, et al. Mesenchymal stem cells expanded in vitro with human serum for the treatment of acute and chronic graft-versus-host disease: results of a phase I/II clinical trial. Haematologica 2011;96(7):1072–6.

[33] Prasad VK, Lucas KG, Kleiner GI, Talano JA, Jacobsohn D, Broadwater G, et al. Efficacy and safety of ex vivo cultured adult human mesenchymal stem cells (Prochymal) in pediatric patients with severe refractory acute graft-versus-host disease in a compassionate use study. Biol Blood Marrow Transplant 2011;17(4):534–41.

[34] Wernicke CM, Grunewald TG, Juenger H, Kuci S, Kuci Z, Koehl U, et al. Mesenchymal stromal cells for treatment of steroid-refractory GvHD: a review of the literature and two pediatric cases. Int Arch Med 2011;4(1):27.

[35] Wu KH, Chan CK, Tsai C, Chang YH, Sieber M, Chiu TH, et al. Effective treatment of severe steroid-resistant acute graft-versus-host disease with umbilical cord-derived mesenchymal stem cells. Transplantation 2011;91(12):1412–6.

[36] Dander E, Lucchini G, Vinci P, Introna M, Masciocchi F, Perseghin P, et al. Mesenchymal stromal cells for the treatment of graft-versus-host disease: understanding the in vivo biological effect through patient immune monitoring. Leukemia 2012;26(7):1681–4.

[37] Herrmann R, Sturm M, Shaw K, Purtill D, Cooney J, Wright M, et al. Mesenchymal stromal cell therapy for steroid-refractory acute and chronic graft versus host disease: a phase 1 study. Int J Hematol 2012;95(2):182–8.

[38] Ball LM, Bernardo ME, Roelofs H, van Tol MJ, Contoli B, Zwaginga JJ, et al. Multiple infusions of mesenchymal stromal cells induce sustained remission in children with steroid-refractory, grade III-IV acute graft-versus-host disease. Brit J Haematol 2013;163(4):501–9.

[39] Muroi K, Miyamura K, Ohashi K, Murata M, Eto T, Kobayashi N, et al. Unrelated allogeneic bone marrow-derived mesenchymal stem cells for steroid-refractory acute graft-versus-host disease: a phase I/II study. Int J Hematol 2013;98(2):206–13.

[40] Resnick IB, Barkats C, Shapira MY, Stepensky P, Bloom AI, Shimoni A, et al. Treatment of severe steroid resistant acute GVHD with mesenchymal stromal cells (MSC). Am J Blood Res 2013;3(3):225–38.

[41] Introna M, Lucchini G, Dander E, Galimberti S, Rovelli A, Balduzzi A, et al. Treatment of graft versus host disease with mesenchymal stromal cells: a phase I study on 40 adult and pediatric patients. Biol Blood Marrow Transplant 2014;20(3):375–81.

[42] Kurtzberg J, Prockop S, Teira P, Bittencourt H, Lewis V, Chan KW, et al. Allogeneic human mesenchymal stem cell therapy (remestemcel-L, Prochymal) as a rescue agent for severe refractory acute graft-versus-host disease in pediatric patients. Biol Blood Marrow Transplant 2014;20(2):229–35.

[43] Sanchez-Guijo F, Caballero-Velazquez T, Lopez-Villar O, Redondo A, Parody R, Martinez C, et al. Sequential third-party mesenchymal stromal cell therapy for refractory acute graft-versus-host disease. Biol Blood Marrow Transplant 2014;20(10):1580–5.
[44] Silla L, Valim V, Amorin B, Alegretti AP, Dos Santos de Oliveira F, Lima da Silva MA, et al. A safety and feasibility study with platelet lysate expanded bone marrow mesenchymal stromal cells for the treatment of acute graft-versus-host disease in Brazil. Leukemia Lymphoma 2014;55(5):1203–5.
[45] Yin F, Battiwalla M, Ito S, Feng X, Chinian F, Melenhorst JJ, et al. Bone marrow mesenchymal stromal cells to treat tissue damage in allogeneic stem cell transplant recipients: correlation of biological markers with clinical responses. Stem Cells 2014;32(5):1278–88.
[46] Zhao K, Lou R, Huang F, Peng Y, Jiang Z, Huang K, et al. Immunomodulation effects of mesenchymal stromal cells on acute graft-versus-host disease after hematopoietic stem cell transplantation. Biol Blood Marrow Transplant 2015;21(1):97–104.
[47] Kebriaei P, Isola L, Bahceci E, Holland K, Rowley S, McGuirk J, et al. Adult human mesenchymal stem cells added to corticosteroid therapy for the treatment of acute graft-versus-host disease. Biol Blood Marrow Transplant 2009;15(7):804–11.
[48] Martin PJ, Uberti JP, Soiffer RJ, Klingemann H, Waller EK, Daly AS, et al. Prochymal improves response rates in patients with steroid-refractory acute graft versus host disease (SR-GVHD) involving the liver and gut: results of a randomized, placebo-controlled, multicenter phase III trial in GVHD. Biol Blood Marrow Transplant 2010;16(2 Supplement 2):S169–70.
[49] Newell LF, Deans RJ, Maziarz RT. Adult adherent stromal cells in the management of graft-versus-host disease. Expert Opin Biol Ther 2014;14(2):231–46.
[50] Cragg L, Blazar BR, Defor T, Kolatker N, Miller W, Kersey J, et al. A randomized trial comparing prednisone with antithymocyte globulin/prednisone as an initial systemic therapy for moderately severe acute graft-versus-host disease. Biol Blood Marrow Transplant 2000;6(4A):441–7.
[51] Ren J, Stroncek DF, Zhao Y, Jin P, Castiello L, Civini S, et al. Intra-subject variability in human bone marrow stromal cell (BMSC) replicative senescence: molecular changes associated with BMSC senescence. Stem Cell Res 2013;11(3):1060–73.
[52] Galipeau J. The mesenchymal stromal cells dilemma—does a negative phase III trial of random donor mesenchymal stromal cells in steroid-resistant graft-versus-host disease represent a death knell or a bump in the road? Cytotherapy 2013;15(1):2–8.
[53] Bartholomew A, Sturgeon C, Siatskas M, Ferrer K, McIntosh K, Patil S, et al. Mesenchymal stem cells suppress lymphocyte proliferation in vitro and prolong skin graft survival in vivo. Exp Hematol 2002;30(1):42–8.
[54] Di Nicola M, Carlo-Stella C, Magni M, Milanesi M, Longoni PD, Matteucci P, et al. Human bone marrow stromal cells suppress T-lymphocyte proliferation induced by cellular or nonspecific mitogenic stimuli. Blood 2002;99(10):3838–43.
[55] Krampera M, Glennie S, Dyson J, Scott D, Laylor R, Simpson E, et al. Bone marrow mesenchymal stem cells inhibit the response of naive and memory antigen-specific T cells to their cognate peptide. Blood 2003;101(9):3722–9.
[56] Uccelli A, Moretta L, Pistoia V. Immunoregulatory function of mesenchymal stem cells. Eur J Immunol 2006;36(10):2566–73.
[57] Bernardo ME, Locatelli F, Fibbe WE. Mesenchymal stromal cells. Ann N Y Acad Sci 2009;1176:101–17.

[58] Barrett A, Melenhorst J. Is human cell therapy research caught in a mousetrap? Mol Ther 2011;19(2):224–7.

[59] Polchert D, Sobinsky J, Douglas G, Kidd M, Moadsiri A, Reina E, et al. IFN-gamma activation of mesenchymal stem cells for treatment and prevention of graft versus host disease. Eur J Immunol 2008;38(6):1745–55.

[60] Caplan AI. Why are MSCs therapeutic? New data: new insight. J Pathol 2009;217(2):318–24.

[61] Le Blanc K, Mougiakakos D. Multipotent mesenchymal stromal cells and the innate immune system. Nat Rev Immunol 2012;12(5):383–96.

[62] Griffin MD, Ryan AE, Alagesan S, Lohan P, Treacy O, Ritter T. Anti-donor immune responses elicited by allogeneic mesenchymal stem cells: what have we learned so far? Immunol Cell Biol 2013;91(1):40–51.

[63] Huang XP, Sun Z, Miyagi Y, McDonald Kinkaid H, Zhang L, Weisel RD, et al. Differentiation of allogeneic mesenchymal stem cells induces immunogenicity and limits their long-term benefits for myocardial repair. Circulation 2010;122(23):2419–29.

[64] Schu S, Nosov M, O'Flynn L, Shaw G, Treacy O, Barry F, et al. Immunogenicity of allogeneic mesenchymal stem cells. J Cell Mol Med 2012;16(9):2094–103.

[65] Battula VL, Treml S, Bareiss PM, Gieseke F, Roelofs H, de Zwart P, et al. Isolation of functionally distinct mesenchymal stem cell subsets using antibodies against CD56, CD271, and mesenchymal stem cell antigen-1. Haematologica 2009;94(2):173–84.

[66] Maijenburg MW, Kleijer M, Vermeul K, Mul EP, van Alphen FP, van der Schoot CE, et al. The composition of the mesenchymal stromal cell compartment in human bone marrow changes during development and aging. Haematologica 2012;97(2):179–83.

[67] Tormin A, Li O, Brune JC, Walsh S, Schutz B, Ehinger M, et al. CD146 expression on primary nonhematopoietic bone marrow stem cells is correlated with in situ localization. Blood 2011;117(19):5067–77.

[68] Haack-Sorensen M, Kastrup J. Cryopreservation and revival of mesenchymal stromal cells. Methods Mol Biol 2011;698:161–74.

[69] Francois M, Copland IB, Yuan S, Romieu-Mourez R, Waller EK, Galipeau J. Cryopreserved mesenchymal stromal cells display impaired immunosuppressive properties as a result of heat-shock response and impaired interferon-gamma licensing. Cytotherapy 2012;14(2):147–52.

[70] Moll G, Alm JJ, Davies LC, von Bahr L, Heldring N, Stenbeck-Funke L, et al. Do cryopreserved mesenchymal stromal cells display impaired immunomodulatory and therapeutic properties? Stem Cells 2014;32(9):2430–42.

[71] Pollock K, Sumstad D, Kadidlo D, McKenna DH, Hubel A. Clinical mesenchymal stromal cell products undergo functional changes in response to freezing. Cytotherapy 2015;17(1):38–45.

[72] Luetzkendorf J, Nerger K, Hering J, Moegel A, Hoffmann K, Hoefers C, et al. Cryopreservation does not alter main characteristics of Good Manufacturing Process-grade human multipotent mesenchymal stromal cells including immunomodulating potential and lack of malignant transformation. Cytotherapy 2015;17(2):186–98.

[73] Dominici M, Le Blanc K, Mueller I, Slaper-Cortenbach I, Marini F, Krause D, et al. Minimal criteria for defining multipotent mesenchymal stromal cells. The International Society for Cellular Therapy position statement. Cytotherapy 2006;8(4):315–7.

[74] Lv FJ, Tuan RS, Cheung KM, Leung VY. Concise review: the surface markers and identity of human mesenchymal stem cells. Stem cells 2014;32(6):1408–19.
[75] Galipeau J, Krampera M. The challenge of defining mesenchymal stromal cell potency assays and their potential use as release criteria. Cytotherapy 2015;17(2):125–7.
[76] Administration UFaD. Guidance for Industry: Potency Tests for Cellular and Gene Therapy Products 2011. Available at: http://www.fda.gov/downloads/BiologicsBloodVaccines/GuidanceComplianceRegulatoryInformation/Guidances/CellularandGeneTherapy/UCM243392.pdf.
[77] Wuchter P, Bieback K, Schrezenmeier H, Bornhauser M, Muller LP, Bonig H, et al. Standardization of Good Manufacturing Practice-compliant production of bone marrow-derived human mesenchymal stromal cells for immunotherapeutic applications. Cytotherapy 2015;17(2):128–39.
[78] Ren J, Jin P, Sabatino M, Balakumaran A, Feng J, Kuznetsov SA, et al. Global transcriptome analysis of human bone marrow stromal cells (BMSC) reveals proliferative, mobile and interactive cells that produce abundant extracellular matrix proteins, some of which may affect BMSC potency. Cytotherapy 2011;13(6):661–74.
[79] Salem B, Miner S, Hensel NF, Battiwalla M, Keyvanfar K, Stroncek DF, et al. Quantitative activation suppression assay to evaluate human bone marrow-derived mesenchymal stromal cell potency. Cytotherapy 2015;17(12):1675–1686.
[80] Bloom DD, Centanni JM, Bhatia N, Emler CA, Drier D, Leverson GE, et al. A reproducible immunopotency assay to measure mesenchymal stromal cell-mediated T-cell suppression. Cytotherapy 2015;17(2):140–51.
[81] Amarnath S, Foley JE, Farthing DE, Gress RE, Laurence A, Eckhaus MA, et al. Bone marrow-derived mesenchymal stromal cells harness purinergenic signaling to tolerize human Th1 cells in vivo. Stem Cells 2015;33(4):1200–12.
[82] Eggenhofer E, Benseler V, Kroemer A, Popp FC, Geissler EK, Schlitt HJ, et al. Mesenchymal stem cells are short-lived and do not migrate beyond the lungs after intravenous infusion. Front Immunol 2012;3:297.
[83] Nystedt J, Anderson H, Tikkanen J, Pietila M, Hirvonen T, Takalo R, et al. Cell surface structures influence lung clearance rate of systemically infused mesenchymal stromal cells. Stem Cells 2013;31(2):317–26.
[84] Plock JA, Schnider JT, Schweizer R, Gorantla VS. Are cultured mesenchymal stromal cells an option for immunomodulation in transplantation? Front Immunol 2013;4:41.
[85] Schlosser S, Dennler C, Schweizer R, Eberli D, Stein JV, Enzmann V, et al. Paracrine effects of mesenchymal stem cells enhance vascular regeneration in ischemic murine skin. Microvasc Res 2012;83(3):267–75.
[86] Romieu-Mourez R, Coutu DL, Galipeau J. The immune plasticity of mesenchymal stromal cells from mice and men: concordances and discrepancies. Front Biosci 2012;4:824–37.
[87] Balakumaran A, Pawelczyk E, Ren J, Sworder B, Chaudhry A, Sabatino M, et al. Superparamagnetic iron oxide nanoparticles labeling of bone marrow stromal (mesenchymal) cells does not affect their "stemness". PLos One 2010;5(7):e11462.
[88] Martin PJ, Bachier CR, Klingemann HG, McCarthy PL, Szabolcs P, Uberti JP, et al. Endpoints for clinical trials testing treatment of acute graft-versus-host disease: a joint statement. Biol Blood Marrow Transplant 2009;15(7):777–84.

[89] Ratanatharathorn V, Nash RA, Przepiorka D, Devine SM, Klein JL, Weisdorf D, et al. Phase III study comparing methotrexate and tacrolimus (prograf, FK506) with methotrexate and cyclosporine for graft-versus-host disease prophylaxis after HLA-identical sibling bone marrow transplantation. Blood 1998;92(7):2303–14.
[90] Vander Lugt MT, Braun TM, Hanash S, Ritz J, Ho VT, Antin JH, et al. ST2 as a marker for risk of therapy-resistant graft-versus-host disease and death. N Eng J Med 2013;369(6):529–39.
[91] Gyorgy B, Szabo TG, Pasztoi M, Pal Z, Misjak P, Aradi B, et al. Membrane vesicles, current state-of-the-art: emerging role of extracellular vesicles. Cell Mol Life Sci 2011;68(16):2667–88.
[92] Akyurekli C, Le Y, Richardson RB, Fergusson D, Tay J, Allan DS. A systematic review of preclinical studies on the therapeutic potential of mesenchymal stromal cell-derived microvesicles. Stem Cell Rev 2015;11(1):150–60.
[93] Kordelas L, Rebmann V, Ludwig AK, Radtke S, Ruesing J, Doeppner TR, et al. MSC-derived exosomes: a novel tool to treat therapy-refractory graft-versus-host disease. Leukemia 2014;28(4):970–3.
[94] Antonioli L, Pacher P, Vizi ES, Hasko G. CD39 and CD73 in immunity and inflammation. Trends Mol Med 2013;19(6):355–67.
[95] Rani S, Ryan AE, Griffin MD, Ritter T. Mesenchymal stem cell-derived extracellular vesicles: toward cell-free therapeutic applications. Mol Ther 2015;23(5):812–23.

Chapter 9

Regulatory Pathway for Mesenchymal Stromal Cell-Based Therapy in the United States

R. Lindblad, A. Fiky, D. Wood and G. Armstrong
The EMMES Corporation, Rockville, MD, United States

CHAPTER OUTLINE

Mesenchymal Stromal Cells. http://dx.doi.org/10.1016/B978-0-12-802826-1.00009-X

9.1 INTRODUCTION

Advancing a Mesenchymal Stromal Cell (MSC) therapy into the clinic can be a slow and arduous task. Knowing the appropriate pathways for communication with the Food and Drug Administration (FDA) can establish a clear framework for the generation and compilation of the required data into an Investigational New Drug Application (IND), as early communication with FDA is key. This is due to the inherent heterologous nature of MSC therapies themselves, which are used across many indications. The identity of MSCs, as described by the International Society of Cellular Therapy (ISCT), worked well to classify MSCs as a distinct cell population, but is lacking specificity when characterizing an MSC cell product for a given indication. As MSCs are a heterogeneous population of cells, knowing the cells types that are identifiable is just as important as knowing the cell types that are not identifiable, as efficacy may rely on more than one component of the cellular therapy. Having a proposed mechanism of action, measured using a potency assay, will allow the optimization of the donor choice, the tissue source within that donor, and the manufacturing techniques for the disease indication proposed.

The heterogeneity of the results of the clinical trials conducted to date using MSCs may be due to the heterogeneity of the cell product used. Selecting from a variable donor pool, the wrong tissue source, or inappropriate processing can have a disastrous effect on the regulatory development pathway, as these factors must be tightly controlled to ensure a consistent MSC product is used in the clinic. However, the isolation and identification of the desired cell population could take months, and the optimization of this process along with the development and validation of potency assays could take years. This work is expensive, labor intensive and requires qualified staff to perform at the highest technical level. The challenges faced in the regulation of MSCs are driven by the challenges in making a consistent, efficacious product. This is not unique to MSCs, but is inherent in using biologic living cell products for disease intervention. The challenges in the clinical translation of MSC-based products include biological variability, characterization, raw materials, adventitious agents, aseptic processing, and cryopreservation [1]. Based on the success of navigating these issues, the regulatory pathway becomes more straightforward.

9.2 REGULATORY PLAN

It is important to understand what critical elements are needed in the development of a cell therapy product.

- Develop a regulatory plan early in the product development process

- Assemble a study team of experts (ie, research scientists, clinicians, quality assurance, regulatory affairs) early to:
  - coordinate the planning and initiation of nonclinical and clinical research studies aimed at bringing the therapeutic candidate to the clinic and,
  - become familiar and understand regulatory expectations for the development of cellular therapy products through a variety of resources and interactions with FDA.

Fig. 9.1 outlines the stages of cell product development and when to engage FDA, that is, pre-IND meeting. Pre-IND meetings are most effective when focused on specific scientific or regulatory issues:

- Clinical trial design issues
- Toxicity profile issues
- Nonstandard or novel formulations
- Dosing limitations
- Species suitability
- Immunogenicity
- Regulatory requirements for demonstrating safety and efficacy to support a new product approval.

These meetings aid in ensuring that necessary studies are designed to provide useful information.

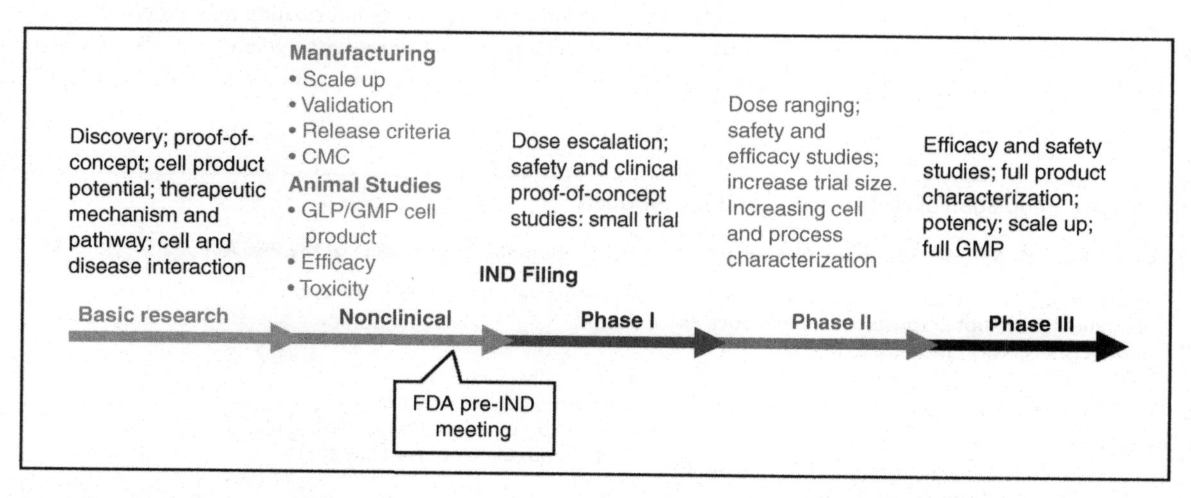

FIGURE 9.1 Cell therapy product development. The product development process is typically divided into three major stages: basic research, nonclinical development, and the clinical trial. The stage of product development serves to determine key issues to address. *CMC*, chemistry, manufacturing, and controls; *GLP*, good laboratory practices; *GMP*, good manufacturing practices; *IND*, Investigational New Drug.

9.3 REGULATORY PATHWAY

The clinical use of Human Cells, Tissues, and Cellular and Tissue-Based Products (HCT/Ps) is regulated under two sections of the Public Health Service (PHS) Act [2,3], Sections 361 and 351. Section 361 includes biologic products that require no premarket approval and are consistent with biologic products defined in 21 CFR 1271.10 [4], that is, products which are minimally manipulated, homologous use, not a combination product, AND either has no cellular or systemic effect, or, if it is active on a cellular or systemic basis, is used in an autologous setting, or in an allogeneic setting in first or second degree blood relatives, or is for reproductive use. Section 351 of the PHS Act [4,5] includes products that require premarket approval, that is, require data with clinical investigations to be collected for FDA review under an Investigational New Drug (IND) application, and do not meet the definition of exempt products described above 361 products. Cell therapies that also have a device or scaffold associated with them can be regulated by the Center for Devices and Radiological Health (CDRH) as a consultant to CBER or as the lead center if the scaffold is the primary mode of action. Table 9.1 summarizes regulation of HCT/Ps as 351 and 361 products.

As MSCs meet the definition for a "biological product" in section 351(i) of the PHS Act [42 U.S.C. 262(i)], prior to their administration to patients in a clinical trial, an IND must be submitted to the Office of Cellular, Tissue, and Gene Therapies (OCTGT) in the Center for Biologics Evaluation and Research (CBER) at the Food and Drug Administration (FDA); herein referred to "FDA." This IND application presents information regarding the source, manipulation, characterization, and bioactivity (potency) of the MSCs to be used in the proposed clinical study, along with the results of nonclinical

Table 9.1 Regulation of HCT/Ps as 351 and 361 Products

HCT/Ps Regulated Under 351 of the PHS Act	HCT/Ps Regulated Under 361 of the PHS Act
Require IND premarket approval; do not meet the definition of exempt products described above 361 products; 21CFR 1271	No premarket approval requirement
More than minimally manipulated	■ Minimally manipulated, and
Not intended for homologous use	■ For homologous use, and
Associated with a device or scaffold	■ Not a combination product, AND either
Cellular or systemic effect or dependent of metabolic activity of the cells for its primary function	❑ has no cellular or systemic effect or ❑ if it is active on a cellular or systemic basis is used in an autologous setting, in an allogeneic setting in first or second degree blood relatives, or is for reproductive use

testing of the product to be used in the clinical trial and the design of the proposed clinical protocol. The IND filing allows FDA to review this information, regarding the risks anticipated to the patients enrolled in the proposed clinical study, against the benefits of their participation and decide if the study may proceed.

9.3.1 Interactions With CBER and OCTGT

Prior to spending time and resources on an extensive nonclinical (safety) program to file as part of the IND, it is advisable to approach the FDA to ask for feedback on the suitability of the program and design of the proposed studies to support the IND, as these studies can be expensive, complex, and time consuming to conduct. Incorporating Agency feedback into study designs will greatly increase the chances of the IND clearing and ensure a quicker path into the clinic.

The appropriate first stage to approach FDA is when the investigator has some efficacy data and is about to embark on your IND-enabling nonclinical studies, for which s/he has synopses available. Approaching FDA with these study designs and short overall summaries of the planned nonclinical model, the MSC manufacturing methods and proposed clinical program (a protocol synopsis), along with background information regarding the indication and therapeutic strategy, will help FDA advise the investigator informally on the appropriateness of the proposed animal model to support the proposed clinical program. The package to provide to FDA, overall, should focus on the nonclinical studies, including their design, justification for species/animal model used, route of administration, dose (exposure), and the biomarkers to be examined. If there are specific issues the investigator needs the FDA to address regarding the nonclinical plan, these should be made clear in the cover letter. This package should be sent to FDA with suggested times for a teleconference within 6–8 weeks of the request. Once data are available from the IND-enabling nonclinical studies, the FDA can be approached once more, this time using the formal pre-IND meeting mechanism (a Type B Meeting) [6].

9.4 PRE-IND MEETING

A pre-IND meeting with OCTGT is highly recommended for new products and can be critical to program success. The pre-IND meeting is a Type B meeting [6] and can be either face to face, via teleconference, or simply be a written response to the questions posed. FDA will usually grant a pre-IND meeting within 60 days of the receipt of a request for a meeting and the meeting will be within 14 days of the requested date. The meeting request

Table 9.2 Sample Pre-IND Meeting Agenda

Item	Time (60 min)	Time (90 min)	Participants
Introduction of participants	5	5	FDA Lead/Company Lead
Goals and objectives	5	5	Company Lead
Discussion of questions	45	70	All
Summary and conclusions	5	10	FDA Lead/Company Lead
Total time	60	90	

should include enough information to allow FDA to have appropriate staff available at the meeting and include a draft of questions to be discussed at the meeting, and follow the FDA suggested format [6]. Important aspects of the meeting request are the objectives and agenda suggested for the meeting, as these provide overall context for the meeting topics; however, the list of questions is the most crucial, as these will focus on the discussion.

Meetings are typically most productive when questions are focused and specific. Questions should be as precise as possible and include a preamble with a brief explanation of the context and purpose of the question. Open-ended questions should be avoided where possible. Questions with yes or no answers are most likely to provide definitive useful feedback. The number of questions should also be kept to a reasonable number which can be answered in the time given. The proposed agenda should include estimated times needed for each agenda item. The length of a pre-IND meeting should allow enough time to address each agenda item adequately. Typically, pre-IND meetings are scheduled for 60 and occasionally 90 min. Table 9.2 is an example of a pre-IND meeting agenda.

Once the pre-IND meeting is granted and scheduled, the meeting briefing package should be prepared. It must be submitted to FDA at least 1 month prior to the meeting. It is generally a good idea to have a first draft of the meeting briefing package at the time of submission of the meeting request, to help drive the questions, as the content is driven by the questions posed.

Preferably the sponsor (Government agency such as NIH, academic center, or a pharmaceutical company), the principal investigator (or sponsor-investigator), the individual responsible for the nonclinical work, and the cell manufacturer should attend the pre-IND meeting. Others, such as a statistician, may also attend depending on the questions posed. The FDA will be represented by reviewers to match these areas. Before the meeting, written feedback will be provided and, if all questions are answered, there is an option to cancel the meeting. If there are still outstanding questions, or clarification is needed, the meeting can be used to discuss these last items. After the pre-IND meeting, formal minutes will be issued by FDA and the reviewers may make themselves available for further discussion and follow-up as needed. This process is a collaborative process with the goal of moving the therapy into the clinical arena as quickly and as safely as possible.

Engaging the FDA early on in the IND submission process is recommended to facilitate the IND's overall success. Within CBER, early meetings are available that are nonbinding, informal scientific discussions between nonclinical (pharmacology/toxicology and manufacturing) reviewers and the sponsor to initiate dialogue at an early stage in the process. This structure reflects CBER's willingness to engage the cell therapy community early in the development process and helps avoid delays due to lack of communication.

9.5 IND SUBMISSION

The submission of the IND will follow the pre-IND meeting and must fully address any issues raised at the pre-IND meeting to be successful. Once the IND is submitted, the FDA has 30 days to respond. Reviewers may have questions, provide nonhold comments, or place the IND on clinical hold within that 30-day time limit. Typically, FDA will have comments and/or questions and will contact the sponsor prior to the 30-day deadline. Being responsive to the agency at this point will help the application clear within the 30-day timeframe. Several key sections in the IND are the clinical protocol with appropriate eligibility, endpoints, stopping rules and dosing justification (starting dose/dose escalation), the CMC section, and the nonclinical section [7].

9.6 IND CONTENT—CHEMISTRY, MANUFACTURING, AND CONTROLS

Prior to clinical use, the identity, quality, purity, and potency of your MSC product must be demonstrated. Current Good Tissue Practices (cGTP) govern the methods and the facilities used for, the manufacturing of HCT/Ps. cGTP focus on the prevention of the introduction, transmission, and spread of communicable diseases or other adverse events while preserving product function and integrity [8]. FDA published a final rule (effective September 15, 2008) in the Federal Register amending the cGMP for human drugs, including biological products to exempt most phase I drugs from complying with the cGMP regulations (21CFR 210/211) [9].

Even though exempt from the requirements of Parts 210/ 211, Phase 1 investigational drugs remain subject to the statutory cGMP requirements of

FD&C Act 501(a)(2)(B). The cGMP requirements in 21 CFR parts 210 and 211 govern the methods to be used in, and the facilities or controls to be used for, the manufacture, processing, packing, or holding of a drug to ensure that such drug meets the requirements of the Act as to safety, identity and strength and meets the quality and purity characteristics that it purports or is represented to possess. These cGMP requirements apply to biological products, regulated under section 351 of the PHS Act, that meet the definition of drug in the Act [10]. The Final Rule for Phase I investigational drugs has not significantly affected current HCT/P manufacturing. The existing risk-based regulatory structure and the progressive application of GMP and GTP regulations remain the same. Several FDA Guidance documents are available [11,12] to assist with compiling this information, which is outlined at a high level below.

Manufacturing Information: Information regarding the components and materials used in the manufacture of the MSCs should be provided, with a summary of the testing performed on these. A listing of all components used in the manufacture should be provided. The cellular source of the MSCs (allogeneic and/or autologous, tissue, and cell type) should be described along with the mobilization protocol, if applicable. The collection protocol or recovery method should be outlined and the location of the collection facility and transport conditions, if the product is transported. For allogeneic cells, information on the donor screening and testing conducted should show that all donors were qualified to donate per protocol and were free from infectious agents [11]. The master cell bank (MCB) section should provide information regarding the history, source, derivation, and characterization, including testing to establish the safety (freedom from contamination), identity, purity, and stability of the cells. The working cell bank (WCB) will need less extensive characterization as it is derived from the MCB, but testing to ensure freedom from contamination and some identity testing is recommended.

All reagents used in cell processing (for cell growth, differentiation, selection, purification) must be listed in the IND no matter if they are present in the final product. These include cytokines, growth factors, antibiotics, and fetal bovine serum. These reagents have an influence on the safety, potency, and purity of the final product and are often a source of adventitious agents, so their source, quality, qualification (if not FDA cleared/approved), and removal from the final product should be demonstrated. Excipients which are present in the final product should similarly be listed.

Manufacturing procedures should be described in detail, with schematics of the collection, production, and purification of the product provided with in-process and final product testing identified within these schematics. Preparation of the cells, process timing and intermediate storage, and the final formulation should be described [11].

Product Testing and Release: Product testing should include microbial (sterility, mycoplasma, adventitious agents), identity (identification of the cells), purity (residual contaminants, pyrogenicity/endotoxin), viability (not less than 70%), and potency testing [11]. Testing should be ongoing throughout the production process to ensure quality and consistency of the product, with specifications set for the intermediate products. Release testing should be performed on each lot of product manufactured; as little as one dose may be considered a lot. Specifications for the final product should be clearly set out in the IND, usually in a tabular format, and final release testing data should preferably be available prior to administration of the product in the clinic, but if this is not possible, then clear procedures to follow in the case of an administered product not meeting release criteria should be in place and described in the IND. All products must be tracked and identified from collection to administration and carry appropriate labeling.

Stability Program: Stability testing must be performed to show that the final cell product is sufficiently stable over the time required in the study (from final release to administration) and an expiration date for the product should be provided to ensure that it is not used past this time [11].

9.7 IND CONTENT—PHARMACOLOGY AND TOXICOLOGY

The nonclinical pharmacology and toxicology section presented in the IND must support the proposed clinical trial [12]. The specific product characteristics, anticipated mechanism of action, target indication, and the route of administration all define the key elements of study design for the nonclinical program. Due to the novel processing techniques, cellular diversity and rapidly evolving scientific landscape of MSCs, the generation of the appropriate nonclinical data can pose a unique challenge and therefore communication with FDA is important. Overall, the objectives of the nonclinical program should be to establish the biological plausibility of the proposed therapeutic product and identify biologically active dose levels. This will allow the identification of a suitable clinical starting dose, drive dose escalation and the dosing regimen. Generally for stem cell-based treatment, a 75-kg patient needs approximately 2×10^6 MSCs per kilogram body weight for transplantation [13]. However, reports suggest that there are variations in optimal or effective doses required for various diseases, although the source of the stem cells is the same. For example, 3–5 × 10^7 cell/kg ex vivo expanded autologous bone marrow-derived mesenchymal stem cells (BM-MSCs) were administrated into each patient with multiple sclerosis [14], whereas in spinal cord injury, 5–10 × 10^6 cells/kg were administered intrathecally in the lumbar region [15]. Clinical trials for graft-versus-host disease (GVHD) using allogeneic MSCs demonstrated beneficial responses in both pediatric and adult patients with varying degrees of GVHD severity when administered (single to multiple times) with different doses [16,17]. A dose escalation phase I clinical trial using different doses of allogeneic MSCs (Prochymal) ranging from 0.5, 1.6, and 5 × 10^6 cells/kg in patients with myocardial infarction reported acceptable safety and efficacy signals [18]. The feasibility and safety of the administration of the product via the proposed route should be established and are used to drive patient eligibility criteria, identify physiologic parameters to guide clinical safety monitoring and identify risks.

9.8 CROSS-REFERENCING

If a cell-processing facility is currently supporting clinical trials and already has information on file at FDA, it is possible to cross-reference that information in multiple INDs. Written authorization must be obtained from the sponsor of the submission that is being cross-referenced and specific details, including the submission and volume number, the heading and page numbers which can be reviewed, should be provided to identify what material is being cross-referenced. This allows FDA reviewers to quickly locate the referenced materials, facilitating the review process.

9.9 FDA ACTION ON AN IND APPLICATION

Once the IND is filed, FDA has 30 days to respond with comments prior to the IND automatically becoming active. FDA, if needed, applies a clinical hold to stop the clinical investigation from proceeding until identified issues are adequately addressed [19]. If the pre-IND process was successful, the chance of a hold will be significantly reduced, as potential hold issues may be sufficiently raised prior to, and addressed in the IND submission.

Some of the examples that FDA put a clinical trial on hold include the following:

- Exposure to unreasonable risk of significant illness or injury
- Clinical investigators are not qualified
- Investigator brochure is misleading, erroneous, or incomplete

- IND does not contain sufficient information to assess risk
- Gender exclusion for a condition that occurs in both men and women

In practice, a clinical hold on cell therapy INDs is applied for several reasons, which include:

- The clinical trial does not provide adequate safety protection, which includes the lack of appropriate dosing, based on the preliminary clinical or nonclinical data; lack of appropriate dose escalation schemes, and lack of appropriate stopping rules for the trial
- The nonclinical data does not support the clinical trial based on product manufacturing or route of delivery
- The manufacturing section has inadequate characterization of the product, inadequate controls over manufacturing and insufficient details

9.10 IND MAINTENANCE

When an IND becomes active, future communication with the FDA outside of specific meeting requests occurs through the submission of IND amendments. Each submission is sequentially numbered and adds to the overall content of the IND. Amendments are submitted to the IND on a rolling basis and include protocol revisions, expedited safety reports, changes to the manufacturing technique or to the facility, key personnel changes, and any other significant changes to the clinical or manufacturing portions of the IND. Additionally, each IND sponsor is required to submit an annual report [20]. In some cases, FDA may require more frequent progress reports. Each annual progress report is an opportunity to submit other details regarding the IND that were not submitted during the year. The annual report is due within 60 days of the anniversary date of the IND becoming active. Table 9.3 is a sample table of contents for an annual report.

9.11 MOVING TO A CLINICAL TRIAL

The ultimate test of a cellular therapy is its performance and demonstration of safety and efficacy in clinical trials. This is the last step of an incredibly complex journey encompassing the nonclinical, manufacturing, and logistical complexities of the development process. The MSC cell population has been identified, along with manufacturing methods and assays to characterize the consistency and key properties of the product. The manufactured product has been tested in animal models for both proof-of-concept data including the route of administration, dosing schemes, and toxicology. An efficient procedure to ensure a seamless transition of the MSC product from bench to bedside depends heavily on a consistent core of facility staff.

Table 9.3 Sample Annual Report Table of Contents

A. Individual Study Information
   - Protocol title
   - Protocol number
   - IND number
   - Study purpose
   - Study population
   - Study status, subjects enrolled to date and study results

B. Summary Information
   1. A narrative or tabular summary showing the most frequent and most serious adverse experiences by body system
   2. A summary of all IND safetyreports submitted during the past year
   3. List of subjects who died during participation in the investigation, with the cause of death
   4. List of subjects who dropped out during the course of the investigation
   5. Brief description of what, if anything, was obtained that is pertinent to an understanding of the drug's actions, including, for example, information about dose response, bioavailability, or relevant information from controlled trials
   6. List of the nonclinical studies (including animal studies) completed or in progress during the past year and a summary of the major nonclinical findings
   7. Summary of any significant manufacturing or microbiological changes made during the last year

C. Update to the General Investigational Plan
   1. Research rationale and objectives
   2. Proposed research plan
      - 2.1. Investigational interventions
      - 2.2. Administration of investigational product
      - 2.3. Phase 1 trial
      - 2.4. Phase 2 trial (if applicable)
      - 2.5. Anticipated risks from study drug

D. Update to Investigator's Brochure

E. Significant Protocol Updates

F. Update on Foreign Marketing Developments

G. A Log of Outstanding Business

9.12 TRIAL DESIGN CONSIDERATIONS

The design of early-phase clinical trials of cell therapeutics must consider clinical safety and CMC portfolios that are unique to these products. During that early stage of trial development, safety assessment encompasses the evaluation of frequency, nature, and attribution of potential adverse events. In addition, a dose escalation protocol to identify the maximum tolerated dose (MTD) may be applicable [21]. As development of the trials continues, clinical trial design can be further subdivided into two broad areas: scientific and operational. The scientific area includes the clinical endpoints to be reached and the statistical methods used to address them. The investigator and the cell processing staff should meet early with statistical staff to determine how to phrase the clinical questions such that they can be answered with the proper endpoints. If a surrogate endpoint is proposed, it is recommended that the sponsor seek regulatory feedback from the FDA on the appropriateness of using the surrogate.

The method of endpoint measurement must be determined. Is there an existing validated assay to measure the endpoint? Is the assay commercially available and is it FDA approved for clinical use or is it a "research-based" assay? The statisticians should calculate the appropriate sample size that is required to answer the questions with sufficient statistical power. Once the sample size is derived, the investigators must determine if the trial will be single or multicenter based. From an operational side an initial Phase I trial, typically conducted at a single center, aims to assess the safety of the new therapy and monitors for possible adverse effects of the product. For example, it may be important to investigate the level of the preexisting antibodies, or that which may develop, as a result of administering the MSC product in enrolled subjects which may cause an adverse reaction, reduce a potential therapeutic effect in the short term or an immunogenic response that may undermine the outcome of potentially beneficial cell therapy or organ transplant in the future.

9.13 DATA AND ADVERSE EVENT MONITORING

In addition to the Quality Control (QC)/Quality Assurance (QA) of the cellular therapeutic, there is also the QC of data collection. Case report forms (CRFs) must be tailored to the study and need to be linked to the source data collected during the manufacture, delivery of the cells and throughout the clinical trial. The forms should be tested for ease of completion so that data submitted through the clinical sites will be easy to interpret. Data coordinators and/or research nurses should be given rigorous training on the completion of these forms and should be aware of the duration needed to complete the CRF, and accordingly generate a reasonable estimate of the number of subjects they can follow within a given time period.

Single center trials are often monitored by an institutional data safety monitoring board (DSMB) while multicenter trials typically have a centralized independent DSMB that is often convened by the sponsor. The Informed Consent Form (ICF) should be tailored to explain the novel and yet to be confirmed outcome of administering MSC along with the possible irreversibility of a cellular transplant that may generate adverse events through the life time of the subject [22]. A monitoring plan must be in place to provide guidelines and procedures for the collection and reporting of adverse events to the DSMB and the FDA. This is another step where good communication is essential. The trial investigator and the cell processing facility must meet regularly to discuss any adverse events and especially events that occur during product administration to determine if they are attributable to the cellular product or the subject's underlying disease. The FDA and the DSMB must be informed of these events to determine if a prespecified stopping boundary has been crossed.

9.14 CONCLUSIONS

Novel cell therapy approaches using MSCs are investigational and require an IND application to be filed with the FDA. Taking an MSC-based therapy into the clinic is a long and complex process. There are many translational steps to be completed before exposure to humans is feasible. These steps range from practical issues, such as the establishment of specifications and compilation of manufacturing information, to the complex design and rationale for nonclinical studies to support the clinical program. The challenges to regulating MSCs are evident with the high degree of variability in regards to MSC sources, manufacturing processes, and in vivo and in vitro characterization. This lack of consensus highlights potential challenges to the clinical translation of MSC-based products [1]. Investigators need to develop good working relationships with the cell-processing facility staff, laboratory and statistical staff, and with FDA to navigate this route and initiate a clinical trial. When this occurs, then it is easier to solve inevitable operational difficulties, attribute adverse events, and resolve recruitment problems which may occur during the course of the trial. The key to a successful IND is early and frequent communication with the FDA.

SUGGESTED READING

- U.S. Food and Drug Administration, Center for Biologics Evaluation and Research "Guidance for Human Somatic Cell Therapy and Gene Therapy". Guidance for Industry, March 1998.
- U.S. Food and Drug Administration, Center for Biologics Evaluation and Research "Eligibility Determination for Donors of Human Cells, Tissues, and Cellular and Tissue-Based Products (HCT/Ps)" Final Guidance, August 2007.
- Bringing safe and effective cell therapies to the bedside. Robert A. Preti. Nature Biotechnology July 2005; 23(7):801–4.
- U.S. Food and Drug Administration, Center for Biologics Evaluation and Research, Center for Drug Evaluation and Research "CGMP for Phase 1 Investigational Drugs" Guidance for Industry, July 2008.
- Cell Therapy cGMP Facilities and Manufacturing. Adrian Gee, chief editor. Springer-Verlag US 2009.

REFERENCES

[1] Mendicino M, Bailey AM, Keith W, Puri RK, Bauer SR. MSC-based product characterization for clinical trials: an FDA perspective. Cell Stem Cell 2014;14(2):141–5.

[2] U.S. Food and Drug Administration, Center for Biologics Evaluation and Research, "References for the regulatory process for the office of cellular, tissue and gene

therapies." Available at: http://www.fda.gov/BiologicsBloodVaccines/GuidanceComplianceRegulatoryInformation/OtherRecommendationsforManufacturers/ucm094338.htm.

[3] U.S. Food and Drug Administration, "FDA History. Available at: http://www.fda.gov/oc/history/.

[4] 21 Code of Federal Regulations Part 1271 Human Cells, Tissues, and Cellular and Tissue-Based Products, 25 May 2005.

[5] U.S. Food and Drug Administration, Center for Biologics Evaluation and Research, "About CBER." Available at: http://www.fda.gov/AboutFDA/CentersOffices/OfficeofMedicalProductsandTobacco/CBER/.

[6] U.S. Food and Drug Administration, Center for Biologics Evaluation and Research, Center for Drug Evaluation and Research, "Formal Meetings Between the FDA and Sponsors or Applicants of PDUFA Products" Guidance for Industry, March 2015.

[7] U.S. Food and Drug Administration, Center for Biologics Evaluation and Research, "Information on submitting an investigational new drug application." Available at: http://www.fda.gov/biologicsbloodvaccines/developmentapprovalprocess/investigationalnewdrugindordeviceexemptionideprocess/ucm094309.htm; August 2011.

[8] 21 Code of Federal Regulations Part 1271 Human Cells, Tissues, and Cellular and Tissue-Based Products, 1 April 2014.

[9] Current Good Manufacturing Practice and Investigational New Drugs Intended for Use in Clinical Trials. 21 CFR210.2 (c) July 2008.

[10] U.S. Food and Drug Administration, Center for Biologics Evaluation and Research, "Current Good Tissue Practice (CGTP) and Additional Requirements for Manufacturers of Human Cells, Tissues, and Cellular and Tissue-Based Products (HCT/Ps)" Draft Guidance for Industry January 2009.

[11] U.S. Food and Drug Administration, Center for Biologics Evaluation and Research, "Guidance for FDA Reviewers and Sponsors: Content and Review of Chemistry, Manufacturing, and Control (CMC) Information for Human Somatic Cell Therapy Investigational New Drug Applications (INDs)" April 2008.

[12] U.S. Food and Drug Administration, Center for Biologics Evaluation and Research, "Preclinical Assessment of Investigational Cellular and Gene Therapy Product" Guidance for Industry, November 2013.

[13] Bartmann C, Rohde E, Schallmoser K, Pürstner P, Lanzer G, Linkesch W, et al. Two steps to functional mesenchymal stromal cells for clinical application. Transfusion 2007;47:1426–35.

[14] Yamout B, Hourani R, Salti H, Barada W, El-Hajj T, Al-Kutoubi A, et al. Bone marrow mesenchymal stem cell transplantation in patients with multiple sclerosis: a pilot study. J Neuroimmunol 2010;227(1–2):185–9.

[15] Kishk NA, Gabr H, Hamdy S, Afifi L, Abokresha N, Mahmoud H, et al. Case control series of intrathecal autologous bone marrow mesenchymal stem cell therapy for chronic spinal cord injury. Neurorehabil Neural Repair 2010;24(8):702–8.

[16] Le Blanc K, Rasmusson I, Sundberg B, Götherström C, Hassan M, Uzunel M, et al. Treatment of severe acute graft-versus-host disease with third party haploidentical mesenchymal stem cells. Lancet 2004;363(9419):1439–41.

[17] Ringdén O, Uzunel M, Rasmusson I, Remberger M, Sundberg B, Lönnies H, et al. Mesenchymal stem cells for treatment of therapy-resistant graft-versus-host disease. Transplantation 2006;81(10):1390–7.

[18] Hare JM, Traverse JH, Henry TD, Dib N, Strumpf RK, Schulman SP, et al. A randomized, double-blind, placebo-controlled, dose-escalation study of intravenous adult human mesenchymal stem cells (Prochymal) after acute myocardial infarction. J Am Coll Cardiol 2009;54(24):2277–86.
[19] U.S. Food and Drug Administration, Center for Biologics Evaluation and Research, SOPP 8201—"Administrative Processing of Clinical Holds for Investigational New Drug Applications" Version 4. 19 May 2014.
[20] 21 Code of Federal Regulations Part 312.33 Annual Reports.
[21] U.S. Food and Drug Administration, Center for Biologics Evaluation and Research, "Considerations for the Design of Early-Phase Clinical Trials of Cellular and Gene Therapy Products" Guidance for Industry, June 2015.
[22] ISSCR Guidelines for the Clinical Translation of Stem Cells, December 3, 2008.

Chapter 10

Global Regulatory Perspective for MSCs

A. Lodge, G. Detela, J. Barry, P. Ginty and N. Mount
Cell and Gene Therapy Catapult, London, United Kingdom

CHAPTER OUTLINE

Mesenchymal Stromal Cells. http://dx.doi.org/10.1016/B978-0-12-802826-1.00010-6

10.1 INTRODUCTION

Mesenchymal stromal cells (MSCs) are being used in the development of cell therapies for a multitude of diseases and conditions. As well as the non-commercial use of unlicensed cell therapies within hospital environments (named-patient use), MSCs are increasingly being used for the development of commercial medicinal products to be placed in the international pharmaceutical markets. These medicinal products are therefore subject to the licensing (or marketing authorization) procedures applicable to traditional small molecule and protein-based (biologics) drugs, which require that quality, safety, and efficacy are demonstrated through appropriately-designed manufacturing strategies, nonclinical studies, and human clinical trials. A search of clinical trial databases such as www.clinicaltrials.gov will identify a few hundred trials involving MSCs—some of which involve the same product for different disease indications—being sponsored by both academic centers and commercial organizations. The disease indications being targeted include, among others, graft versus host disease (GVHD), acute myocardial infarction (AMI), congestive heart failure (CHF), bone disease, Crohn disease, ulcerative colitis, acute respiratory distress syndrome (ARDS), ischemic stroke, and diabetes. The trials span Phases I–III, but well over 80% are currently in Phases I and II, while the MSC-based products may either comprise early multipotent cells (eg, MSC progenitors or MSCs themselves) or MSC-derived cells (eg, osteocyte progenitors or tendon progenitors).

The suitability of MSCs for cell-based medicinal product manufacture is multifaceted [1,2]. First, they can be obtained from a number of adult tissue sources including bone marrow and adipose tissue, and they can also be isolated from perinatal tissues including placenta, umbilical cord, cord blood, and Wharton's jelly. Second, not only can these tissues often be obtained using minimally-invasive techniques, but MSCs can be isolated easily from them and expanded to clinically relevant yields. Third, the therapeutic effect of MSCs is thought to be related to their immunomodulatory and tissue repair-promoting activities, which make them potential candidates for a number of diseases and conditions, particularly those with an inflammatory mechanism. Finally, by virtue of negligible HLA class II expression, allogeneic MSCs appear largely able to escape immune rejection following transplantation into a non-HLA-matched host, meaning that allogeneic products can be manufactured as "off-the-shelf" medicines to overcome the inconvenience of needing to manufacture autologous products to order.

However, despite the efforts of established biopharmaceutical companies such as Mesoblast and Athersys in the traditionally strong pharmaceutical markets of the United States (US), the European Union (EU), and Japan, only a handful of MSC-based medicinal products have received a marketing authorization (Table 10.1), including Prochymal [3] in Canada and New Zealand, TEMCELL HS (which is based on Prochymal) in Japan, and Hearticellgram-AMI, Cupistem, Cartistem, QueenCell, and Ossron in Korea (Table 10.1). Some MSC-containing products classified as medical devices, such as AlloStem, OsteoCel, and Trinity Evolution, have been approved for marketing in the United States, but we will not address the regulation of medical devices in this discussion of medicinal products. The approved products include those manufactured from both autologous and allogeneic cells, demonstrating that autologous products are not just the domain of unlicensed, named-patient therapies. However, the development of cell banks of MSCs with multiple potential indications presents an attractive global commercialization strategy.

In the next few years, a number of MSC-based medicinal products will hope to progress from clinical trials into marketing authorization applications in a number of countries/economic areas, and to gain marketing authorization the manufacturers will need to successfully navigate regulatory pathways to approval. This chapter provides an overview of the regulatory issues that manufacturers of MSC products will need to address to gain approval for the use of their products in pharmaceutical markets globally.

Table 10.1 Mesenchymal Stromal Cell-Based Products with Marketing Authorization

Medicinal Product (Marketing Authorization)	Cell Source	Company	Disease Indication
Prochymal (Canada and New Zealand) [3]	Allogeneic BM MSCs	Mesoblast	Acute graft versus host disease in pediatric patients
TEMCELL HS (Japan)	Allogeneic BM MSCs	JCR Pharmaceuticals	Acute graft versus host disease in pediatric patients
Hearticellgram-AMI (Korea)	Autologous BM MSCs	Pharmicell	Acute myocardial infarction
Cartistem (Korea)	Allogeneic UCB MSCs	Medipost	Osteoarthritis knee cartilage defects
Cupistem (Korea)	Autologous adipose MSCs	Anterogen	Crohn's fistula
QueenCell (Korea)	Autologous adipose MSCs	Anterogen	Regeneration of subcutaneous adipose tissue
Ossron (Korea)	Autologous BM MSCs	RMS	Bone regeneration
Allostem (US medical device)	Allogeneic adipose MSCs	AlloSource	Bone regeneration
OsteoCel (US medical device)	Allogeneic MSCs	NuVasive	Spinal bone regeneration
Trinity Evolution (US medical device)	Allogeneic MSCs	Orthofix	Bone regeneration

10.2 MEDICINAL PRODUCT LEGISLATION APPLICABLE TO MESENCHYMAL STROMAL CELL PRODUCTS

The development of all medicinal products for human use is subject to legal requirements with which the developer must comply. Although there are differences in how the laws around medicinal products are written and implemented in different countries or economic areas, the basic premise is the same in that medicinal products must be demonstrated to be safe, efficacious, and of suitable quality (ie, meet defined and validated specifications that correlate with safety and efficacy) for use in humans. Safety and efficacy are demonstrated through nonclinical studies (eg, in animal models of disease) and human clinical trials, while the quality criteria applied to a medicinal product are elaborated as the manufacturing process evolves throughout the course of the nonclinical studies and clinical trials (including postmarketing studies). Satisfactory demonstration that a medicinal product meets the required quality, safety, and efficacy criteria will enable the developer to receive authorization from one or more regulatory agencies (the bodies responsible for ensuring that medicines have been developed in accordance with pharmaceutical law) to place the medicinal product on the market in designated countries. We will refer to such successful licensing of a medicinal product as "marketing authorization" (MA) throughout this text, although certain countries/economic areas may have their own specific terminology.

A key element of medicinal product legislation is that it defines what a medicinal product is. For example, in the European Union, *Directive 2001/83/EC* (the Medicines Directive) [4] defines a medicinal product as:

1. Any substance or combination of substances presented as having properties for treating or preventing disease in human beings; or
2. Any substance or combination of substances which may be used in or administered to human beings either with a view to restoring, correcting or modifying physiological functions by exerting a pharmacological, immunological or metabolic action, or to making a medical diagnosis.

These EU definitions are descriptive of all medicinal products globally. With respect to medicinal products manufactured from human cells or tissues, their recognition as actual medicinal products is largely a recent development driven by the realization that the quality, safety, and efficacy of these products need to be appropriately controlled in a manner similar to other medicinal products (ie, small molecules and biologics) as the science around their development has evolved. So, while it is true that medical practice involving human cells and tissues (eg, hematopoietic stem cell transplantation and solid organ transplantation) has been established for several decades, a new field has been developed in which cells and tissues are either the *starting materials* for the *manufacture* of a product in a process that involves their *substantial manipulation*, or processed cells are used for a purpose other than their normal (*same essential*) function (ie, their use is *nonhomologous*) (Table 10.2). These concepts of starting materials, manufacture, substantial manipulation, and nonhomologous use (or equivalent terms, as defined in medicinal product legislation) are key to understanding when human cells and tissues transition from being subject to transplantation/named-patient (homologous use) regulations to being subject to medicinal product regulations. Essentially, when cells undergo only "minimal manipulation" and are used for the "same essential function" that they perform in their natural environment within the body, their administration to humans is considered a transplantation or homologous use.

The development of legislation around medicinal products manufactured from human starting materials has also led to the implementation of specific nomenclature to describe them (Table 10.2). At the generic level, "(somatic) cell therapies," "gene therapies," and "engineered tissues" are useful descriptions that are meaningful globally, but different countries or economic areas may also have specific nomenclature for these products. This is

Table 10.2 Key Concepts in the Regulation of Cell Therapies

Cell-based product nomenclature		*Minimal manipulation*[a]	
ATMPs | European Union Cellular therapy products | United States Regenerative medicinal products | Japan Biologicals | Australia Biological drugs | Canada Cell therapy products | Korea		Cutting, grinding, shaping of tissues Soaking in antibiotic or antimicrobial solutions Cell separation, centrifugation, concentration, or purification Sterilization Filtration Cryopreservation	
Quality	*Safety*	*Efficacy*	*Homologous use*[b]
Starting materials Raw materials Manufacturing process Product specifications Stability GMP	Nonclinical studies GLP	Clinical trials GCP	The repair, reconstruction, replacement or supplementation of a recipient's cells or tissues with a cell-based medicinal product that performs the same essential function in the recipient as in the donor

[a] *Any manipulations other than minimal manipulations are considered substantial manipulations, processing steps, or manufacturing steps. Any products manufactured using substantial manipulations become subject to medicinal product legislation, and must meet specified quality, safety, and efficacy criteria to achieve marketing authorization.*
[b] *Cell therapies for homologous use (ie, when the cells perform their same essential function) are only subject to medicinal product legislation if they are substantially manipulated. Cell therapies containing cells for nonhomologous use are always subject to medicinal product legislation.*

discussed in further detail, together with their legislative frameworks, in the following subsections on medicinal product regulation in the United States, the European Union, Japan, Australia, Canada, and Korea—regions in which the development of cell-based products is particularly active and where relevant legislation is established.

MSC-based medicinal products, together with all medicinal products derived from human cells, genes, or tissues, are subject to additional legislation governing the procurement and use of the human tissue starting material. Such legislation is principally in place to minimize the transmission of communicable diseases including HIV, hepatitis B, hepatitis C, syphilis, and West Nile virus (and also transmissible spongiform encephalopathies in some countries) by setting appropriate standards for donor screening, selection, and testing. In some cases, this legislation overlaps with, or is the same as, legislation governing the use of human cells and tissues for same essential function/transplantation purposes or in the manufacture of unlicensed products through processes involving only minimal manipulation.

10.2.1 United States

In the United States, the legislation governing the regulation of medicinal products for human use originates from the *Federal Food, Drug and Cosmetic Act*, which was first passed by the US Congress in 1938. This act still forms the basis of most food and drug laws but has subsequently been updated by both the 1997 *Food and Drug Administration Modernization Act* (FDAMA) and the 2007 *Food and Drug Administration Amendment Act* (FDAAA). The original act of 1938 established the Food and Drug Administration (FDA), which is currently an agency of the US Department of Health and Human Services (DHHS) and remains the body responsible for enforcing the current FDAAA (the Act). The Act is enforced by *Regulations*, which are codified in the *US Code of Federal Regulations* (CFR). The FDA Regulations applicable to medicinal products are published in Title 21 (Food and Drugs) [5] of the CFR and are updated annually in April. The role of the FDA in evaluating medicinal products for human use is therefore to ensure that their development is performed in compliance with the appropriate Regulations in Title 21 of the CFR. For medicinal products, the FDA has established a number of offices for this purpose, each with specific responsibility for different product types. These offices are the Center for Drug Evaluation and Research (CDER), the Center for Biologics Evaluation and Research (CBER), and the Center for Devices and Radiological Health (CDRH). MSC (and other cell)-based medicinal products (that are not classified as medical devices) fall under the remit of CBER, specifically the Office of Cellular, Tissue and Gene Therapies (OCTGT), who regulate such products as "human cells, tissue, and cellular and tissue-based products" (HCT/Ps) under Part 1271 of CFR Title 21. OCTGT also has responsibility for gene therapy products. HCT/Ps include both "cellular therapy products" manufactured using processes involving substantial manipulation (Table 10.2), and cells and tissues intended for implantation, transplantation, infusion, or transfer into a human recipient. The former are regulated as biological drugs under Section 351 of the *Public Health Services (PHS) Act*, which describes the requirements for marketing authorization via the New Drug Application (NDA) and Biologics License Application (BLA) procedures. The latter are regulated under Section 361 of *PHS Act* as therapies undergoing minimal manipulation or for homologous use (Table 10.2), and in contrast to "351 products" are not eligible for marketing authorization. Examples of Section 361 HCT/Ps include bone, skin, corneas, ligaments, tendons, oocytes, and semen. Guidance (see Section 10.3.1) is provided by CBER for the manufacture of both 351 and 361 HCT/Ps in accordance with the relevant legislation. The majority of MSC-based therapies are likely to be classified as cellular therapy products in the United States, while some may be medical devices (Table 10.1).

For all HCT/Ps, 21CFR1271 (Human Cells, Tissue, and Cellular and Tissue-Based Products) describes the regulations in place to prevent the introduction, transmission, and spread of communicable diseases by such products. In 21CFR1271.50, the requirements for donor eligibility determination based on the results of donor screening (21CFR1271.75) and testing (21CFR1271.80, 21CFR1271.85) are described.

10.2.2 **European Union**

EU legislation on cell therapies, gene therapies, and engineered tissues is set out in a number of *Directives* and *Regulations*. A Directive is a legislative act that sets out a goal that all EU countries must achieve, and it is up to the individual countries to decide how. A Regulation instead is a binding legislative act that must be applied in its entirety across the European Union. Directives and Regulations are published by the European Commission in the Official Journal of the European Union [6].

In the EU, all medicinal products are regulated in accordance with *Directive 2001/83/EC* [4]. The European Medicines Agency (EMA) is responsible for ensuring that this Directive is enacted through the scientific evaluation of data generated during the development of medicinal products. *Directive 2001/83/EC* outlines the legal obligations and procedural aspects with which manufacturers of medicinal products must comply to enable their products to be granted an EU Marketing Authorisation.

With regard to medicinal products comprising human cells, genes, or tissues, *Directive 2001/83/EC* was amended in 2007 by *Regulation (EC) No. 1394/2007* on "advanced therapy medicinal products" [7]. *Regulation (EC) No. 1394/2007* (the ATMP Regulation) defines advanced therapy medicinal products (ATMPs) as specific types of medicinal products including "gene therapy medicinal products" (GTMPs), "somatic cell therapy medicinal products" (SCTMPs) and "tissue engineered products" (TEPs) (see Glossary for full definitions). In addition, "combined ATMPs" are those that contain a medical device, as defined in *Directive 93/42/EEC* [8], as an integral part of a viable cell- or tissue-containing product. Of the three types of ATMPs (other than combined ATMPs), MSC products would fall within the definition of somatic cell therapy medicinal products when prepared as cell suspensions utilizing their immuno-modulatory mechanism, as gene therapy medicinal products if they were genetically-modified, and as tissue-engineered products if used for the regeneration, repair, or replacement of tissues, for example, when combined with a scaffold or matrix for bone repair.

The human tissues and cells used as the starting materials for ATMP manufacture are governed by the "*EU Tissues and Cells Directives*" (EUTCD) and the "*Blood Directive*" (2002/98/EC), which were developed to regulate the donation, procurement, testing, and processing (including for named-patient use) of human tissues and cells, and blood, respectively. The EUTCD comprise the parent directive, *2004/23/EC* [9], and the three technical directives, *2006/17/EC* [10], *2006/86/EC* [11] and *2012/39/EU* [12]. The *Blood Directive, 2002/98/EC* [13], is the equivalent of the EUTCD parent directive when the donated tissue is blood. The Blood Directive also has three technical directives: *2004/33/EC* [14], *2005/61/EC* [15], and *2005/62/EC* [16]. In cases where human cells, tissues, or blood are not being used for commercial ATMP manufacture, the EUTCD and Blood Directive are implemented by member state regulations rather than by the ATMP regulation.

A key concept in the regulation of ATMPs is the optional "risk-based approach" (introduced through *Directive 2009/120/EC* [17], which amends *Directive 2001/83/EC* [4] with respect to ATMPs) to demonstrate quality, safety, and efficacy. The purpose of the risk-based approach is to determine the extent of quality, safety, and efficacy data to be included in the EU Marketing Authorisation Application (MAA) on an individual product basis. The output should be a risk analysis—which may cover any or all parts of the product development strategy—that is included in the MAA. Use of the risk-based approach may additionally be advantageous in identifying opportunities for pursuing an alternative licensing route to the traditional EU Marketing Authorisation procedure (eg, Conditional Approval or Adaptive Pathways), as described in Section 10.7.2.

Within EMA, the Committee for Advanced Therapies (CAT) is responsible for reviewing the quality, safety, and efficacy of ATMPs. However, the CAT does not grant a marketing authorization for an ATMP but rather makes a recommendation to the Committee for Medicinal Products for Human Use (CHMP), who may then issue a "positive opinion" to the European Commission for adoption. The CHMP is the key committee within EMA for the regulation of all types of medicinal products other than herbal medicinal products and veterinary medicinal products. The CHMP has representation from all of the EU member states, and the representatives are also members of the "national competent authority" (NCA) in each member state. The NCAs have authority devolved from EMA to approve MAAs in their own member state for some medicines, and also to approve Clinical Trial Authorisation (CTA) applications for Investigational Medicinal Products (IMPs) before an MAA is submitted. However, in the case of ATMPs, an MAA may not be submitted to an individual member state, but rather the Centralised Procedure described in *Regulation (EC) No. 726/2004* [18] must be followed. The Centralised Procedure involves all EU member states and is required for ATMPs (as well as products manufactured using recombinant DNA technology) because the EMA considers it essential that broad expertise is necessary for the review of these complex products. The Centralised Procedure results in a single marketing authorization that is valid in all EU countries, as well as in the European Economic Area (EEA) countries of Iceland, Liechtenstein, and Norway. However, in the EU, *Regulation (EC) No 1394/2007* [6] enables ATMPs to be supplied within an individual member state under the Hospital Exemption scheme if they are manufactured on a *nonroutine basis*. Under this scheme, ATMPs are exempt from the Centralised Procedure if they are for use in a hospital under the exclusive responsibility of a medical practitioner and comply with an individual medical prescription for a custom-made product for an individual patient.

10.2.3 Japan

In Japan, the Pharmaceuticals and Medical Devices Agency (PMDA) is responsible for evaluating medicinal products for use in clinical trials or for marketing authorization. Since Nov. 2014, medicinal products manufactured from human cells, genes, or tissues have been regulated under the *Pharmaceuticals and Medical Devices Act* (PMD Act) [19], which was introduced to replace the previously established *Pharmaceutical Affairs Law* (PAL) and create a new regulatory pathway for such products. The *PMD Act* is enforced via a number of Ministerial Ordinances, Notifications, and Administrative Notices (regulations) [20] (note that, at the time of writing, English translations of the *PMD Act* and the relevant regulations are not available from PMDA). Under the PAL, cell-, gene-, and tissue-based products with a marketing authorization were regulated as either drugs or medical devices depending on their mode of action, while unlicensed products were managed within hospitals in accordance with the principles of medical practice. Under the *PMD Act*, medicinal products comprising human cells and tissues that will be marketed are now regulated generically as "regenerative medicinal products" (RMPs; which also include gene therapy products), specifically "Cellular and Tissue-based Products" (CTBPs), by the Office of Cellular and Tissue-Based Products within PMDA. CTBPs are defined as processed (substantially manipulated) human cells that are intended to be used either for (1) the reconstruction, repair, or formation of structures or functions of the human body, or (2) the treatment or prevention of human diseases.

The *PMD Act* introduced a new pathway for the approval of RMPs that is distinct from the traditional pathway for small molecule drugs. Under the *PMD Act*, a RMP can obtain expedited conditional marketing authorization on the basis of safety and predicted probable efficacy demonstrated in early stage clinical trials. The specific requirements for authorization of an individual RMP will need to be established on a case-by-case basis in consultation with the PMDA, and the legislation offers the possibility of expedited approval of RMPs to fulfil currently unmet medical needs. However, the quality development must be sufficiently well advanced and the benefit–risk balance must justify the use of the product for marketing authorization to be granted. The first MSC-based RMP given conditional marketing authorization under the PMD Act, in September 2015, was TEMCELL HS, which was developed by JCR Pharmaceuticals in collaboration with Mesoblast, on whose product, Prochymal, TEMCELL HS is based. This was the first stem cell-based RMP of any kind approved under the PMD Act.

Conditional marketing authorization of a RMP under the *PMD Act* lasts for a maximum of 7 years, during which time the applicant is required to perform the later stage trials that will be required for subsequent full marketing authorization. If these trials are not performed, or if the data from them are considered inadequate to support full marketing authorization, the product must be withdrawn from the market at the end of the 7-year conditional authorization period.

The introduction of the *PMD Act* to replace the *PAL* was a response to the passing of the *Regenerative Medicine Promotion Law* as part of the National Diet of Japan (Japan's bicameral legislature) in May 2013. In addition to the *PMD Act*, the *Regenerative Medicine Promotion Law* also introduced the *Act on the Safety of Regenerative Medicine* [19]. This latter act establishes stricter regulation of unlicensed CBTPs in Japan for use in hospitals under a scheme similar to the EU Hospital Exemption scheme, in which clinical studies are classified as "clinical research" because they are not being performed as Good Clinical Practice (GCP)-compliant clinical trials (see Section 10.6) with the aim of submitting a MAA.

In Japan, marketing authorization of RMPs is granted in the absence of comprehensive efficacy data under the *PMD Act*. This act therefore requires that clinicians obtain informed consent from patients for the use of such products, and the marketing authorization holder (MAH) must perform postmarketing safety surveillance (pharmacovigilance) and provide regular reports on the incidence of infectious diseases. In addition, the Relief Services for Adverse Health Effects has been established within PMDA to compensate patients who experience an adverse reaction to the product, or an infection caused by its use.

10.2.4 **Australia**

The primary legislation governing the regulation of medicinal products in Australia is the *Therapeutic Goods Act 1989*, which is implemented by the *Therapeutic Goods Regulations 1990* [21]. The *Therapeutic Goods Act 1989* provides a uniform framework for import, export, manufacture, and supply of therapeutic goods in Australia. All therapeutic goods must be entered on the *Australian Register of Therapeutic Goods* (ARTG) before they can be imported, exported, or supplied for use in Australia. The Therapeutic Goods Administration (TGA) within the Department of Health is the government body responsible for the regulation of medicinal products according to the *Therapeutic Goods Act 1989* and ARTG.

Human cell- and tissue-derived medicinal products are defined as "biologicals," regulated according to the *Australian Regulatory Guidelines for Biologicals* (ARGB) [22], with a biological being defined in Part 3-2A of the *Therapeutic Goods Act 1989* "as a thing made from, or that contains, human cells or human tissues, and that is used to treat or prevent disease, ailment, defect or injury; diagnose a condition of a person; alter the physiological processes of a person; test the susceptibility of a person to disease; or replace or modify a person's body parts." The ARGB came into force on May 31, 2011 following a recommendation from Australian Commonwealth, State and Territory Health Ministers to improve the regulation of human tissues and cell-based therapies, which were previously regulated as medicines or medical devices. Products that are now regulated as biologicals under the Biologicals Regulatory Framework defined by the ARGB include human stem cell products, tissue-based products (skin, bone, ocular, cardiovascular), cell-based products (including genetically-modified cells), and combined cell and tissue products. Note that blood and bone marrow transplants are outside of this framework because they were already regulated as medicines under the *Therapeutic Goods Act 1989*. Other cell- and tissue-based therapies *not* regulated as biologicals include organs for transplantation, some cell-based therapies (including skin grafts, pancreatic islet cells and hematopoietic stem cells) prepared for use on a named-patient basis under medical practice, and autologous cell therapies used under the supervision of a medical practitioner for a single indication in a single course of treatment. Such autologous therapies (including some MSC therapies) are therefore not covered by the *Therapeutic Goods Act 1989* but their use is becoming more widespread in a range of therapeutic applications and they are therefore currently subject to review by TGA following a public consultation [23] to determine whether or not they should be regulated as biologicals to better ensure their quality, safety, and efficacy. Finally, according to the *Therapeutic Goods (Things that are not Biologicals) Determination No.1 of 2011*, some products of human origin are regulated by the TGA as therapeutic goods but not as biologicals. These include vaccines, plasma derivatives, recombinant products, labile blood and blood components, and hematopoietic progenitor cells used for hematopoietic reconstitution (nonfresh transplants).

Biologicals are subject to different levels of regulation based on the specific risks associated with their manufacture (minimal or substantial manipulation), the source of the starting material (autologous or allogeneic), and whether they are for homologous or nonhomologous use. Biologicals are therefore classified according to these risks within the Biologicals Regulatory Framework as Class 1–4 biologicals (Table 10.3). The classification

Table 10.3 Classification of Biologicals Under the Australian Regulatory Guidelines for Biologicals (ARGB)

Classification	Definition	Risk
Class 1	A biological that is included in Schedule 16 of the TG Regulations based on a justification that it is not a Class 2, 3, or 4 biological	Low
Class 2	A biological processed using (1) only one or more of the actions of minimal manipulation, and (2) is for homologous use	Low The product should be manufactured in accordance with the Australian Code of GMP for human blood and tissues
Class 3	A biological processed (1) using a method in addition to any of the actions of minimal manipulation; and (2) in a way that does not change an inherent biochemical, physiological, or immunological property	Medium To be included on the ARTG for supply in Australia, Class 3 biologicals must be evaluated by the TGA for safety, quality and efficacy.
Class 4	A biological processed (1) using a method in addition to any of the actions of minimal manipulation; and (2) in a way that changes an inherent biochemical, physiological, or immunological property	High To be included on the ARTG for supply in Australia, Class 4 biologicals will require additional supporting data, and must be evaluated by the TGA for safety, efficacy and quality (as for Class 3), but with further assessment and analysis of the supporting data.

of biologicals is performed on a product-specific basis in consultation with TGA.

The Biologicals Regulatory Framework sets new product standards and manufacturing principles with direct relevance and application for biologicals (ie, distinct from those for medicines and medical devices), and which have been harmonized with international standards (including ICH and EMA guidelines, see Section 10.3.1) where possible and relevant. This framework enables the approval of a biological based on advice from an independent expert committee, the Advisory Committee on Biologicals. After biologicals have been approved for marketing, ongoing controls are implemented including manufacturing surveillance, targeted review and laboratory testing, adverse events reporting, investigations, and recalls. The Biologicals Regulatory Framework additionally includes provisions for biologicals to be exempt from the TGA's usual requirements for inclusion on the ARTG to allow legal supply under certain circumstances, for example, for clinical trials, emergency situations, or use by individual patients.

The legislation implemented by TGA to regulate starting materials of human origin with regards to donor selection and testing is *Therapeutic Goods Order No. 88* [24]. This legislation is applicable to the starting materials used to manufacture biologicals.

10.2.5 **Canada**

The legal framework governing the approval of medicinal products for marketing authorization in Canada is defined by the *Food and Drugs Act of Canada* [25]. Medicinal products comprising human cells, genes, or tissues are regulated as "biological drugs" under *Schedule D of the Food and Drugs Act* by the Biologics and Genetic Therapies Directorate (BGTD). The *Food and Drugs Act* is implemented by a number of *Regulations*. The Regulations applicable to medicinal products manufactured from human cells, genes, or tissues are described in the *Food and Drug Regulations* (codified as C.R.C., c. 870), specifically Division 4 Part C (C.04.001) [26].

BGTD is a department within the Health Products and Food Branch of Health Canada, the federal agency responsible for promoting health. BGTD's definition of biological drugs includes, in addition to protein therapeutics, blood and blood products, gene and cell therapies, tissues, and organs. Within BGTD, The Centre for Evaluation of Radiopharmaceuticals and Biotherapeutics (CERB) is responsible for gene therapies, and The Centre for Blood and Tissues Evaluation (CBTE) is responsible for cell-based medicines, tissues, and organs. The BGTD also oversees the activities of the blood establishments.

The BGTD evaluates marketing authorization applications (New Drug Submissions), which are submitted to the Office of Regulatory Affairs (ORA), to ensure that all medicinal products meet the required quality, safety, and efficacy criteria specified by the legal framework of the *Food and Drugs Act* [25]. If the product is considered to meet the criteria, a Notice of Compliance (NOC) will be issued. In addition, the applicant is required to obtain a Drug Identification Number (DIN) to place the product on the market.

After being placed on the market, biological drugs are monitored according to a lot release schedule specific to their potential risk, and according to their manufacturing, testing, and inspection history. Each lot of higher risk biological drugs must be tested before being released for sale in Canada. Moderate risk biological drugs are periodically tested at the discretion of Health Canada, while low risk biological drugs are typically monitored as lots sold via certification of complete and satisfactory testing. However, the lot release schedule for an individual product may be amended at any time by Health Canada based on the available data.

In addition to the *Food and Drug Regulations*, other regulations implemented by the *Food and Drugs Act* to control the quality and safety of donated human cells and tissues used as starting materials or for named-patient use are the *Blood Regulations (SOR/2013-178)* [27,28], and the *Safety of Human Cells, Tissues and Organs for Transplantation Regulations (SOR/2007-118)* [29,30].

10.2.6 **Korea**

In Korea, cell-, tissue-, and gene-based products are regulated as "cell and gene therapy products" under the *Pharmaceutical Affairs Act* (PAA) and the *Enforcement Rule of Medicinal Product Safety* by the Ministry of Food and Drug Safety (MFDS), which, prior to 2013, was known as the Korea Food and Drug Administration. Cell and gene therapy products are a class of "biologics" as defined in the MFDS Notification, *Regulation on Review and Authorization of Biological Products*. In common with medicinal product legislation in other countries/economic areas, cell and gene therapy products are considered to be those manufactured using substantial manipulation techniques, and also minimally manipulated products manufactured by commercial organizations (in contrast, minimal manipulation of unlicensed products within hospitals for named-patient use is regulated as medical practice under the *Medical Service Act*). MFDS also regulates human tissues derived from living or cadaveric donors for use as medicinal product starting materials under the *Human Tissue Safety and Control Act*.

The PAA and the Enforcement Rule of Medicinal Product Safety are enforced by MFDS Notifications (regulations), and guidance (see Section 10.3.1) on implementing the regulations to achieve the required quality, safety, and efficacy for marketing authorization of cell and gene therapy products is available from MFDS. In recognition of the fact that the development of cell-based medicinal products poses specific challenges such as the availability of autologous tissue for CMC studies (see Section 10.4), the availability of appropriate nonclinical animal models (see Section 10.5), and the problems associated with running large-scale clinical trials with such products (see Section 10.6), MFDS, in common with EMA in Europe, operates a risk-based approach to the development and review of these products. For the cell therapy products, including MSC-based products, approved in Korea under the PAA to date (Table 10.1) [31], marketing authorization has been granted with the recommendation to perform long-term follow-up of patients enrolled in clinical trials, thus enabling the initial approval of products in smaller trial populations than would be required for small molecule and biologic drugs. The follow-up measures include periodic safety update reports (PSURs) post-MA, and a 5-year product license renewal system based on the marketing history and safety data. In addition, all medicinal products in Korea are reexamined to identify any adverse events not previously recorded, and product labels may be updated accordingly. Product labels are also periodically reevaluated based on the new scientific information from published study results. There is an additional system specific for cell therapy products, which requires mandatory safety reporting for every use of an approved cell therapy product for a specified time period. Early approval is also possible for medicines being developed in Korea which meet the definition of an orphan drug (see Section 10.7), for which approval can be granted on condition that a confirmatory Phase III study is carried out after marketing authorization.

At the time of writing, Korean legislation, regulations, and guidance documents are not widely available in English, and therefore while certain publications [31,32] can provide a valuable overview, the use of an intermediary with specialist expertise in Korean regulatory requirements is recommended for product developers based outside of Korea.

10.2.7 Use of Medicinal Products Prior to Marketing Authorization

The use of a medicinal product destined for commercialization outside of a clinical trial typically requires that it is approved for marketing authorization. When used in a clinical trial prior to authorization, the product is often referred to as an "investigational medicinal product" (IMP) to indicate its developmental, unapproved status. However, when justified based on medical need and risk–benefit balance, IMPs may be used outside of clinical trials under regulatory agency oversight. Such schemes are operative in the European Union (Compassionate Use; enacted by *Regulation (EC) No. 726/2004*) [18], United States (Expanded Access) [33], Canada (Special Access Programme; SAP)

[34], Australia (Special Access Scheme) [35] and Korea (Treatment Use of an Investigational New Drug) [36]. These schemes enable certain patients to benefit from effective treatments before they are made available for general use.

10.3 PRODUCT DEVELOPMENT: FROM NONCLINICAL STUDIES TO CLINICAL TRIALS TO MARKETING AUTHORIZATION

In a review of a MAA for a medicinal product under development, the regulatory agency responsible for licensing the product will thoroughly examine the quality, safety, and efficacy data presented in the application dossier as a key part of their decision making process. The terms "quality," "safety," and "efficacy" correspond to specific elements of the product development process, these being (1) chemistry, manufacturing and controls (CMC), (2) nonclinical testing, and (3) human clinical trials, respectively. CMC data should demonstrate that a manufacturing process can consistently produce a medicinal product with controlled specifications (sometimes called critical quality attributes or CQAs) and a stability profile that correlate with safe and efficacious use. CMC is an ongoing activity throughout all stages of product development. Nonclinical testing is performed to demonstrate that the medicinal product, with specifications comparable to or representative of those of the product manufactured for use in humans, is predicted to be safe for human use through studies in in vivo or in vitro model systems of disease. Safety data from nonclinical studies are required to support the use of a medicinal product in human clinical trials, which are performed to show that the product is safe and efficacious (ie, demonstrates the ability to produce a beneficial effect) when administered to the target human population. Data from clinical trials are required to support a MAA for a medicinal product, at which stage a regulatory agency's requirements for quality, safety, and efficacy must be met for marketing authorization to be granted.

10.3.1 Guidance on Achieving Quality, Safety, and Efficacy When Developing Mesenchymal Stromal Cell Products

To help product developers meet the required standards of quality, safety, and efficacy for the licensing of their product, guidance on specific elements of the product development process has been developed by regulators, which is both globally applicable and territory-specific.

Globally-relevant guidance is provided by the *International Council for Harmonisation of Technical Requirements for Registration of Pharmaceuticals for Human Use* (ICH) [37]. ICH was established to standardize, where possible, the approach to product development such that the requirements for medicinal product registration in different countries are common. This reduces the need to perform multiple, different studies for medicinal products that will be marketed in different countries/economic areas, thus improving patient access to the treatments they need. The ICH guidelines provide detailed information on how key aspects of the quality, safety, and efficacy of a medicinal product should be demonstrated. These guidelines are officially adopted by the United States, the European Union, and Japan (the founding members and signatories of ICH), by the European Free Trade Association (EFTA) states of Iceland, Lichtenstein, Norway, Switzerland, and also by Canada, which is an observer. They are broadly used in other countries/economic areas too. It should be noted that none of the ICH guidelines are specific to products derived from human cells, genes, or tissues, but are generically applicable to the key elements of the product development process for all medicinal products. At the time of writing, there are 12 guidelines on Quality (ICH Q1–Q12), 11 guidelines on Safety (ICH S1–S11), and 18 guidelines on Efficacy (ICH E1–E18).

Territory-specific guidance is provided to help the product developer understand how the requirements of the appropriate legislation can be met, and the inference is that if, for example, a medicinal product is being developed with the intention of applying for marketing authorization in the European Union, following the EMA's guidance documents is more likely to enable a positive opinion to be issued than if the guidance is not followed. It is also recognized by the regulators that sometimes certain guidance cannot be followed as intended. In this situation, dialog should be entered into with the regulators (using the European Union again as an example, EMA operates a "Scientific Advice" procedure partly for this purpose; and similar procedures are available elsewhere) and recommendations on the product development strategy proposed should always be followed. The EMA publishes a number of "Scientific Guidelines" [38] to help organizations who are developing products for marketing in the EU understand how to comply with the regulations (these guidelines are also recommended by the TGA in Australia). MSC-based medicinal products are classified as ATMPs in the EU, and a number of ATMP-specific, as well as "biological/biotechnology-derived product"-specific, scientific guidelines are available and applicable. A key document is *Guideline On Human Cell-Based Medicinal Products* [39], which is a multidisciplinary guideline that addresses quality, safety, and efficacy aspects of cell-based ATMP development. The FDA publishes a number of "Guidance for Industry" documents [40] to help organizations who are developing products for marketing in the United States understand how to comply with the regulations. These guidance documents are specific to the quality, safety, and efficacy aspects of the product development process, rather than being multidisciplinary. In Japan, the PMDA publishes several "Guidelines on Cellular and Gene Therapy Products." These documents are currently published in Japanese only, and so nonnative Japanese speakers will need to enter into Consultation Meetings (scientific advice) with PMDA via a Japan-based intermediary. In common with the EMA *Guideline On Human Cell-Based Medicinal Products*, there are two multidisciplinary *PMDA Guidances for Ensuring the Quality and Safety of Processed Human Somatic Stem Cell Products*, one for autologous products [41] and one for allogeneic products [42]. Finally, Health Canada publishes a number of "Guidance Documents" covering Canada-specific procedures, including some applicable to MSC products [43].

Because the ICH guidelines are generic, not all are applicable to MSC (or other cell-based) products. Those that are of relevance, plus the applicable territory-specific guidance documents additional to those described above, are discussed in the sections below on quality development (Section 10.4), nonclinical development (Section 10.5), and clinical development (Section 10.6).

10.3.2 The Common Technical Document

The importance of quality, safety, and efficacy as specific elements of the product development process is reflected in the organization of the dossiers used to present such data in applications for clinical trial or marketing authorization. In the ICH regions of European Union, United States, and Japan, an application for marketing authorization is made using the (electronic) "Common Technical Document" (CTD) format [44]. The CTD is composed of five modules. Module 1 is region specific while Modules 2, 3, 4, and 5 are intended to be common for all regions, which describe, respectively, the Quality, Nonclinical and Clinical Summaries, the Quality data and information, the Non-clinical Study Reports, and the Clinical Study Reports. In Jul. 2003, the CTD became the mandatory format for Marketing Authorisation Applications in the EU, New Drug Applications in Japan, and the strongly recommended format of choice in the United States for NDAs and BLAs.

The CTD format is actually used more broadly than just for dossiers submitted with MAAs, and it is used as a template for presenting quality, safety, and efficacy data in IMP dossiers [eg, the EU IMP Dossier (IMPD) and the US IND] submitted with clinical trial applications. Therefore, in effect, the structure of the CTD provides a plan for how the product development process should be designed and executed. Further details on the nonclinical, clinical, and CMC elements of the product development process are discussed in the following sections.

10.4 CMC AND QUALITY DEVELOPMENT

The quality development of a medicinal product creates a manufacturing process validated to consistently deliver a product with a defined composition and *specification* that correlate with safe use and clinical efficacy in human patients. Current territory-specific guidance [39–42] for cell-based medicinal products typically requires that the release specification is based on parameters or CQAs including cell viability, cell number, cell identity, purity, impurities, and potency. In addition, microbiological sterility and absence of mycoplasma and bacterial endotoxins will also need to be shown.

Manufacturing of an MSC-based product in accordance with CTD requirements will involve, first, the manufacture of the *Drug Substance* (DS) from the *starting materials* (tissues or cells) and *raw materials* (eg, cell culture medium and supplements), and second, the manufacture and/or final formulation of the *Drug Product* (DP) from the DS and any necessary *excipients* (nonmedicinal ingredients, eg, cryopreservation medium). For example, for the MSC product Prochymal, the DS is "Ex vivo Cultured Adult Human Mesenchymal Stem Cells" and the excipients are dimethyl sulfoxide (10%) and human serum albumin (5%) in Plasma-Lyte A [3]. According to the CTD format for application dossiers, the quality information for a medicinal product under development is presented primarily in two separate sections in Module 3, one on DS manufacture and one on DP manufacture (Table 10.4). The requirements of these sections are discussed later, full guidance on which is provided in the ICH guideline, M4Q(R1): Quality [45].

A typical allogeneic MSC-based product will be manufactured by first expanding the cells from starting material in culture to manufacture the DS, before mixing the DS cells with a cryopreservation medium to create aliquots of DP, each representing, if possible, a single dose. Specifications and sterility tests will need to be developed for both DS and DP. To enable the results

Table 10.4 Quality Section (Module 3) of the CTD

Section 3.2.S Drug Substance	Section 3.2.P Drug Product
3.2.S.1 General Information	3.2.P.1 Description and Composition of the Drug Product
3.2.S.2 Manufacture	3.2.P.2 Pharmaceutical Development
3.2.S.3 Characterization	3.2.P.3 Manufacture
3.2.S.4 Control of Drug Substance	3.2.P.4 Control of Excipients
3.2.S.5 Reference Standards or Materials	3.2.P.5 Control of Drug Product
3.2.S.6 Container Closure System	3.2.P.6 Reference Standards or Materials
3.2.S.7 Stability	3.2.P.7 Container Closure System
	3.2.P.8 Stability

of all tests on DS and DP to be obtained before the product is released, this may require that both DS and DP are cryopreserved, possibly with the DS cells being further expanded in culture prior to cryopreservation as the DP.

The starting material for most *allogeneic* MSC-based products will be a bank of cells prepared from a donated tissue source (although other strategies such as repeated manufactured from bone marrow aspirate starting material are feasible), and in the ICH regions, the cell bank will need to be manufactured and tested in accordance with ICH Q5D [46] and Q5A [47]. For MSCs, the typical approach of creating a two-tiered cell bank—in which a Master Cell Bank (MCB) is initially produced, from which one or more Working Cell Banks (WCBs) are then prepared to guarantee the supply of the starting material (ie, the WCB) in sufficient capacity to meet the needs for commercial manufacture—may not be possible with all MSCs because of their finite *in vitro* lifespan. Therefore, continued manufacturing of multiple MCBs as the starting material will be required, and comparability studies with products manufactured from different MCBs will be necessary to ensure product consistency. Note that such cell banking would not be applicable to *autologous* products under most circumstances, unless repeated administration to the patient is required.

All raw materials and excipients should, wherever possible, be of "compendial quality." Compendial quality indicates that the material has a defined specification that is accepted for use in medicinal product manufacture, and that this specification can be tested for using specific analytical methods. These specifications and analytical procedures are published in a *monograph* in a *pharmacopeia*, for example, European Pharmacopeia (Ph. Eur.), United States Pharmacopeia (USP) and Japanese Pharmacopoeia (JP) (see ICH Guideline Q4 [48] for more information). For noncompendial materials, their suitability should be determined by the developer of the medicinal product, and guidance is available on this (eg, USP Chapter 1043) [49]. Animal-derived materials (eg, bovine serum, porcine trypsin) should be qualified to ensure their suitability with respect to adventitious agent contamination. Additionally, in some countries, materials of bovine origin may need to be demonstrated to be from cattle free of bovine spongiform encephalopathy (BSE). Guidance is available on determining the suitability of animal-derived materials [50–53].

When developing specifications for the DS and DP of an MSC-based product, a minimum data set will include information on cell viability, cell number, cell identity, cell purity, any impurities introduced by the manufacturing process (eg, residual cell culture reagents of animal origin), and potency [39–42,54]. A characterization program should be initiated to collect data for all of these parameters from the start of the product development program, such that those data which correlate with nonclinical and clinical safety and efficacy can be identified and refined to propose final specifications for DS and DP. Given the intrinsic heterogeneity of MSCs isolated from different tissues [55,56] and the diverse tissue sources from which different MSC starting materials may be obtained, characterization of cell identity and potency, in particular, require a number of molecular markers and bioassays to be evaluated to ensure that correlations of cell identity and potency with safety and efficacy are valid. These markers will include the standard basic panel of MSC markers (>95% population positive expression of CD73, CD90, CD105, and <2% population positive expression of CD19, CD14 or CD11b, CD34, CD45, C79α, or CD19, and HLA-DR) [57] plus any other markers that may be important to the product functionality or specific for the particular MSC product under development and its starting material tissue source (eg, CD13, CD44, CD166, CD271, STRO-1 for bone marrow MSCs [58,59] or CD44, CD54, CD166 for peripheral blood MSCs [58]). These markers should be expressed in a consistent pattern throughout the manufacturing process to ensure that the product has not undergone phenotypic change from the defining MSC phenotype. Such phenotypic changes may be induced by specific culture conditions, differentiation and/or in vitro aging (MSCs are known to have a finite population doubling level in in vitro culture before undergoing phenotypic change and replicative senescence [60–62]). Morphological and genetic analyses may also help in this respect, as well as the characteristic differentiation potential of MSCs by which they are also defined, that is, their ability to become osteoblasts, adipocytes and chondroblasts in vitro [57,63]. Mechanisms of action of MSC products are still largely debated, although the most widely held consensus on how MSCs exert their therapeutic functions relates to their paracrine activity [64]. Therefore, regarding potency, suitable release assays will reflect the known (or hypothesized) mechanism of action (eg, growth factor and cytokine release) of the MSC being used, and this will be informed by nonclinical studies (see Section 10.5). The proposed specifications should also be suitable to support comprehensive stability studies which can be used to determine shelf lives of the DS and DP.

The manufacturing process itself needs to evolve to meet the increasing demands of late-stage clinical trials and commercial supply. For a Phase III trial to support a subsequent marketing authorization application, the manufacturing process must be validated to produce medicinal product with the specifications required for safe and efficacious use in the desired volume. Manufacturing process development will also need to ensure comparability of product manufactured at different scales and/or when changes are introduced, for example, when changing from one batch of serum to another.

Guidance relevant to the CMC aspects of a product development strategy, in addition to that provided in the territory-specific multidisciplinary guidance [39–42], includes the documents listed in Table 10.5.

Table 10.5 Guidance Documents Relevant to the Quality Development of an MSC-Based Medicinal Product

Organization	Guidance Document	Applicability to CTD Module 3
ICH	ICH Q2(R1) Validation of Analytical Procedures: Text and Methodology	3.2.S.4 Control of Drug Substance 3.2.P.5 Control of Drug Product
	ICH Q4–Q4B Pharmacopoeias	3.2.S.2 Manufacture 3.2.P.4 Control of Excipients
	ICH Q5A(R1) Viral Safety Evaluation of Biotechnology Products Derived from Cell Lines of Human or Animal Origin	3.2.S.2 Manufacture
	ICH Q5C Stability Testing Of Biotechnological/Biological Products	3.2.S.7 Stability 3.2.P.8 Stability
	ICH Q5D Derivation and Characterization of Cell Substrates Used for Production of Biotechnological/Biological Products	3.2.S.2 Manufacture
	ICH Q5E Comparability Of Biotechnological/Biological Products Subject To Changes In Their Manufacturing Process	3.2.S.4 Control of Drug Substance 3.2.P.5 Control of Drug Product
	ICH Q6B Specifications: Test Procedures And Acceptance Criteria For Biotechnological/Biological Products	3.2.S.4 Control of Drug Substance 3.2.P.5 Control of Drug Product
	ICH M4Q(R1) The Common Technical Document for the Registration of Pharmaceuticals for Human Use: Quality	Broadly applicable
EMA	EMEA/CHMP/410869/2006 Guideline On Human Cell-Based Medicinal Products	Broadly applicable
	CHMP/CAT/BWP/353632/2010 CHMP/CAT position statement on Creutzfeldt-Jakob disease and advanced therapy medicinal products	3.2.S.4 Control of Drug Substance
	CHMP/BWP/271475/06 Potency testing of cell-based immunotherapy medicinal products for the treatment of cancer	3.2.S.4 Control of Drug Substance 3.2.P.5 Control of Drug Product
	EMA/CHMP/BWP/187338/2014 Process validation for the manufacture of biotechnology-derived active substances and data to be provided in the regulatory submission	3.2.S.2 Manufacture 3.2.P.3 Manufacture
	CHMP/BWP/457920/2012 Rev. 1 Use of bovine serum in the manufacture of human biological medicinal products	3.2.S.2 Manufacture
	EMA/CHMP/BWP/814397/2011 Use of porcine trypsin used in the manufacture of human biological medicinal products	3.2.S.2 Manufacture
	EMEA/410/01 Rev. 3 Minimizing the Risk of Transmitting Animal Spongiform Encephalopathy Agents via Human and Veterinary Medicinal Products	3.2.S.2 Manufacture
	EMA/CAT/CPWP/686637/2011 Draft guideline on the risk-based approach according to Annex I, part IV of Directive 2001/83/EC applied to Advanced Therapy Medicinal Products	Broadly applicable
	EMA/CAT/571134/2009 Reflection paper on stem cell-based medicinal products	Broadly applicable

Table 10.5 Guidance Documents Relevant to the Quality Development of an MSC-Based Medicinal Product (*cont.*)

Organization	Guidance Document	Applicability to CTD Module 3
FDA	Guidance for Industry Potency Tests for Cellular and Gene Therapy Products 2011	3.2.S.4 Control of Drug Substance
	Guidance for FDA Reviewers and Sponsors Content and Review of Chemistry, Manufacturing, and Control (CMC) Information for Human Somatic Cell Therapy Investigational New Drug Applications (INDs) 2008	Broadly applicable
	Guidance for Industry Eligibility Determination for Donors of Human Cells, Tissues, and Cellular and Tissue-Based Products 2007	3.2.S.2 Manufacture
	Guidance for Industry Guidance for Human Somatic Cell Therapy and Gene Therapy 1998	Broadly applicable
	Guidance for Industry Guidance for the Submission of Chemistry, Manufacturing, and Controls Information and Establishment Description for Autologous Somatic Cell Therapy Products 1997	Broadly applicable
	USP Chapter 1043 Ancillary Materials for Cell, Gene, and Tissue-Engineered Products	3.2.S.2 Manufacture
PMDA	Notification No. 1314 of the PFSB/MHLW, Dec. 26, 2000 Guidance for ensuring the quality and safety of products manufactured from human or animal derived components	Broadly applicable
	Notification No. 0208003 of the PFSB/MHLW, Feb. 8, 2008 Guidance for ensuring the quality and safety of processed autologous human derived cells/tissues	Broadly applicable
	Notification No. 0912006 of the PFSB/MHLW, Sep. 12, 2008 Guidance for ensuring the quality and safety of processed allogeneic human derived cell/tissue products	Broadly applicable
	Notification No. 0907-2 of the PFSB/MHLW, Sep. 7, 2012 Guidance for ensuring the quality and safety of processed autologous human somatic stem cell products	Broadly applicable
	Notification No. 0907-3 of the PFSB/MHLW, Sep. 7, 2012 Guidance for ensuring the quality and safety of processed allogeneic human somatic stem cell products	Broadly applicable

10.4.1 Good Manufacturing Practice

In addition to the product-specific "quality parameters" that are determined and controlled by the CMC element of a product development strategy, all medicinal products must be manufactured in accordance with controlled procedures which ensure that the production processes are subject to appropriate "quality management." This is the basis of Good Manufacturing Practice (GMP), which is a quality assurance activity that is implemented through the use of a quality management system for controlling and documenting the procedures used to manufacture products and track their use, and to maintain the required level of hygiene within manufacturing facilities.

One of the key elements of GMP with respect to medicinal products manufactured using human cells, genes, or tissues is the lack of sterility of these starting materials. Because it is not possible to sterilize the products manufactured from such starting materials by methods such as sterile filtration or irradiation used for other medicinal products, aseptic processing must be performed to prevent contamination with adventitious agents from sources including operators, raw materials, and equipment. Facilities should therefore be designed to enable aseptic processing to be performed, and operators should be appropriately trained.

GMP is implemented through territory-specific legislation. In the EU, *Directive 2003/94/EC* [65] is implemented via the European Commission's GMP Guidelines [66], compliance with which is mandatory within the European Economic Area. In the United States, where GMP is referred to as "current Good Manufacturing Practices (cGMPs)," the regulations are codified in 21CFR210 and 21CFR211, for which Guidance for Industry documents are available [67,68]. Health Canada publishes their GMP legislation in C.02.001 of the Food and Drug Regulations. ICH GMP guidelines (ICH Q7 [69]) are also available where territory-specific guidance is not provided.

10.5 SAFETY STUDIES AND NONCLINICAL DEVELOPMENT

Nonclinical development involves the safety and preliminary efficacy testing of developmental medicinal products in animal models of disease or in in vitro assays that reflect the disease pathway. Such studies can also provide insight into the mechanism of action. Guidance on nonclinical development for medicinal products in general is provided via the ICH Safety guidelines [70]. However, although there are currently eleven ICH guidelines on safety, few are considered by ICH to be directly applicable to medicinal products manufactured from human cells, genes, or tissues.

Because of the limitations of the ICH Safety guidelines for such products, specific guidance has been developed in the European Union, United States, and Japan. For the development of cell-based ATMPs in the EU, the CHMP *Guideline On Human Cell-Based Medicinal Products* [39] acknowledges that "conventional nonclinical pharmacology and toxicology studies may not be appropriate for cell-based medicinal products. Therefore, the guideline addresses which nonclinical studies are necessary to demonstrate proof-of-principle and to define the pharmacological and toxicological effects predictive of the human response." In the United States, CBER has published a Guidance for Industry document entitled *Preclinical Assessment of Investigational Cellular and Gene Therapy Products* [71] for similar reasons, while the PMDA guidelines on ensuring quality and safety of autologous [41] and allogeneic [43] human cell-based products in Japan should be implemented in consultation with PMDA. These documents should be considered as the most appropriate sources of guidance for the type, extent, and design of nonclinical studies to be performed for cell-based medicinal products, although dialog with the appropriate regulatory agency (ie, through Scientific Advice or consultation-type meetings) is encouraged on a product-specific basis.

Nonclinical safety studies are categorized according to the CTD format into Pharmacology studies (Module 4.2.1), Pharmacokinetics studies (Module 4.2.2), and Toxicology studies (Module 4.2.3), the purposes of which are discussed next.

Pharmacology studies can be divided into primary pharmacodynamics, secondary pharmacology, and safety pharmacology studies. Primary pharmacodynamics studies are those which address the mode of action and/or effects of a substance in relation to its desired therapeutic target. Secondary pharmacology studies are those which address the mode of action and/or effects of a substance not related to its desired therapeutic target. Safety pharmacology studies are defined as those studies which investigate the potential undesirable pharmacodynamic effects of a substance on physiological functions, particularly the cardiovascular, central nervous, and respiratory systems, in relation to exposure in the therapeutic range and above. Safety pharmacology studies should generally be conducted before human exposure. Using Prochymal as an example of an MSC-based product that has undergone nonclinical safety testing, in vitro and in vivo pharmacodynamics studies indicated a mechanism of action involving inhibition of immune and inflammatory responses at target sites, protection of inflamed tissue from collateral damage, and facilitation of damaged tissue repair [3]. Safety pharmacology studies showed that high doses of MSCs caused death from pulmonary/respiratory distress in rats, probably because of cell clumping in the vascular system, while no effects on cardiac safety were observed in pigs [3].

Pharmacokinetics studies are performed to investigate the absorption, distribution, metabolism, and excretion of a developmental medicinal product. These studies are typically referred to as "ADME" studies. For a mesenchymal stromal cell-based product, only distribution studies are likely to be relevant. Such studies with Prochymal in a number of animal species provided consistent evidence of how the MSCs are cleared from the blood to initially accumulate in the lungs within minutes of infusion before moving into the liver, kidneys, and spleen at later time points, thus giving valuable insight into how the cells would behave when administered to humans [3].

Toxicology studies investigate the toxicity of a developmental medicinal product in single-dose (acute toxicity) and repeat-dose (chronic toxicity) studies. For cell-based therapies such as MSCs, consideration needs to be given to whether the human MSC product will be administered to animals under appropriate immunosuppression or whether creating an animal species specific analog of the product would provide useful data. Traditional safety endpoints such as clinical signs, physical examination, food consumption, body weight measurements, clinical pathology and hematology, organ weights, gross pathology, and histopathology should be monitored to identify potential targets for the toxic effects of the medicinal product. Using the principles of reduction, refinement, and replacement to reduce animal use, it may be possible to combine safety and pharmacokinetic evaluations with preclinical efficacy models under appropriate conditions. Traditional toxicology studies also address genotoxicity, carcinogenicity, reproductive and developmental toxicity, tumorigenicity, and immunogenicity, but whether all are appropriate for an MSC-based product will need to be determined on a case-by-case basis, and in the European Union in particular, a risk-based approach is encouraged. For an MSC product, tumorigenicity will need to be evaluated, but it may be possible to combine tumorigenicity studies with the toxicology and/or biodistribution studies. The immunogenicity of MSCs has been extensively studied and allogeneic MSCs provide a lower risk of immunogenicity than most other allogeneic cell types, which is one of the reasons underlying their potential broad utility and suitability for large-scale use. Toxicology (single- and repeat-dose) and tumorigenicity studies with Prochymal demonstrated no adverse effects [3].

10.5.1 Good Laboratory Practice

Nonclinical studies should typically be conducted in accordance with the principles of Good Laboratory Practice (GLP), and this requirement is enforced by legislation. GLP is implemented to ensure that studies are designed, executed, monitored, recorded, and reported in line with a quality management system such that they meet the standards required for supporting clinical use of an IMP in humans. In the United States, GLP regulations are codified in 21CFR58. In the European Union, the legislation on GLP is provided in *Directives 2004/9/EC* [72] and *2004/10/EC* [73]. In Japan, the GLP regulations are described in *MHLW Ordinance No. 114* [20], which partially revises the original GLP regulations described in *MHLW Ordinance No. 21*.

For all IMPs entering into clinical trials, pivotal nonclinical safety studies are required to be performed according to GLP, and specialist contract research organizations (CROs) are often used to perform these studies with traditional medicinal products. However, CHMP and FDA guidance [39,68] recognizes that some or even all toxicology studies with cell-based medicinal products cannot comply fully with GLP regulations. For example, key safety studies may only be possible in animal models of disease or injury which have been developed in academic or specialist research facilities and may not be transferred easily to a CRO's GLP facility. In addition, certain specialized assays (eg, biodistribution assays) required for an MSC-based product may not be available from a CRO. Indeed, the biodistribution studies for Prochymal were non-GLP [3]. Early dialog with regulatory agencies is therefore important when designing nonclinical studies that will generate data supportive of a clinical trial with an MSC-based product.

10.6 CLINICAL DEVELOPMENT

Clinical development of medicinal products traditionally involves the testing of IMPs in clinical trials with human volunteers after they have been shown to be safe in nonclinical studies. Clinical trials are progressive in their design, starting out with the Phase I, or "first-in-human," trials that provide an initial indication of safety. For small molecule or biologic drugs, Phase I studies are usually performed in healthy volunteers. However, for MSC and other cell-based medicinal products, Phase I trials are conducted in a small target patient population for ethical reasons, and evaluation of safety is often combined with an early evaluation of efficacy in a Phase I/II transitional study type design. In this respect, early dialog with regulatory authorities is essential when designing the clinical trial protocol. Subsequent Phase II and III trials will continue to gather safety data and also study a number of efficacy endpoints designed to show that the IMP has a beneficial therapeutic effect in increasing patient numbers. MSCs are being studied in a range of disease indications and the clinical efficacy data required to achieve a marketing authorization for an MSC product in a particular indication will be dependent on the rarity of the indication, the urgency of the unmet medical need, and the magnitude of benefit observed with the MSC product. Again, how the safety and efficacy endpoints are staged, and the size and demographics of the patient population in which they will be tested, will need to be agreed with the regulatory authority responsible for approving the trial for each individual MSC-based IMP. This is acknowledged in the specific guidance on cell-based medicinal products developed in the European Union, United States, and Japan [39–42]. For example, the CHMP *Guideline on Cell-Based Medicinal Products* [39] states: "Special problems might be associated with the clinical development of human cell-based medicinal products. Guidance is therefore provided on the conduct of pharmacodynamic/pharmacokinetic studies, dose finding and clinical efficacy and safety studies. The guideline describes the special consideration that should be given to pharmacovigilance aspects and the risk management plan for these products". "Pharmacovigilance" is defined by EMA as "Science and activities relating to the detection, assessment, understanding, and prevention of adverse effects or any other medicine-related problem." A "risk management plan" describes how these activities will be performed once a medicinal product is marketed. These activities constitute the so-called Phase IV, that is, postmarketing, clinical trials.

Guidance on clinical trials and GCP (see later) is thoroughly documented, and is too extensive to list here. Instead, the reader is referred to the resources listed in Refs. [74–81], in addition to the guidelines specific to cell-based products provided by CHMP [39], CBER [40], and PMDA [41,42]. The results of clinical trials are presented in Module 5 of the CTD, in sections on Biopharmaceutic Studies (Module 5.3.1), Studies Pertinent to Pharmacokinetics Using Human Biomaterials (Module 5.3.2), Human Pharmacokinetic Studies (Module 5.3.3), Human Pharmacodynamic Studies (Module 5.3.4), Efficacy and Safety Studies (Module 5.3.5), Post-Marketing Experience (Module 5.3.6), and Case Report Forms and Individual Patient Listings (Module 5.3.7).

10.6.1 Good Clinical Practice

Clinical trials worldwide are performed in compliance with GCP, which was initially introduced via ICH guidelines [71] but has been elaborated by specific legislation in certain countries/economic areas, and this legislation supersedes the ICH guidance where available. For example, clinical trials and GCP in the European Union are governed by *Directive 2001/20/EC* [82] (however, this Directive is due to be replaced by *Regulation EU No. 536/2014* [83] in late 2016). In the United States, a number of FDA regulations [74] cover clinical trials and GCP, including 21CFR312, 21CFR50 and 21CFR56. In Japan, clinical trials legislation is published in the *Ministerial Ordinance on Good Clinical Practice for Drugs* [20]. The aim of all GCP legislation and guidance is to protect the safety and dignity of human volunteers who enter into a trial of an IMP by defining ethical and scientific quality standards for the design, conduct and recording of the studies. In this respect, obtaining informed patient consent and independent ethical approval are key elements underlying the approval of clinical trials. In addition, in some territories specific regulations for pediatric and geriatric populations have also been developed in recognition of the fact that clinical protocols need to be tailored to these patients rather than them being treated according to an adult trial protocol. For example, the US Pediatric Study Plan [84] and the EU Paediatric Investigation Plan [85,86] procedures have been introduced specifically to address clinical trials in children and adolescent patients.

10.7 INCENTIVES AND ACCELERATED APPROVAL SCHEMES APPLICABLE TO MESENCHYMAL STROMAL CELL PRODUCTS

The ultimate aim when developing a medicinal product is to obtain marketing authorization through the demonstration of quality, safety, and efficacy. This is a lengthy and often very expensive process. In some cases, the time it takes to bring a product to market can mean that some patients who urgently require an alternative product to those available do not receive it in a timely manner. In other cases, developing medicinal products for small target patient populations (that is, those with rare diseases or conditions) may not justify a return on investment for the manufacturer, even though the patients' needs are significant. There are several ways in which these issues have been addressed by the regulatory authorities, and this discussion will focus on those adopted in the ICH regions of the United States, Europe, and Japan.

For rare diseases, "orphan designation" provides developers of medicinal products with a number of incentives to justify the financial investment required to bring a product to market for a small patient population. Of the currently approved MSC products (Table 10.1), Cupistem is an orphan drug in Korea, and Prochymal has received orphan designation for acute GVHD from regulatory agencies including EMA and FDA. However, the majority of MSC-based products currently in development do not target rare diseases. Meanwhile, "accelerated approval/early access" schemes enable patients with an urgent unmet medical need to obtain access to medicinal products earlier than the traditional licensing routes would allow. In Europe in particular, there are also incentives specifically aimed at "small- to medium-sized enterprises" (SMEs), who have only limited resources to dedicate to product development. How these procedures operate in each of the core ICH regions is discussed further. Note that orphan designation procedures also exist in Korea, Australia, and Singapore, while Health Canada operates an Innovative Drugs program which provides an 8-year period of market exclusivity for qualifying products. Innovative Drug status was assigned to Prochymal, together with an additional 6-month extension because it is for pediatric use.

10.7.1 United States

10.7.1.1 Orphan drug designation

The United States was the first country to introduce an orphan drug designation procedure (the US Orphan Drug Designation program) in 1983 via the *Orphan Drug Act* (ODA) [87]. For a medicinal product to qualify for US orphan designation, the disease or condition must meet certain criteria specified in the ODA and in the implementing regulations, 21CFR316. These criteria are that either the condition for which the drug is intended affects less than 200,000 people in the United States, or that it affects more than 200,000 people but the development and marketing costs are unlikely to be recovered by the US sales.

Certain incentives are available to developers of medicinal products with a US orphan designation, including tax credits worth up to 50% of the development costs, a 7-year period of market exclusivity, and waiving of MAA (in this case, a BLA) review fees and annual FDA product fees.

10.7.1.2 Expedited programs for serious conditions

The FDA has implemented a number of "expedited programs," and these are described in the Guidance for Industry document, "Expedited Programs for Serious Conditions—Drugs and Biologics" [88]. Four expedited programs are available, which meet different needs, including *Fast Track Designation*, *Breakthrough Therapy Designation*, *Accelerated Approval*, and *Priority Review Designation*.

These programs have been introduced to facilitate and expedite the development and review of new medicinal products which address unmet medical need in the treatment of a serious or life-threatening disease or condition. More specifically, they help ensure that products for serious conditions are approved and available to patients as soon as it can be concluded that their benefits justify the potential risks associated with their use. The specific aims of, and differences between, these schemes are shown in Table 10.6. Note that the schemes are not mutually exclusive, and a drug development program may qualify for more than one expedited program. Each program has its own specific "qualifying criteria" which are fully described in the Guidance for Industry document [88].

As discussed in the Guidance for Industry document on expedited programs [88], communication with the FDA is a critical aspect of these schemes. In this respect, there will need to be discussion around the CMC, nonclinical, and clinical aspects of the medicinal product development program to give the FDA assurance that the requirements for marketing authorization can be

Table 10.6 Overview of FDA Expedited Programs for Serious Conditions

Expedited Program	Program Aims	Qualifying Criteria
Fast track designation	A process designed to facilitate the development and expedite the review of drugs to treat serious conditions and fill an unmet medical need	A drug that is intended to treat a serious condition AND nonclinical or clinical data demonstrate the potential to address unmet medical need
Breakthrough therapy designation	A process designed to expedite the development and review of drugs which may demonstrate substantial improvement over available therapy	A drug that is intended to treat a serious condition AND preliminary clinical evidence indicates that the drug may demonstrate substantial improvement on a clinically significant endpoint(s) over available therapies
Accelerated approval	A process to allow drugs for serious conditions that fill an unmet medical need to be approved based on a surrogate endpoint	A drug that treats a serious condition AND generally provides a meaningful advantage over available therapies AND demonstrates an effect on a surrogate endpoint that is reasonably likely to predict clinical benefit or on a clinical endpoint that can be measured earlier than irreversible morbidity or mortality (IMM) that is reasonably likely to predict an effect on IMM or other clinical benefit (ie, an intermediate clinical endpoint)
Priority review designation	A process which obligates the FDA to take action on an application within 6 months	An application (original or efficacy supplement) for a drug that treats a serious condition AND, if approved, would provide a significant improvement in safety or effectiveness

met on time and that the product development program will continue to an agreed postmarketing plan where necessary.

The FDA has defined certain concepts which apply to expedited programs, including the definitions of "Serious Condition," "Available Therapy" and "Unmet Medical Need," and these are described fully in the Guidance for Industry document [88].

10.7.2 European Union

10.7.2.1 *Orphan designation*

An orphan designation procedure was introduced in Europe in 2000 with the implementation of *Regulation (EC) 141/2000* on Orphan Medicinal Products [89]. According to Article 3[1](a) of this Regulation, for a medicinal product to qualify for EU orphan designation, the product must be "...intended for the diagnosis, prevention or treatment of a life-threatening or chronically debilitating condition affecting not more than 5 in 10,000 persons." Further guidance is provided in the EMA publications, ENTR/6283/00 [90] and 2003/C 178/02 [91]. It should be noted that there are differences in qualifying criteria for orphan designation between, for example, the United States and the European Union, and therefore developers should go through the procedures in each territory where they plan to develop/market. In common with the US procedure, incentives are provided to developers of orphan medicinal products, including market exclusivity for 10 years following the granting of the MA (plus a further 2 years if a pediatric investigation plan is also approved), and Protocol Assistance, a specific form of Scientific Advice that is charged at a reduced rate compared with standard Scientific Advice.

10.7.2.2 ***Alternative routes to marketing authorization***

10.7.2.2.1 **Conditional marketing authorisation**

A "conditional marketing authorisation" may be granted by the CHMP on the basis of less complete clinical data than would ordinarily be required for a MAA. Conditional MAs are granted in the interest of public health for medicinal products which fill an unmet medical need, including life-threatening diseases and emergency situations, and are valid for one year on a renewable basis. The MA is conditional because it is subject to certain obligations which the applicant must meet following approval. Such obligations may typically include further clinical trials, which must either be ongoing, or planned and approved, as part of the conditional MA review process. The legal framework for conditional MA is provided by *Regulation (EC) No. 726/2004* [18], and the provisions for the granting of such an authorization are laid down in *Regulation (EC) No. 507/2006* [92], which require that the risk–benefit balance of the medicinal product is positive and it is likely that the applicant will be in a position to provide the comprehensive clinical data. Agreement from the CHMP on whether the medicinal product is eligible for the conditional MA route is typically addressed during Scientific Advice or Protocol Assistance during the product development stage.

It is worth noting here that Health Canada provides a similar conditional marketing authorization pathway via the granting of a Notice of Compliance with conditions (NOC/c). Indeed, Prochymal was initially licensed in this manner because only preliminary evidence was available to support its therapeutic effect. However, its use was supported by positive safety data, and the NOC/c therefore provided the opportunity, based on a positive risk–benefit balance, for patients with an unmet medical need to be treated while further efficacy data are collected.

10.7.2.2.2 **Marketing authorisation under exceptional circumstances**

Marketing authorisation under exceptional circumstances may be considered in the absence of complete efficacy data when the indications for which

Table 10.7 Differences Between Conditional MA and MA under Exceptional Circumstances

Conditional Marketing Authorisation	Marketing Authorisation under Exceptional Circumstances
Demonstrate positive benefit–risk balance, based on scientific data, pending confirmation	Comprehensive data cannot be provided (specific reasons foreseen in the legislation)
Authorisation valid for one year, on a renewable basis	Reviewed annually to reassess the risk-benefit balance, in an annual reassessment procedure
Once the pending studies are provided, it can become a "normal" marketing authorisation	Will normally not lead to the completion of a full dossier and become a "normal" marketing authorisation

the product in question is intended are encountered so rarely that the applicant cannot reasonably be expected to provide comprehensive data supporting clinical benefit, including for scientific or ethical reasons. Granting of an MA under exceptional circumstances requires the applicant to further address the safety of the product in patients on an ongoing basis.

There are a number of differences between conditional MA and MA under exceptional circumstances, as shown in Table 10.7. In additional, while designated orphan medicinal products are automatically eligible for conditional MA, designated orphan products are only eligible for approval under exceptional circumstances if the criteria considered for the approval under exceptional circumstances are fulfilled.

In common with all routes to EU marketing authorisation, the legal framework is specified in *Regulation (EC) No. 726/2004* [18], and the applicability of the exceptional circumstances route should be addressed with the CHMP during Scientific Advice/Protocol Assistance in the product development stage. In this respect, a MA under exceptional circumstances will not be granted when a conditional MA is more appropriate, that is, when comprehensive clinical data will likely be provided by the applicant within a short timeframe. Rather, MA under exceptional circumstances will normally not lead to the completion of a full dossier because of the qualifying criteria specified earlier.

10.7.2.3 Accelerated assessment

In the European Union, assessment of the MAA by the CAT will ordinarily be performed to a schedule with a maximum timeframe of 210 days (excluding clock stops, which are periods during the 210 days when the applicant is given the opportunity to respond to any issues raised by the CAT). *Regulation (EC) No. 726/2004* [18] also makes provision for an applicant to request an "Accelerated Assessment" of their MAA which, if granted, will take a maximum of 150 days excluding clock stops. According to *Regulation (EC) No. 726/2004*, Accelerated Assessment is for "… medicinal products of major interest from the point of view of public health and in particular from the view point of therapeutic innovation." Accelerated Assessment can be considered similar to FDA's Accelerated Approval scheme, and Health Canada operates a similar scheme called Priority Review of Drug Submissions [93].

10.7.2.4 **Adaptive Pathways**

Adaptive Pathways and PRIME EMA has recently introduced two schemes to make better use of Conditional Marketing Authorisation and Accelerated Assessment by including them in frameworks that utilize more dedicated agency support through early and ongoing formal dialog, including Scientific Advice at defined timepoints throughout the product development phase. These schemes are known as Adaptive Pathways [94] and PRIME [95], and are being implemented to provide patients with early access to innovative medicines that address unmet needs. Adaptive Pathways primarily utilises the Conditional Marketing Authorisation procedure, with the concept being based on either an initial approval in a well-defined patient subgroup with a high medical need and subsequent widening of the indication to a larger patient population, or a planned conditional approval with post-approval monitoring of the product's use in patients. In addition, a commitment to meeting the CMC requirements for market supply should be achievable. Early Scientific Advice also involving Health Technology Assessment bodies (who are responsible for the reimbursement of medicines in the EU) and NCAs is a key element of the Adaptive Pathways approach, with the aim of providing patients with timely access to a medicinal product based on an evolving risk–benefit profile. The PRIME scheme is primarily intended to prospectively plan Accelerated Assessment for 'priority medicines', which are defined as medicinal products that may offer a major therapeutic advantage over existing treatments, or benefit patients with no treatment options. In common with Adaptive Pathways, a strategic regulatory plan implemented through agency support is a key aspect of the PRIME scheme. Medicinal products are accepted onto the scheme if they demonstrate the potential to benefit patients with unmet medical needs through early clinical data or compelling nonclinical data. Although it would not be accurate to perform a direct comparison of the EMA and FDA schemes, they present a number of elements in common. For instance, FDA Breakthrough Therapy Designation is most similar to the EMA PRIME scheme because of the shared need to demonstrate substantial improvement over available therapies (therapeutic innovation), while FDA Priority Review Designation shows analogies with EMA Accelerated Assessment because they both are intended to reduce the marketing authorization application review timeline. Furthermore, similarities can be drawn between the FDA Fast Track Designation and Accelerated Approval schemes and the EMA Conditional Marketing Authorisation and Adaptive Pathways schemes, because for all of them complete clinical data are not needed and a surrogate endpoint may be used instead to demonstrate initial product efficacy. Additionally, non-clinical data can be considered for the granting of a Fast Track Designation by FDA, which is similar to the EMA PRIME scheme.

10.7.3 Japan

10.7.3.1 ***Orphan drugs***

Orphan drug designation in Japan was originally enabled via Article 77-2 of the former Pharmaceutical Affairs Law (now replaced by the *PMD Act*), which stipulates that such products are intended for use in less than 50,000 patients in Japan and there is a high medical need for their use. Orphan drug designation is provided by the Minister of Health, Labour and Welfare (MHLW) based on the opinion of the Pharmaceutical Affairs and Food Sanitation Council (PAFSC).

In common with the United States and European Union orphan designation programs, a number of incentives are available for developers of orphan medicinal products in Japan, including subsidy payments for product development activities, fee reductions for PMDA consultations and MAAs, and priority review of MAAs.

10.7.3.2 ***Expedited conditional approval of regenerative medicinal products***

In Japan, regenerative medicinal products have been regulated under the *PMD Act* since Nov. 2014, which allows for expedited conditional approval of RMPs based on safety and (predicted) efficacy data from early stage (Phase I and II) clinical trials (see Section 10.2.3). Expedited conditional marketing authorization is therefore a way of supporting developers of RMPs, who typically are not large pharma or biotech companies who can sponsor their own large, late stage clinical trials, to treat patients with promising early stage products whose further development can be enabled by obtaining data from their use in human subjects with an unmet medical need.

10.8 CONCLUSIONS

The global regulatory agencies have established systems that enable cell-based medicinal products to be approved for marketing, thus potentially making cell-based medicinal products—including those manufactured using MSCs—available to large numbers of patients. The regulatory routes to market for such products are based on medicinal product legislation for small molecule and biologic drugs, which requires that the quality, safety, and efficacy of the products are demonstrated by evidence from controlled manufacturing strategies, nonclinical safety studies, and clinical trials in human patients. In this respect, previously established medicinal product legislation has been amended to make it relevant to cell-based products, and specific guidance has been developed to enable the requirements of the legislation to be met through appropriately-informed product development strategies. The importance of following this guidance should not be underestimated by developers who ultimately wish their product to achieve marketing authorization. In addition, product developers should engage in dialog with the regulatory agencies through scientific advice/consultation-type meetings, not only to ensure that the product development program is on track but also to identify the correct regulatory pathway for the product in question. For example, incentive schemes, such as those offered by orphan designation, or accelerated approval may be applicable. Furthermore, in countries such as Japan and Korea, it is possible to bring cell-based medicinal products to market on the basis of limited, but positive, clinical efficacy data, which should be investigated further postmarketing. The approval of several MSC-based products in Korea, and also of Prochymal in Canada and New Zealand, by such pathways is a positive move for the field, and further clinical development can be expected to result in the progression of these and other MSC-based products toward marketing authorization and patient access in pharmaceutical markets globally.

GLOSSARY

1. EMA definitions of Advanced Therapy Medicinal Products (ATMPs), as provided in Regulation (EC) No. 1394/2007.

   ATMPs are specific types of medicinal products including "gene therapy medicinal products" (GTMPs), "somatic cell therapy medicinal products" (SCTMPs) and "tissue engineered products" (TEPs), where:

   - GTMP means a biological medicinal product which has the following characteristics (but is not a vaccine against an infectious disease):
     - It contains an active substance which contains or consists of a recombinant nucleic acid used in or administered to human

beings with a view to regulating, repairing, replacing, adding, or deleting a genetic sequence;
  - Its therapeutic, prophylactic, or diagnostic effect relates directly to the recombinant nucleic acid sequence it contains, or to the product of genetic expression of this sequence.
- SCTMP means a biological medicinal product which has the following characteristics:
  - Contains or consists of cells or tissues that have been subject to substantial manipulation so that biological characteristics, physiological functions, or structural properties relevant for the intended clinical use have been altered, or of cells or tissues that are not intended to be used for the same essential function(s) in the recipient and the donor;
  - Is presented as having properties for, or is used in or administered to human beings with a view to treating, preventing or diagnosing a disease through the pharmacological, immunological or metabolic action of its cells or tissues.
- TEP means a product that:
  - Contains or consists of engineered cells or tissues, and
  - Presented as having properties for, or is used in or administered to human beings with a view to regenerating, repairing, or replacing a human tissue.
  - It is noted that the cells in a TEP may be of human or animal origin, and may be a mixture of viable and non-viable cells but cannot be exclusively nonviable. In addition, cells or tissues shall be considered "engineered" if they fulfil at least one of the following conditions:
    - The cells or tissues have been subject to substantial manipulation, so that biological characteristics, physiological functions or structural properties relevant for the intended regeneration, repair or replacement are achieved.
    - The cells or tissues are not intended to be used for the same essential function or functions in the recipient as in the donor.
- Finally, "combined ATMPs" are those that contain a medical device [as defined in *Directive 93/42/EEC* (REF)] as an integral part of a viable cell- or tissue-containing product.

ABBREVIATIONS

ADME Absorption, distribution, metabolism, and excretion
AMI Acute myocardial infarction
ARDS Acute respiratory distress syndrome
ARGB Australian Regulatory Guidelines for Biologicals

ATMP	Advanced therapy medicinal product (EU)
BGTD	Biologics and Genetic Therapies Directorate (Canada)
BLA	Biologics License Applications (US)
BSE	Bovine spongiform encephalopathy
CAT	Committee for Advanced Therapies (EU)
CBER	Center for Biologics Evaluation and Research (US)
CBTE	Centre for Blood and Tissues Evaluation (Canada)
CDER	Center for Drug Evaluation and Research (US)
CDRH	Center for Devices and Radiological Health (US)
CERB	Centre for Evaluation of Radiopharmaceuticals and Biotherapeutics (Canada)
CFR	Code of Federal Regulations (US)
cGMPs	Current good manufacturing practices
CHF	Congestive heart failure
CHMP	Committee for Medicinal Products for Human Use (EU)
CMC	Chemistry, manufacturing and controls
CQAs	Critical quality attributes
CRO	Contract Research Organizations
CTA	Clinical trial application (Canada)
CTA	Clinical trial authorization (EU)
CTBP	Cellular and tissue-based products (Japan)
CTD	Common technical document (ICH)
DIN	Drug identification number (Canada)
DP	Drug product
DS	Drug substance
EFTA	European Free Trade Association
EMA	European Medicines Agency
Ph. Eur.	European pharmacopeia
EU	European Union
EUTCD	EU Tissues and cells directives
FDA	Food and Drug Administration (US)
FDAAA	Food and Drug Administration Amendment Act 2007 (US)
FDAMA	Food and Drug Administration Modernization Act (US)
GCP	Good clinical practice
GLP	Good laboratory practice
GMP	Good manufacturing practice
GTMP	Gene therapy medicinal product (EU)
GVHD	Graft versus host disease
HCT/Ps	Human cells, tissues, and cellular and tissue-based products (US)
ICH	International Conference on Harmonisation of Technical Requirements for Registration of Pharmaceuticals for Human Use
IMP	Investigational medicinal product
JP	Japanese Pharmacopoeia
MA	Marketing Authorization
MAA	Marketing Authorisation application (EU)
MAH	Marketing Authorization Holder
MCB	Master cell bank

MFDS	Ministry of Food and Drug Safety (Korea)
MHLW	Minister of Health, Labour and Welfare (Japan)
MSCs	Mesenchymal stromal cells
NCA	National Competent Authority (EU)
NOC	Notice of compliance (Canada)
NOC/c	Notice of compliance with conditions (Canada)
OCTGT	Office of Cellular, Tissue and Gene Therapies (US)
ODA	Orphan Drug Act (US)
OOPD	Office of Orphan Products Development (US)
ORA	Office of Regulatory Affairs (Canada)
PAA	Pharmaceutical Affairs Act (Korea)
PAFSC	Pharmaceutical Affairs and Food Sanitation Council (Japan)
PHS Act	Public Health Services Act (US)
PMDA	Pharmaceuticals and Medical Devices Agency (Japan)
PMD Act	Pharmaceuticals and Medical Devices Act (Japan)
PSUR	Periodic Safety Update Report
RMPs	Regenerative medicinal products (Japan)
SCTMP	Somatic cell therapy medicinal product (EU)
SME	Small- to medium-sized enterprise (EU)
TEP	Tissue engineered product (EU)
TGA	Therapeutic Goods Administration (Australia)
US	United States
USP	United States Pharmacopeia
WCB	Working Cell Bank

REFERENCES

[1] Hematti P, Keating A. Mesenchymal Stromal Cells: Biology and Clinical Applications. New York: Springer Science & Business Media; 2013.

[2] Weyand B, Dominici M, Hass R, Jacobs R, Kasper C. Mesenchymal Stem Cells—Basics and Clinical Application II. Berlin Heidelberg: Springer; 2014.

[3] Prochymal® Summary Basis of Decision (Health Canada). Available from: http://www.hc-sc.gc.ca/dhp-mps/prodpharma/sbd-smd/drug-med/sbd_smd_2012_Prochymal®_150026-eng.php.

[4] Directive 2001/83/EC of the European Parliament and of the Council of 6 November 2001 on the Community code relating to medicinal products for human use.

[5] US Code of Federal Regulations Title 21. Available from: http://www.accessdata.fda.gov/scripts/cdrh/cfdocs/cfcfr/cfrsearch.cfm.

[6] Official Journal of the European Union. Available from: http://eur-lex.europa.eu/homepage.html.

[7] Regulation (EC) No 1394/2007 of the European Parliament and of the Council of 13 November 2007 on advanced therapy medicinal products and amending Directive 2001/83/EC and Regulation (EC) No 726/2004.

[8] Council Directive 93/42/EEC of 14 June 1993 concerning medical devices.

[9] Directive 2004/23/EC of the European Parliament and of the Council of 31 March 2004 on setting standards of quality and safety for the donation, procurement, testing, processing, preservation, storage and distribution of human tissues and cells.

[10] Commission Directive 2006/17/EC of 8 February 2006 implementing Directive 2004/23/EC of the European Parliament and of the Council as regards certain technical requirements for the donation, procurement and testing of human tissues and cells.
[11] Commission Directive 2006/86/EC of 24 October 2006 implementing Directive 2004/23/EC of the European Parliament and of the Council as regards traceability requirements, notification of serious adverse reactions and events and certain technical requirements for the coding, processing, preservation, storage and distribution of human tissues and cells.
[12] Commission Directive 2012/39/EU of 26 November 2012 amending Directive 2006/17/EC as regards certain technical requirements for the testing of human tissues and cells.
[13] Directive 2002/98/EC of the European Parliament and of the Council of 27 January 2003 setting standards of quality and safety for the collection, testing, processing, storage and distribution of human blood and blood components and amending Directive 2001/83/EC.
[14] Commission Directive 2004/33/EC of 22 March 2004 implementing Directive 2002/98/EC of the European Parliament and of the Council as regards certain technical requirements for blood and blood components.
[15] Commission Directive 2005/61/EC of 30 September 2005 implementing Directive 2002/98/EC of the European Parliament and of the Council as regards traceability requirements and notification of serious adverse reactions and events.
[16] Commission Directive 2005/62/EC of 30 September 2005 implementing Directive 2002/98/EC of the European Parliament and of the Council as regards Community standards and specifications relating to a quality system for blood establishments.
[17] Commission Directive 2009/120/EC of 14 September 2009 amending Directive 2001/83/EC of the European Parliament and of the Council on the Community code relating to medicinal products for human use as regards advanced therapy medicinal products.
[18] Regulation (EC) No 726/2004 of the European Parliament and of the Council of 31 March 2004 laying down Community procedures for the authorisation and supervision of medicinal products for human and veterinary use and establishing a European Medicines Agency.
[19] Hara A, Sato D, Sahara Y. New governmental regulatory system for stem cell-based therapies in Japan. Ther Innov Regul Sci 2014;48:681–8.
[20] PMDA Ministerial Ordinances. Available from: http://www.pmda.go.jp/english/review-services/regulatory-info/0002.html.
[21] Therapeutic Goods Administration Legislation and Legislative Instruments. Available from: https://www.tga.gov.au/legislation-legislative-instruments.
[22] Therapeutic Goods Administration Australian Regulatory Guidelines for Biologicals (ARGB). Available from: https://www.tga.gov.au/publication/australian-regulatory-guidelines-biologicals-argb.
[23] TGA Consultation: Regulation of autologous stem cell therapies. Available from: https://www.tga.gov.au/consultation/consultation-regulation-autologous-stem-cell-therapies.
[24] Australian Government Therapeutic Goods Order No. 88. Standards for donor selection, testing, and minimising infectious disease transmission via therapeutic

goods that are human blood and blood components, human tissues and human cellular therapy products. Available from: https://www.comlaw.gov.au/Details/F2013L00854.

[25] Government of Canada Food and Drugs Act. Available from: http://laws-lois.justice.gc.ca/eng/acts/F-27/index.html.

[26] Government of Canada Food and Drug Regulations. Available from: http://laws-lois.justice.gc.ca/eng/regulations/C.R.C.%2C%5Fc.%5F870/.

[27] Government of Canada Blood Regulations. Available from: http://laws-lois.justice.gc.ca/eng/regulations/SOR-2013-178/.

[28] Health Canada Guidance Document: Blood Regulations. Available from: http://www.hc-sc.gc.ca/dhp-mps/alt_formats/pdf/brgtherap/applic-demande/guides/blood-reg-sang/blood-guid-sang-ligne-2014-10-23-eng.pdf.

[29] Government of Canada Safety of Human Cells, Tissues and Organs for Transplantation Regulations (SOR/2007-118). Available from: http://laws-lois.justice.gc.ca/eng/regulations/SOR-2007-118/.

[30] Health Canada Guidance on Cells, Tissues, and Organs for Transplantation and Assisted Human Reproduction. Available from: http://www.hc-sc.gc.ca/dhp-mps/brgtherap/reg-init/cell/index-eng.php.

[31] Han E, Shin W. Regulation of cell therapy products in Korea. ISBT Sci Series 2015;10(Suppl. 1):129–33.

[32] Oh I-H. Regulatory issues in stem cell therapeutics in Korea: efficacy or efficiency? Korean J Hematol 2012;47:87–9.

[33] FDA Expanded Access Program. Available from: http://www.fda.gov/NewsEvents/PublicHealthFocus/ExpandedAccessCompassionateUse/ucm429687.htm.

[34] Health Canada Special Access Programme. Available from: http://www.hc-sc.gc.ca/dhp-mps/acces/drugs-drogues/index-eng.php.

[35] Therapeutic Goods Administration Special Access Scheme. Available from: https://www.tga.gov.au/form/special-access-scheme.

[36] Korea Food & Drug Administration Notification No. 2002-65: Guidelines to Clinical Study Authorization for Drugs. Available from: http://www.asiacroalliance.com/pdf/KFDAguidelines.pdf.

[37] International Conference on Harmonisation of Technical Requirements for Registration of Pharmaceuticals for Human Use (ICH). Available from: http://www.ich.org/home.html.

[38] EMA Scientific Guidelines. Available from: http://www.ema.europa.eu/ema/index.jsp?curl(pages/regulation/general/general_content_000043.jsp&mid(WC0b01ac05800240cb.

[39] EMA Guideline On Human Cell-Based Medicinal Products (EMEA/CHMP/410869/2006). Available from: http://www.ema.europa.eu/docs/en_GB/document_library/Scientific_guideline/2009/09/WC500003894.pdf.

[40] FDA Cellular & Gene Therapy Guidances. Available from: http://www.fda.gov/BiologicsBloodVaccines/GuidanceComplianceRegulatoryInformation/Guidances/CellularandGeneTherapy/default.htm.

[41] MHLW Notification No. 0907-2. Guidance for ensuring the quality and safety of processed autologous human somatic stem cell products. 2012.

[42] MHLW Notification No. 0907-3.Guidance for ensuring the quality and safety of processed allogenic human somatic stem cell products. 2012.

[43] Health Canada Guidance Documents. Available from: http://www.hc-sc.gc.ca/dhp-mps/brgtherap/applic-demande/guides/index-eng.php.

[44] ICH Common Technical Document. Available from: http://www.ich.org/products/ctd.html.

[45] ICH M4Q(R1). The Common Technical Document for the Registration of Pharmaceuticals for Human Use: Quality.

[46] ICH Q5D. Derivation and Characterisation of Cell Substrates Used for Production of Biotechnological/Biological Products.

[47] ICH Q5A(R1). Viral Safety Evaluation of Biotechnology Products Derived from Cell Lines of Human or Animal Origin.

[48] ICH Q4. Pharmacopoeias.

[49] USP Chapter 1043. Ancillary Materials for Cell, Gene, and Tissue-Engineered Products. Available from: http://www.pharmacopeia.cn/v29240/usp29nf24s0_c1043.html.

[50] EMA Guideline on Xenogeneic Cell-based Medicinal Products. EMEA/CHMP/CPWP/83508/2009. Available from: http://www.ema.europa.eu/docs/en_GB/ document_library/Scientific_guideline/2009/12/WC500016936.pdf.

[51] EMA Guideline on the use of bovine serum in the manufacture of human biological medicinal products. CHMP/BWP/457920/2012 Rev. 1. Available from: http://www.ema.europa.eu/docs/en_GB/document_library/Scientific_guideline/2013/06/WC500143930.pdf.

[52] EMA Guideline on the use of porcine trypsin used in the manufacture of human biological medicinal products. EMA/CHMP/BWP/814397/2011. Available from: http://www.ema.europa.eu/docs/en_GB/document_library/Scientific_guideline/2014/02/WC500162147.pdf.

[53] MHLW Notification No. 1314. Guidance for ensuring the quality and safety of products manufactured from human or animal derived components. 2000.

[54] Carmen J, Burger SR, McCaman M, Rowley JA. Developing assays to address identity, potency, purity and safety: cell characterization in cell therapy process development. Regen Med 2012;7:85–100.

[55] Pevsner-Fischer M, Levin S, Zipori D. The origins of mesenchymal stromal cell heterogeneity. Stem Cell Rev 2011;7:560–8.

[56] Bianco P, Cao X, Frenette PS, Mao JJ, Robey PG, Simmons PJ, Wang C-Y. The meaning, the sense and the significantd: translating the science of mesenchymal stem cells into medicine. Nat Med 2013;19:35–42.

[57] Dominici M, Le Blanc K, Mueller I, Slaper-Cortenbach I, Marini FC, Krause DS, Deans RJ, Keating A, Prockop DJ, Horwitz EM. Minimal criteria for defining multipotent mesenchymal stromal cells. The International Society for Cellular Therapy position statement. Cytotherapy 2006;8:315–7.

[58] Hass R, Kasper C, Böhm S, Jacobs R. Different populations and sources of human mesenchymal stem cells (MSC): a comparison of adult and neonatal tissue-derived MSC. Cell Commun Signal 2011;9:12–25.

[59] Alvarez-Viejo M, Menendez-Menendez Y, Blanco-Gelaz MA, Ferrero-Gutierrez A, Fernandez-Rodríguez MA, Perez-Basterrechea M, Gracia-Gala JM, Perez-Lopez S, Otero-Hernandez J. LNGFR (CD271) as marker to identify mesenchymal stem cells from different human sources: umbilical cord blood, Wharton's Jelly and bone marrow. J Bone Marrow Res 2013;1:4–9.

[60] Banfi A, Muraglia A, Dozin B, Mastrogiacomo M, Cancedda R, Quarto R. Proliferation kinetics and differentiation potential of ex vivo expanded human bone marrow stromal cells: Implications for their use in cell therapy. Exp Hematol 2000;8:707–15.
[61] Banfi A, Bianchi G, Notaro R, Luzzatto L, Cancedda R, Quarto R. Replicative aging and gene expression in long-term cultures of human bone marrow stromal cells. Tissue Eng 2002;8:901–10.
[62] Wagner W, Horn P, Castoldi M, Diehlmann A, Bork S, Saffrich R, Benes V, Blake J, Pfister S, Eckstein V, Ho AD. Replicative senescence of mesenchymal stem cells: a continuous and organized process. PLoS One 2008;3:e2213.
[63] Menicanin D, Bartold PM, Zannettino ACW, Gronthos S. Genomic Profiling of Mesenchymal Stem Cells. Stem Cell Rev 2009;1:36–50.
[64] Murphy MB, Moncivais K, Caplan AI. Mesenchymal stem cells: environmentally responsive therapeutics for regenerative medicine. Exp Mol Med 2013;45:e54.
[65] Commission Directive 2003/94/EC of 8 October 2003 laying down the principles and guidelines of good manufacturing practice in respect of medicinal products for human use and investigational medicinal products for human use.
[66] EudraLex Volume 4. Good Manufacturing Practice (GMP) Guidelines. Available from: http://ec.europa.eu/health/documents/eudralex/vol-4/index_en.htm.
[67] FDA Guidance for Industry. Sterile Drug Products Produced by Aseptic Processing—Current Good Manufacturing Practice. Available from: http://www.fda.gov/downloads/Drugs/GuidanceComplianceRegulatoryInformation/Guidances/UCM070342.pdf.
[68] FDA Guidance for Industry CGMP for Phase 1 Investigational Drugs. Available from: http://www.fda.gov/downloads/Drugs/GuidanceComplianceRegulatoryInformation/Guidances/UCM070273.pdf.
[69] ICH Q7: Good Manufacturing Practice Guide for Active Pharmaceutical Ingredients. Available from: http://www.ich.org/fileadmin/Public_Web_Site/ICH_Products/Guidelines/Quality/Q7/Step4/Q7_Guideline.pdf
[70] ICH Safety Guidelines. Available from: http://www.ich.org/products/guidelines/safety/article/safety-guidelines.html.
[71] FDA Guidance for Industry Preclinical Assessment of Investigational Cellular and Gene Therapy Products. Available from: http://www.fda.gov/downloads/BiologicsBloodVaccines/GuidanceComplianceRegulatoryInformation/Guidances/CellularandGeneTherapy/UCM376521.pdf.
[72] Directive 2004/9/EC of the European Parliament and of the Council of 11 February 2004 on the inspection and verification of good laboratory practice (GLP).
[73] Directive 2004/10/EC of the European Parliament and of the Council of 11 February 2004 on the harmonisation of laws, regulations and administrative provisions relating to the application of the principles of good laboratory practice and the verification of their applications for tests on chemical substances.
[74] ICH Efficacy Guidelines. Available from: http://www.ich.org/products/guidelines/efficacy/article/efficacy-guidelines.html.
[75] FDA GCP/Clinical Trial Regulations. http://www.fda.gov/ScienceResearch/SpecialTopics/RunningClinicalTrials/ucm155713.htm.
[76] FDA Guidance for Industry. Considerations for the Design of Early-Phase Clinical Trials of Cellular and Gene Therapy Products. Available from: http://www.fda.gov/downloads/BiologicsBloodVaccines/GuidanceComplianceRegulatoryInformation/Guidances/CellularandGeneTherapy/UCM359073.pdf.

[77] EudraLex Volume 10. Clinical Trials Guidelines. Available from: http://ec.europa.eu/health/documents/eudralex/vol-10/index_en.htm.

[78] EMA Clinical Efficacy and Safety Guideline. Available from: http://www.ema.europa.eu/ema/index.jsp?curl(pages/regulation/general/general_content_000085.jsp&mid(WC0b01ac0580027549.

[79] MHLW Notification No. 0812-26. Regarding Clinical Trail Notification for Cell Derived Products. 2014.

[80] MHLW Notification No. 0812-1. Regarding Handling of Clinical Trail Notification for Cell Derived Products. 2014.

[81] MHLW Notification No. 425. Guidelines for Clinical Studies using Human Stem Cells. 2006.

[82] Directive 2001/20/EC of the European Parliament and of the Council of 4 April 2001 on the approximation of the laws, regulations and administrative provisions of the Member States relating to the implementation of good clinical practice in the conduct of clinical trials on medicinal products for human use.

[83] Regulation (EU) No 536/2014 of the European Parliament and of the Council of 16 April 2014 on clinical trials on medicinal products for human use, and repealing Directive 2001/20/EC.

[84] FDA Guidance for Industry. Pediatric Study Plans: Content of and Process for Submitting Initial Pediatric Study Plans and Amended Pediatric Study Plans. Available from: http://www.fda.gov/downloads/Drugs/GuidanceComplianceRegulatoryInformation/Guidances/UCM360507.pdf.

[85] Regulation (EC) No 1901/2006 of the European Parliament and of the Council of 12 December 2006 on medicinal products for paediatric use and amending Regulation (EEC) No 1768/92, Directive 2001/20/EC, Directive 2001/83/EC and Regulation (EC) No 726/2004.

[86] Regulation (EC) No 1902/2006 of the European Parliament and of the Council of 20 December 2006 amending Regulation 1901/2006 on medicinal products for paediatric use.

[87] FDA Orphan Drug Designation. Available from: http://www.fda.gov/forindustry/developingproductsforrarediseasesconditions/howtoapplyfororphanproductdesignation/default.htm.

[88] FDA Guidance for Industry Expedited Programs for Serious Conditions—Drugs and Biologics. Available from: http://www.fda.gov/downloads/drugs/guidancecomplianceregulatoryinformation/guidances/ucm358301.pdf.

[89] Regulation (EC) No 141/2000 of the European Parliament and of the Council of 16 December 1999 on orphan medicinal products.

[90] European Commission Guideline on the format and content of applications for designation as orphan medicinal products and on the transfer of designations from one sponsor to another, 9 July 2007. ENTR/6283/00 Rev 3. Available from: http://ec.europa.eu/health/files/orphanmp/doc/2007_07/format_content_orphan_applications_rev3_200707_en.pdf.

[91] Communication from the Commission on Regulation (EC) No 141/2000 of the European Parliament and of the Council on orphan medicinal products. Official Journal of the European Union. (2003/C 178/02). Available from: http://eur-lex.europa.eu/LexUriServ/LexUriServ.do?uri(OJ:C: 2003:178:0002:0008:en:PDF.

[92] Commission Regulation (EC) No 507/2006 of 29 March 2006 on the conditional marketing authorisation for medicinal products for human use falling within the scope of Regulation (EC) No 726/2004 of the European Parliament and of the Council.

[93] Health Canada Guidance Document. Priority Review of Drug Submissions 09-101263-867. Available from: http://www.hc-sc.gc.ca/dhp-mps/alt_formats/hpfb-dgpsa/pdf/prodpharma/priordr-eng.pdf.

[94] EMA Pilot project on adaptive licensing. EMA/254350/2012. Available from: http://www.ema.europa.eu/docs/en_GB/document_library/Other/2014/03/WC500163409.pdf.

[95] EMA PRIME (Priority Medicines) scheme. http://www.ema.europa.eu/ema/index.jsp?curl=pages/regulation/general/general_content_000660.jsp&mid=WC0b01ac05809f8439.

Chapter 11

The Health Economics for Regenerative Medicine: How Payers Think and What That Means for Developers

C. McCabe* and T. Bubela**

*Department of Emergency Medicine Research, Department of Economics, School of Public Health, University of Alberta, Alberta, Canada; Academic Unit of Health Economics, University of Leeds, Leeds, United Kingdom; **Public Health, Alberta School of Business, Alberta, Canada

CHAPTER OUTLINE

11.1 INTRODUCTION

In this chapter, we outline the standard framework for considering the value of new health technologies in the presence of limited health care budgets, including how health care payers take account of uncertainty in the evidence for the value of those technologies. We then outline some implications of these insights for the development and adoption of regenerative medicine technologies.

Mesenchymal Stromal Cells. http://dx.doi.org/10.1016/B978-0-12-802826-1.00011-8

The chapter is structured as follows. Section 11.2 provides a brief overview of health technology assessment and appraisal and the role of cost effectiveness analysis. Section 11.3 outlines the key concepts of cost effectiveness analysis; and Section 11.4 then explains the conceptual framework for the utilization of cost effectiveness analysis in health technology assessment and appraisal processes. Section 11.5 describes how it is possible to characterize the value of research in terms of its expected impact on obtaining a positive reimbursement decision using the evidence on the expected cost effectiveness of a technology. In Section 11.6, we described the Value Engineered Translation framework that can support developers in the prospective design of research and development processes to meet the evidence needs of health care payers and hence optimize the likelihood of positive reimbursement decisions leading to the necessary returns on investment. Section 11.7 provides some specific observations on the challenges that regenerative medicine technologies may face in achieving reimbursement.

11.2 HEALTH TECHNOLOGY ASSESSMENT AND APPRAISAL

Health Technology Assessment (HTA) is "a multidisciplinary process that summarizes information about the medical, social, economic, and ethical issues related to the use of a health technology in a systematic transparent unbiased and robust manner." [1] The aim of HTA is to "inform the formulation of safe effective health policies that are patient focused and seek to achieve best value."

Health Technology Appraisal is the process of considering the synthesized evidence produced by HTA to arrive at a decision regarding whether a specific technology should be included in the portfolio of technologies provided by a specific health care system or covered by a specific health care payer. While HTA focuses on the evidence to inform a policy decision, appraisal focuses on the decision itself. As the rationale for the former is the desire to promote high-quality decision making in the latter, there is limited meaning to considering technology assessment processes divorced from the decision(s) that they are designed to inform.

11.3 COST EFFECTIVENESS ANALYSIS

Cost effectiveness analysis (CEA) lies at the heart of the vast majority of HTA processes. Bending and Smith identified 12 markets where a CEA is a mandated component of all initial submissions to HTA bodies (see Table 11.1). In a further nine markets CEAs are mandated for certain categories of technology, and in another four markets the HTA bodies can choose to require a CEA as part of a submission [2].

Table 11.1 Status of CEA as a Component of HTA Submissions by Jurisdictions [6]

CEA Required as Part of All HTA Submission	CEA Required as Part of Some HTA Submissions	CEA Can Be Required as Part of an HTA Submission
Australia (PBAC)	Austria (HEK)	Denmark (DMA)
Canada (CADTH CDR and pCODR)	Belgium (NIHDI)	Italy (AIFA)
England and Wales (NICE)	Ireland (CPU)	Switzerland (FOPH)
Finland (PPB)	Netherlands (CVZ)	Germany (GBA)
Israel (MoH)	New Zealand (PHARMAC)	
Hungary (OHTA)	Norway (NOMA)	
South Korea (HIRA)	Spain (MOH)	
Mexico (GHC)	France (HAS)	
Poland (AHTOPol)		
Portugal (INFARMed)		
Scotland (SMC)		
Sweden (TLV)		

CEAs are a form of economic evaluation that compare the expected costs and outcomes of alternative management strategies for an identified condition/patient group to establish the incremental costs and benefits of changing from the current practice to one or more alternatives [3]. In CEA the benefits of treatment are increasingly measured using Quality Adjusted Life Years (QALYs), as these can capture the impact of alternatives on both life expectancy and health-related quality of life (HrQol). When decision makers are concerned with allocating resources between different clinical specialties such as cardiovascular, musculoskeletal, and oncology, the use of QALYs as the measure of benefit allows the direct comparison of the value of alternative uses of health care dollars across these specialties. It therefore promotes transparency and horizontal equity in the decision-making process [4,5].

The output of a CEA is the Expected Incremental Cost Effectiveness Ratio (ICER). The ICER is calculated as the difference between the expected costs of current practice and the expected costs with the alternative technology, divided by the difference in the expected benefits (QALYs), assuming current practice, and the expected QALYs with the alternative technology [see Eq. 11.1].

$$ICER = \frac{C_2 - C_1}{Q_2 - Q_1} \tag{11.1}$$

where

C_2 = cost of the alternative technology
C_1 = cost of current practice

Q_2 = expected quality adjusted life years produced by the alternative technology
Q_1 = expected quality adjusted life years produced by current practice

In the health technology appraisal process, decision makers will frequently decide whether to adopt a new technology by comparing the ICER produced by the HTA process to a reference value. This reference value is referred to as the Cost Effectiveness Threshold or the Willingness to Pay for a QALY. Substantial debate exists as to both the conceptual and empirical basis for identifying the reference value because of its pivotal role in determining whether a new technology is made available to patients who might benefit from it. Broadly, there are two competing approaches; the first can be thought of as the demand side definition of the Willingness to Pay for health (care); and the second as the supply side definition. For health care decision makers operating with a fixed or semifixed budget, whether funded from public or private finance, the supply side definition is the appropriate framework for specifying the cost effectiveness threshold. Since almost all health care systems fall within this category, the next section expands on the supply side conceptual model of the cost effectiveness threshold and how it is used in HTA decision making [7].

Before moving on to consider the cost effectiveness threshold in more detail, we need to understand which factors are considered in well-conducted CEA, as there is often confusion on this issue among communities that are new to CEA. Here we discuss five key factors:

1. As the impact of effective health care will often include differences in life expectancy, cost effectiveness analyses will often adopt a life time horizon to identify all relevant differences in health outcomes and health care resource utilization. Hence, treatments that have large upfront costs that lead to substantial longer term costs savings by eliminating the need for future health care utilization will not be disadvantaged compared to treatments where the costs are spread more evenly over a patient's life time. Similarly, treatments that have a large benefit in the distant future (eg, preventive therapies) will not be disadvantaged compared to ones that provide a smaller but immediate benefit.
2. Which costs and health outcomes are captured in a CEA depend quite heavily upon the perspective that is adopted by the CEA. If the perspective adopted is that of the health care payer/system then only costs that fall on the payer's budget should be included in the analysis. Typically out-of-pocket costs that fall on the patients will also be included in the analysis, even when the perspective is that of the health care payer, to avoid creating an incentive to shift costs from the health care system on to the patients affected by the condition.

3. The outcomes that are appropriate for consideration will depend upon what the payer is trying to achieve through funding health care. If the payer is only concerned with health rather than the broader effects of health care such as labor force activity and productivity, then changes in life expectancy and health-related quality of life are appropriate. However, if broader impacts are of interest, then including productivity costs is appropriate [3,8]. Even when the focus of the payer is limited to health benefits, the impact of improvements in the treated individual's health on the health of family and friends providing informal care is appropriately included in the analysis [9].
4. The description of the impact of a therapy on patients' health-related quality of life is typically obtained from the patients who have the condition. The value attached to those health effects in the calculation of Quality Adjusted Life Years may come from either patients or from the general public. They do not come from the individuals who decide whether a treatment should be paid for.
5. Given that CEAs typically evaluate the long-term costs and outcomes of new and existing therapies, extrapolation from short and medium-term evidence is unavoidable. These extrapolations are typically done using a suite of statistical methods called decision analytic modeling [10]. These models are not designed to test hypotheses in the way that clinical experiments are, but rather to make a prediction of the expected value of long-term costs and outcomes. One of the implications of this difference in intent is that the methods for analyzing uncertainty are different from conventional (frequentist) tests of statistical significance. Decision analytic models produce expected values and characterize the probability that a decision based upon the expected value of the costs and outcomes will prove to be wrong. This is called the "decision uncertainty." When the decision uncertainty is large it may be efficient to delay funding a new treatment to allow for further research to reduce the decision uncertainty. This question can be formally evaluated through the use of Value of Information Analyses, discussed later [11,12].

11.4 COST EFFECTIVENESS ANALYSIS IN HEALTH TECHNOLOGY APPRAISAL

In a series of articles McCabe and colleagues have described how the use of CEA and a cost effectiveness threshold decision rule is consistent with improving population health [5,6,13,14]. Fig. 11.1 illustrates the conceptual model. The model assumes that the health system has a finite budget and that the budget is fully allocated, funding a portfolio of technologies

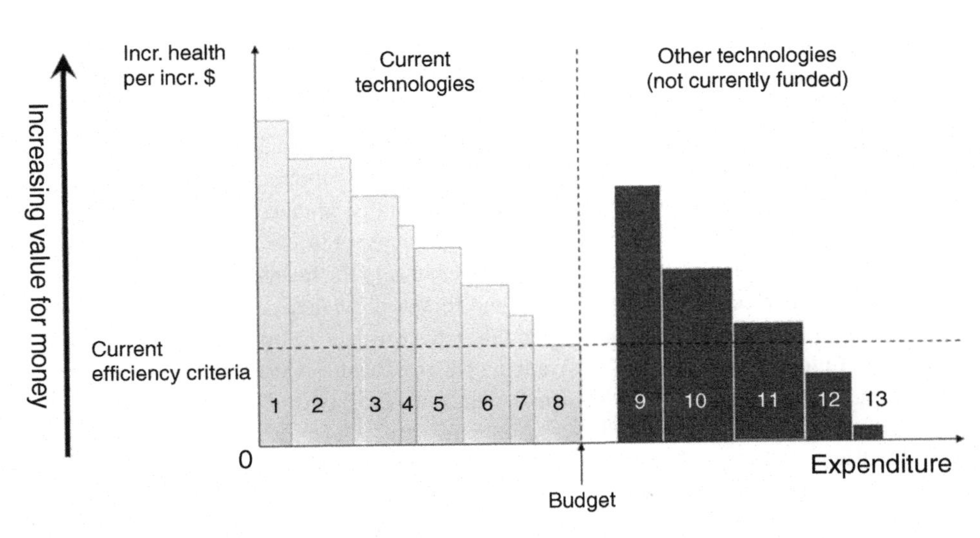

■ **FIGURE 11.1** Value and health technology adoption with fixed budget.

(numbers 1 to 8). There exists a set of currently unfunded technologies that manufacturers, clinicians, and patients who might benefit would like to see funded by the health care payer. These are technologies 9 to 13.

The payer wishes to maximize the value of the health produced by the expenditure of the health care budget, and therefore will fund those technologies that produce the most value per $ spent before funding technologies that produce less value per $ spent. This is reflected in Fig. 11.1 by ordering the technologies according to how much value they produce. Technology 1 produces the most value and technology 8 the least, among the currently funded technologies. It is worth noting that the value produced by a technology may be constrained to its impact on health, or it can be modified to reflect factors that modify the value health gained such as the severity of the disease, the availability of alternative treatments, and the characteristics of the patients who would benefit. What is important is that the same measure of value is applied to all the technologies: both the currently funded and currently unfunded. The width of the bar for each technology represents the budget impact of adopting the technology.

For a new technology to be funded, at least one of the currently funded technologies will have to be displaced, to free up the necessary resources. For a payer who wishes to increase the value of the health produced by consuming the health care budget, the decision rule for adopting a new technology is that it produces more value (health) per $ expended than the technology it displaces. Further, the payer will wish to give up the least valuable currently funded technology first. The dotted horizontal line (Current Efficiency Criteria), in Fig. 11.1, illustrates this decision rule.

We can see that technologies 12 and 13 produce less value per $ spent than technology 8 and therefore, adopting either of them at the expense of technology 8 would reduce the value of the health produced from the health care budget. However, technologies 9, 10, and 11 produce more value than technology 8, and thus are candidates for adoption by the health care system. Under the decision rule of adopting the most valuable technologies first, we would adopt technology 9 first. It has a similar budget impact to technology 8, and it produces more value than all of the currently funded technologies except for 1 and 2. Once Technology 7 is the new "least valuable" technology and as it is less valuable than technology 10, hence it looks like technology 10 should be adopted. Unfortunately, the budget impact of technology 10 is greater than that of technology 7, hence to fully adopt technology 10, will require disinvesting from technology 6 also. Examination of Fig. 11.1 shows that the value per $ spent for technology 10 is slightly higher than for technology 6, and disinvesting from technologies 6 and 7 to fund technology 10 will increase the total value of the health produced by the expenditure of the health care budget.

Fig. 11.2 shows the position after the adoption of the technologies 9 and 10. While technology 11 was better value for money than technology 8, as we adopt the most valuable new technology first, technology 11 should

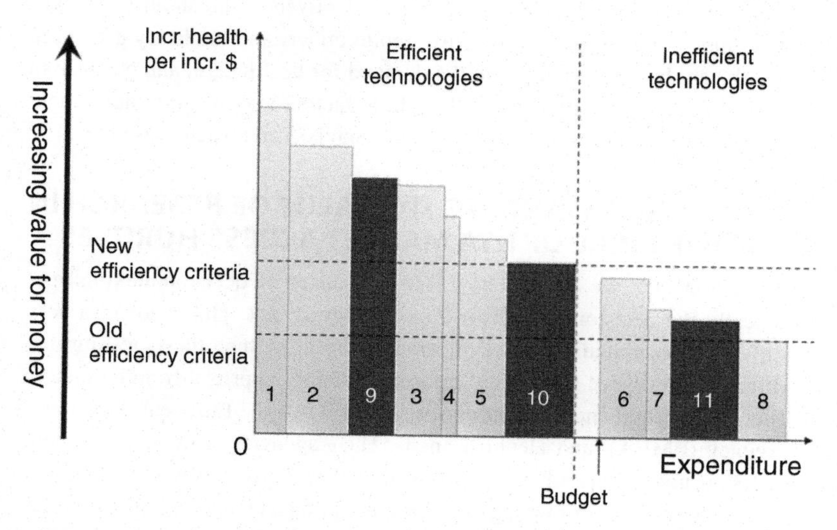

FIGURE 11.2 Value after adoption of new technologies using Cost Effectiveness Threshold Decision Rule.

not be adopted as it is less valuable than the new least valuable technology, technology 10.

The new efficiency criteria deciding whether to adopt a currently unfunded technology, often called the Cost Effectiveness Threshold, is the cost per unit of health produced by technology 10; the inverse of the incremental health per $ spent that is plotted in Fig. 11.2.

As well as illustrating the rationale for using a cost effectiveness threshold to decide which technologies should be funded, this figure illustrates two other important characteristics of the health technology appraisal decision. First, new technologies are adopted at the expense of other people's health care. In our illustration the patients who would benefit from technologies 6, 7, and 8 no longer receive these therapies at the health care payer's expense. Thus saying "no" to the patients who would benefit from a new technology is also saying "yes" to the patients who benefit from the currently funded technologies; and saying "yes" to a new technology involves saying "no" to the patients who would benefit from a current technology.

The second point is that as health care systems adopt more and more new technologies, unless the budget is increased to completely cover the additional cost of these technologies, the threshold for adopting the next new technology becomes increasingly challenging; that is the cost effectiveness threshold above which technologies are not paid for will be lower. Hence, for investors considering technologies that are 5–10 years from market access, consideration of the most likely cost effectiveness threshold(s) that will determine commercial success will need to understand the likely growth in health care budgets, changes in the demand for health care, and changes in health care system efficiency. All of these factors may mean a much more challenging reimbursement environment than is currently the case [13].

11.5 UNDERSTANDING THE VALUE OF RESEARCH IN THE PRESENCE OF HTA MARKET ACCESS HURDLES

In Section 11.2, we outlined how HTA is focused on decision making, specifically the decision to pay for a technology or not. This leads to a very different conceptualization of uncertainty from the hypothesis testing that drives the design of research to inform regulatory approval from regulators such as the Food and Drug Administration (FDA), the European Medicines Agency (EMEA), and Health Canada. HTA agencies are concerned with decision uncertainty; that is the risk that they make the wrong decision. There are two types of wrong decisions: adopting a technology that is not as valuable as those it displaces; and failing to adopt a technology that is more valuable than the technologies it would have displaced.

In this decision-making framework, research is valuable to the extent that it reduces the risk of making the wrong decision. This insight is operationalized as a quantitative tool for assessing the value of investing in research using a suite of methods called value of information analysis (VoI). The detailed algebra for value of information analysis is beyond the scope of this chapter, but in this section, we illustrate the conceptual model of VoI, and the value of sample information, drawing on material published by Edlin et al. [15]

When we make decisions, rather than test hypotheses, it is appropriate to compare the expected value of alternative courses of action in deciding which option is better [16]. Unlike hypothesis testing, we do not have the option of not making a decision. The choice of maintaining the status quo is still a decision with implications for the costs borne by the health care system and patients' health outcomes. The expected value of the two alternatives may be too close to be statistically significantly different, such as technology 10 and technology 6 in Fig. 11.2, but the choice which maximizes the expected health produced from the health care budget is to adopt technology 10 at the expense of technology 6.

Uncertainty in the evidence base means that there is some risk that technology 6 will turn out to be better than technology 10, and we have made the wrong decision. However, only an omniscient decision maker would know this. Such an omniscient decision maker would have perfect information about how the world was actually going to turn out and thus would always make the right choice. In Table 11.2, we have characterized the uncertainty facing a decision maker who has to choose between two alternative technologies: Technology A and Technology B.

In this much simplified state of the world, the decision maker knows that there are five possible future states of the world. In each of these states of the world the net value of the two technologies differs, and because there is no way of knowing which state of the world will turn out to be the true state

Table 11.2 Choosing Between Technologies Under Uncertainty

	Treatment Health Benefit (Quality Adjusted Life Years)	
	Technology A	**Technology B**
State of the world 1	9	12
State of the world 2	12	10
State of the world 3	14	20
State of the world 4	11	10
State of the world 5	14	13
Expectation	12	13

of the world, the choice of which technology to reimburse is made of the basis of the expected value (average) for each of the technologies over all the possible states of the world. Hence, Technology B is chosen over Technology A because it is expected to provide 13 Quality Adjusted Life Years (QALYs) compared to 12 QALY from Technology A.

However, an omniscient decision maker would always choose the superior technology for the state of the world that they know would be realized. Sometimes, this would be Technology A and others it would be Technology B. Table 11.3 shows which technology would be chosen in each state of the world and the expected health gain from always making the right choice.

We can see that for State of the World 1, the omniscient decision maker would choose Technology B; the same as the decision maker without perfect information, and thus there is no health loss due to the uncertainty, in this state of the world. In contrast in State of the World 2, perfect information allows the decision maker to choose Technology A, and we can see that compared to the decision based upon uncertain information, we have lost two QALYs. States of the World 4 and 5 impose a health loss of 1 QALY each, while State of the World 3 leads to the same choice as using the expected value and hence there is no cost of uncertainty (Table 11.3).

If the omniscient decision maker, working with perfect information, made the decisions, then the expected health benefit would be 13.8 QALYs; 0.8 QALYs higher, than the 13 QALYs delivered by the uncertain decision maker who chooses the Technology B for all states of the world. No decision maker could do better than making the best choice on every occasion and hence the Value of Perfect Information is the difference in the expected health gain under uncertainty (13 QALYs) and the expected health gain under conditions of certainty (13.8 QALYs)—0.8 QALYs. When the objective of research to inform reimbursement decisions is to eliminate the risk of

Table 11.3 Decision Making With Perfect Information and the Health Loss Due to Uncertainty

	Treatment Health Benefit		Optimal Choice	Maximum Health Benefit	Health Loss
	A	B			
State of the world 1	9	12	B	12	0
State of the world 2	12	10	A	12	2
State of the world 3	14	20	B	20	0
State of the world 4	11	10	A	11	1
State of the world 5	14	13	A	14	1
Expectation	12	13		13.8	0.8

making the wrong decision, this 0.8 is the maximum value of any future research; hence we call it the Expected Value of Perfect Information.

Of course no research can produce perfect information, and therefore to attach a value to a specific piece of research we need to consider the degree to which it reduces the risk of making the wrong decision. In Table 5.2, we can see that the largest single contributor to the expected health loss due to uncertainty is State of the World 2.

Fig. 11.3 is a tri-plot. It summarizes the evidence for the effectiveness of Technology B compared to Technology A as follows: The Green Probability Density Function (PDF) represents the synthesis of all existing evidence. Hence, the data that underlies the description of the 5 states of the world in Table 11.2; formally referred to as the Prior. The Red PDF represents the data produced from additional research, a trial comparing Technologies A and B. This is called the Likelihood. The Blue PDF (the Posterior) describes the synthesized evidence for the Performance of Technologies A and B, when the data from the additional research (the Likelihood), is added to the preexisting data (the Prior).

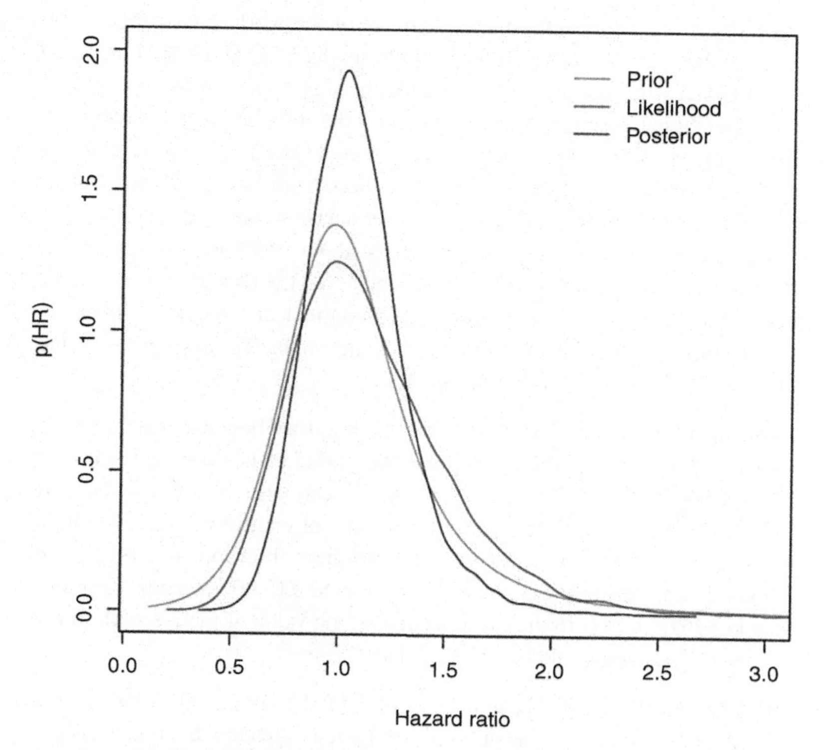

FIGURE 11.3 Tri-plot of the evidence for the relative effectiveness of Technologies A and B.

Table 11.4 Impact of Additional Information on Decision Uncertainty and Value of Information

	Treatment Health Benefit		Optimal Choice	Maximum Health Benefit	Health Loss
	A	B			
State of the world 1	9	12	B		0
State of the world 2	10	11	B	11	0
State of the world 3	14	20	B	20	0
State of the world 4	11	10	A	11	1
State of the world 5	14	13	A	14	1
Expectation	12	13.2		13.6	0.4

We can see that the additional evidence reduces the uncertainty about the relative performance of the technologies. Hence, a smaller range of possible states of the world need to be considered. Table 11.4 illustrates the potential impact of this additional information on the evidence presented to a decision maker.

With the new evidence, the range of states of the world has changed, and State of the World 2 has Technology B producing 11 QALYs and compared to 10 QALYs for Technology A's. Thus, there is no health loss associated with the choice based upon the expected value (Technology B). With the incorporation of the new evidence, the expected QALYs produced by Technology B are now 13.2 and the expected maximum health benefit; that is, what would be achieved by the omniscient decision maker is 13.6 QALYs. The expected health loss due to uncertainty about which technology actually turns out to be the better value is reduced to 0.4 QALYs, from 0.8 QALYs before the additional research reported. From the decision maker's perspective, the value of that additional research is the change in the expected value of perfect information, that is, 0.4 QALYs.

Remember that in the previous section we described how the cost effectiveness threshold provides an estimate of the value of health that is implicit in the budget of the health care system. We can use this value to convert the "health loss" characterization of the value of additional research into a monetary valuation. Thus, the cost effectiveness threshold for the United Kingdom was recently calculated to be around £13,000 sterling, approximately $26,000. [17] In the example above, the value of the research to the health care system would be calculated as:

Expected change in the Health Loss (0.4 QALYs) times the Value of a QALY ($34,000) times the number of patient affected by the decision.

Thus, if Technologies A and B were to treat a disease that affected 5000 people, the maximum population value of the additional research would be $52,000,000.

The value of any specific piece of research is dependent upon, *inter alia,* the sample size, the cost of sampling, how long it takes the research to report, and the difference in the expected difference in the health trajectory of patients on alternative therapies while the research is on-going [18].

A further consideration from the commercial perspective is that the cost of making the wrong decision is not the health lost, as it is for the health system, but differences in the expected revenue to the company from the use of the technology. Hence from a commercial perspective the value of additional research is a function of the change in the likelihood of a positive decision given additional research, the net revenue per utilization of the technology, and the expected volume of utilization of the technology. Thus, the payoff function for research is different from a commercial perspective, but the principles of characterizing the impact of research on the expected cost of making the wrong decision are the same.

11.6 CONSIDERING VALUE ASSESSMENT IN TRANSLATIONAL RESEARCH: THE VALUE ENGINEERED TRANSLATION FRAMEWORK

The increasing utilization of cost effectiveness analyses in reimbursement processes in North America and Europe means that investors will increasingly wish to consider the likelihood of positive reimbursement when deciding where to put their scarce translational dollars. Innovators will need to provide evidence that there is sufficient scope for the production of health compared to likely alternative technologies and at a cost that will be attractive and financially sustainable for health care payers. The Value Engineered Translation (VET) Framework has been developed as a structured approach to the initial assessment of the scope for a cost-effective technology and a prospective approach to the design of an efficient translational research strategy for those technologies that pass the initial scoping triage [19].

Fig. 11.4 illustrates the main components of the VET framework. It consists of three Phases. The first phase consists of a set of headroom analyses. The assessment of effectiveness headroom considers the total health gain available for a curative technology compared to the lifetime health trajectory for patients treated with (1) currently available technologies; and (2) technologies that are in development for the indication of interest and might be on the market by the time the candidate innovation received marketing approval. This component of the headroom analysis is concerned with the

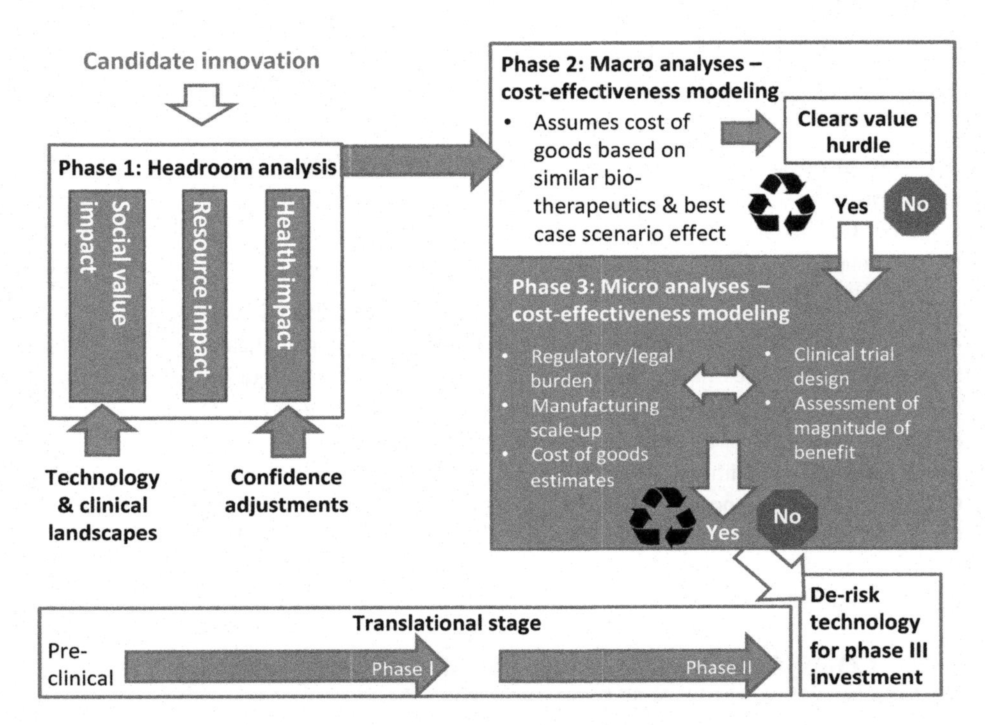

■ **FIGURE 11.4** The value engineered translation framework.

question "Is there sufficient room for improvement in health outcomes to make the indication an attractive commercial proposition?".

The second component of the headroom analysis considers whether there is sufficient opportunity for resource savings to the health system budget from an innovative technology to contribute to the commercial attractiveness of the candidate technology. Unlike conventional regulatory approval processes that will trade of (within reason) safety and efficacy, health technology appraisal processes also consider cost savings, primarily because cost savings can be reallocated to fund health care for others. Hence, it is appropriate to add the resource impact potential to the conventional effectiveness headroom analysis.

The final component of the headroom analysis is the consideration of social values. Policy documents from HTA organizations make it clear that that the value attached to health gain can be affected by a range of considerations such as the severity of the disease, the socioeconomic characteristics of the recipients and broader public policy objectives. As a result, the value headroom associated with a given clinical effectiveness may vary. A comprehensive assessment of the scope for a commercially successful innovation in the context of HTA should systematically assess whether any specific value premia are relevant for a specific technology/indication dyad.

Given the rapidly changing nature of health care, the headroom assessments draw upon clinical and technology landscaping [20,21] to construct likely scenarios at the time of regulatory approval, rather than assuming that the future is well represented by current practice. The outputs from the Headroom Analyses are a mixture of quantitative and qualitative, which while they may be able to rule out investments in certain technologies, will primarily be structured informational inputs to investment decision-making processes.

The second phase of the VET process involves the development of a "first pass" estimate of the expected cost effectiveness of the candidate technology compared to the technologies that may be on the market at the time of regulatory approval. Given the early stage of development, these analyses will assume that the innovative technology can be delivered at similar costs to current technologies and will use formally elicited expert judgements regarding the likely effectiveness of the new and sometimes the expected comparator technologies. The estimates of the cost effectiveness will then be compared to current, and credible estimates of future reimbursement thresholds in the major target markets such as the United Kingdom, Europe, Japan, and Australasia. The function of the second phase is again to provide information that may help investors to triage out technologies that are unlikely to be good value in enough markets (health care payers) to generate the volume of sales required for commercial viability.

The third phase entails the detailed assessment of multiple components of the translational process, including the regulatory pathway for both safety and efficacy and product manufacture; the proposed method and organization of manufacturing capacity (eg, scale up to a few large manufacturing facilities vs scale out of a network of local manufacturing facilities); the target magnitude of benefit to be demonstrated by the clinical trial programme and the associated design of that programme. All this information is then synthesized to characterize the likely scale of investment required by the later stages of the translational programme and the time horizon over which those resources will be required.

Particular attention should be paid to the design of the clinical trial programme and how this feeds into the likelihood of obtaining positive reimbursement decisions in major markets at specific prices, and whether it is likely to be conditional or open. This information can enable investors to consider the tradeoffs between duration and cost of premarket research, time on market-on-patent, price and scope of market access.

The overall objective of the framework is to provide investors with a comprehensive account of the risks and expected returns from investing in the later stages of translation, when commercial success is dependent upon meeting value based reimbursement thresholds.

11.7 POTENTIAL CHALLENGES TO NOVEL CELL THERAPY TECHNOLOGIES IN ACHIEVING REIMBURSEMENT

While recognizing that reimbursement processes and decisions vary between member countries, lessons may be taken from the early experience of some reimbursement authorities. In its review of Provenge, NICE indicated Dendreon (now Valeant) was unable to show that its immunotherapy performed better than already-available treatments, and unlike current treatments, it did not delay the progression of the disease [22]. In other words, it did not perform at a level to justify a premium price when compared to existing therapies. Nevertheless, despite coming to a similar conclusion initially, Germany's Institute for Quality and Efficiency in Health Care (IQWiG) reversed this conclusion, but did so with the proviso that the added benefit of Provenge could not be quantified [23]. While we cannot know what these unquantifiable added benefits may be, the policy literature would indicate that support for innovation may well have been what IQWiG had in mind. Historically, NICE has explicitly endorsed a value premium for innovation [24] but recent empirical research found that the UK public did not support such a premium [25]. Developers of regenerative medicine technologies may be well advised to seek prices that do not rely on such a premium in the UK market, which will have implications for the price they can obtain in up to 25% of the global market due to the UK's role as a reference price market [26]. This may have implications for the value, to manufacturers, of investing in later stage research studies and their choice of trial outcome measures.

While these observations apply to most innovative health technologies, there may be specific challenges, for regenerative medicine and specially cellular therapies such as MSCs. The current high cost of manufacture will unavoidably have a negative impact upon the cost effectiveness of regenerative technologies. While we would expect these costs to reduce over time, in the same way that the costs of goods for biologics have, first mover technologies in this space would be ill-advised to target indications where there are already effective therapies, as this will make it harder to generate the magnitude of QALY gains required to justify the expenditure.

At the time of regulatory approval cellular therapies are unlikely to have substantial evidence of medium and long-term clinical effectiveness and safety [27]. As a result, the value proposition that will be presented to reimbursement authorities will be subject to substantial uncertainty. Under these circumstances, reimbursement authorities are increasingly using patient access schemes and conventional price discounts to manage the risks associated with adopting uncertain but potentially high value technologies [28]. A rational evaluation of the investment opportunities presented by regenerative medicine technologies should consider alternative revenue flow models, including lower initial prices, reduced patient populations, and conditional payment agreements. These analyses should include scenarios that take account of the "spill over" effects of reimbursement decisions by sophisticated payers such as NICE, on the decisions of other payers.

Some cellular therapy products are aiming to be curative. For these products, the challenge of short-term data to support long-term value claims is particularly acute. The early examples of such regenerative therapies are likely to be for extremely rare genetic diseases. In these circumstances, the reimbursement challenge will be a double whammy of high cost of goods and a value proposition based upon extensive extrapolation of short-term data, for example, 12 months or less follow-up. The challenging nature of adopting these therapies is now recognized by both licensing and reimbursement authorities, leading to discussions around conditional licensing and the coordination of regulatory and reimbursement decision making [29,30]. The success of these developments is likely to be pivotal to the long-term success of curative regenerative therapies; hence, investment appraisal of investment opportunities in regenerative medicine would be advised to monitor developments closely and incorporate these novel models of technology adoption into their analyses.

11.8 SUMMARY

The utilization of HTA processes is increasingly prevalent in all major health care systems. As a result, high-cost innovative therapies—such as those being pursued by many in the regenerative medicine space—are inevitably seen as high-risk investments by the venture capital communities. Innovators wishing to attract investment need to address the issue of evidence of value that goes beyond clinical utility if they are to attract the scale of investment their technologies required. To do this, they must understand the nature of the value assessment that their technologies will be subject to prior to market access and adopt a systematic approach to addressing the evidence requirements for those assessments. In this chapter, we have described the intellectual framework underlying HTA as a process for making reimbursement decisions in the presence of limited health care budgets, and described a formal process for the early assessment of candidate innovations and the design of research and development processes that meet the needs of HTA organizations.

REFERENCES

[1] European Network on Health Technology Assessment. What is HTA? http://www.eunethta.eu/faq/Category%201-0#t287n73.

[2] Bending MW, Smith TA. Value and Multiple Criteria used in OECD Countries' Medicine Reimbursement Decision-making Processes using Health Economic Evidence. ISPOR Berlin. November 2012.

[3] Drummond M, Sculpher M, Torrance G, O'Brien BJ, Stoddart GL. Methods for the economic evaluation of health care programmes. Oxford: OUP; 2005.

[4] Paulden M, Stafinski T, Menon D, McCabe C. Value based reimbursement decisions for orphan drugs: a scoping review and decision framework. PharmacoEconomics 2015;33(3):255–69.

[5] Paulden M, O'Mahoney JF, Culyer AJ, McCabe C. Some inconsistencies in NICE's consideration of social values. PharmacoEconomics 2014;32(11):1043–53.

[6] Bending MW. The application of the cost effectiveness threshold in five-countries. *UCL Cost effectiveness threshold workshop MAPI* December 2014 London.

[7] McCabe C, Claxton K, Culyer AJ. The NICE cost effectiveness threshold: what it is and what that means. PharmacoEconomics 2008;26(9):733–44.

[8] Krol M, Brouwer W, Rutten F. Productivity costs: past, present and future. PharmacoEconomics 2013;31(7):537–49.

[9] Dixon S, Walker M, Salek S. Incorporating carer effects into economic evaluation. PharmacoEconomcs 2006;24(1):43–53.

[10] Caro J, Briggs AH, Siebert U, Kuntz KM. Modelling Good Research Practices Overview: a report of the ISPOR-SMDM modeling good research practices task force. Med Decision Making 2012;32(5):667–77.

[11] Claxton K, Ginnelly L, Sculpher M, Philips Z, Palmer S. A pilot study of the use of decision theory and value of information analysis as part of the NHS health technology assessment programme. Health Technol Assessment 2004;8(31):1–103.

[12] Wilson E. A practical guide to value of information analysis. PharmacoEconomics 2015;33(2):105–21.

[13] Culyer AJ, McCabe C, Briggs A, Claxton K, Buxton M, Akehurst R, et al. Searching for a threshold, not setting one: the role of the National Institute for Clinical Excellence. J Health Services Res Policy 2007;12(1):56–8.

[14] Paulden M, O'Mahoney JF, McCabe C. Determinants of changes in the cost effectiveness threshold; PWP 2015_01. PACEOMICS 2015 Available at: http://paceomics.org/wp-content/uploads/2015/10/PACEOMICS-Working-Paper-PWP-2015_01.pdf.

[15] Edlin R, McCabe C, Hulme C, Hall P, Wright J. Cost effectiveness analysis for health technology assessment: a practical course. Auckland: Springer; 2015.

[16] Claxton K. The irrelevance of inferentd: a decision making approach to the stochastic evaluation of health care technologies. JHE 1999;18(3):341–64.

[17] Claxton K, Martin S, Soares M, Rice N, Spackman E, Hinde S, et al. Methods for the estimation of the National Institute for Health and Clinical Excellence cost effectiveness threshold. Health Technol Assessment 2015;19(14):1–503.

[18] Hall PS, Edlin R, Kharroubi S, Gregory W, McCabe C. Expected net present value of sample information: form burden to investment. Med Decision Making 2012;32:E11–21.

[19] Bubela T, McCabe C. Value engineered translation for regenerative medicine: meeting the needs of health systems. Stem Cells Dev 2013;22(Suppl. 1):89–93.
[20] Euroscan International Network. Euroscan international network. 2015 (April 7). 2015.
[21] Packer C. Regenerative medicine, horizon scanning and the NHS. Presented at: Regenerative Medicine Translation – A Pathway into Clinical Practice. London (2014).
[22] NICE. Sipuleucel-T for treating asymptomatic or minimally symptomatic metastatic hormone-relapsed prostate cancer. 2015.
[23] Al Idrus A. Germany's cost watchdog decides Valeant's Provenge has added benefit after all. FierceVaccines, 2015.
[24] NICE. Guide to the methods of health technology appraisal. 3rd ed. London: NICE; 2008.
[25] UK Houses of Parliament: Parliamentary Office of Science and Technology. Value based assessment of drugs. (4), 2015, 1–4.
[26] Minhas R, Moon JC. The office of fair trading report: a prescription for value-based drug pricing. J R Soc Med 2007;100(5):216–8.
[27] Crabb N. Evaluation and reimbursement of regenerative medicines and cell therapies. Presented at: Regenerative Medicine Translation – A Pathway into Clinical Practice. London. 2014.
[28] Stafinski T, McCabe C, Menon D. Funding the unfundable. PharmacoEconomics 2010.
[29] EMA. European Medicines Agency launches adaptive licensing pilot project. 2014.
[30] Sipp D. Conditional approval for regenerative medicine products: Japan lowers the bar. Cell Stem Cell 2015;16(4):353–6.

Chapter 12

Mesenchymal Stromal Cells: Clinical Experience, Challenges, and Future Directions

P. Hematti

Department of Medicine, University of Wisconsin-Madison, School of Medicine and Public Health, Madison, WI, United States; University of Wisconsin Carbone Cancer Center, Madison, WI, United States

CHAPTER OUTLINE

Mesenchymal Stromal Cells. http://dx.doi.org/10.1016/B978-0-12-802826-1.00012-X

12.1 INTRODUCTION

The cells currently known as mesenchymal stromal cells (MSCs), were originally described by Friedenstein et al. more than four decades ago as a fibroblast-like population of cells that could be cultured from the bone marrow (BM) of rodents and when implanted had the ability to generate rudimentary bone tissues that were capable of supporting hematopoiesis [1]. Later, similar cells were derived from human BM [2] and shown to be capable of supporting hematopoiesis in long-term culture assays [3]. Thus, not surprisingly one of the earliest clinical uses of MSCs had been for their observed hematopoietic supportive activities. Later discoveries indicating the potential role of MSCs in the regeneration of many other tissues paved the way for their widespread use to treat many different types of human diseases [4]. These cells are thought to migrate to sites of tissue damage and inflammation and contribute to the healing process [5]. While there are many different proposed mechanisms of action for the regenerative potential of MSCs, this process, most probably, works by the secretory activity of these cells to generate many bioactive molecules that can promote tissue repair, and by their capability to modulate many different types of immune cells to suppress inflammation. Other attributes of these cells that have made them suitable candidates for cell therapies include their relative ease of production from small amounts of (different) tissue sources, and their presumed lack of immunogenicity obviating the need for HLA-matching before their use. While the characteristics and potentials of MSC therapeutics are very attractive many questions remain on the basic research, translational aspect, and clinical applicability of these cells. This chapter will summarize our current clinical experience with using these cells and the issues that need to be considered and resolved for the field to progress.

Since the first reported clinical experience with MSCs in 1995, hundreds of clinical trials have been conducted and the interest in their use in both clinical and veterinary medicine has been exploded. While we have learned a great deal about the clinical safety and potential efficacy of these cells, there are still a lot of works that need to be done to determine the ultimate place of these cells in the field of regenerative medicine. For example, since these cells exist only in very small numbers within the BM, clinical MSC production usually begins by density gradient separation of BM and plating the isolated mononuclear cells in specially formulated MSC medium allowing for their growth [6]. However, still there is no reliable and widely accepted cell surface marker for the identity or direct isolation of them from the BM aspirates or other tissues; and although different groups have proposed certain markers for isolation of these cells, such as Stro-1, CD271, and CD146 marker, none of them have been investigated in depth or widely accepted in the field [7]. Another issue deals with the heterogeneity of the MSCs for clinical use, because to obtain sufficient number of MSCs for therapeutic use they have to be expanded ex vivo to a significant degree. However, since the propagation methods vary widely between groups, the resulting final end product could be biologically very different. To provide some guidelines, in 2006 the International Society for Cellular Therapy (ISCT) recommended a set of criteria for the characterization of MSCs. These include expression of a certain set of cluster of differentiation (CD) markers (ie, CD105, CD73, CD90) and lack of expression of certain other markers that are usually associated with hematopoietic lineages (ie, CD45, CD34, CD14, CD19, and HLA-DR) [8]. It should be noted that although these criteria define a seemingly homogenous population of cells, the population of MSCs characterized this way are indeed a heterogeneous population of cells with different functional capabilities [9]. All these fundamental uncertainties could have major implications in the design, implementation, and interpretation of our clinical experience with MSCs. This chapter provides an overview of the experience in using ex vivo culture-expanded MSCs in the clinical medicine, and the biological and practical issues that could impact their ultimate utilization.

12.2 CLINICAL EXPERIENCE WITH MSCs

Hematopoietic stem cell (HSC) transplantation has been the oldest and most established form of stem cell therapy and has been the standard of care for many diseases such as hematologic malignancies or immunodeficiency disorders [10]. Since MSCs had long been considered to be an important cellular constituent of the BM stromal microenvironment that presumably supports hematopoiesis [11] Dr Lazarus and his team of HSC transplant physicians were the first to test MSCs in a clinical setting in 1995. Specifically they tested the safety of MSCs by infusing autologous BM-derived MSCs that were culture-expanded ex vivo over several weeks [12]. This phase-I clinical study did not show any infusional toxicity or any other adverse reactions after infusions of cells that had undergone significant ex vivo expansion, and thus paved the way for further examination of MSCs for their potential beneficial therapeutic effect. The hope that the cotransplantation of MSCs with HSCs would lead to faster engraftment, and immune reconstitution led this same group to investigate the use of ex vivo culture-expanded autologous MSCs for accelerating the recovery in breast cancer patients after autologous HSC transplantation [13], and subsequently, the safety and efficacy of using allogeneic BM-MSCs from human leukocyte antigen (HLA)-identical sibling donors into their respective allogeneic HSC transplant recipients [14]. Although these studies showed the safety of the infusion of culture-expanded, either autologous or allogeneic, MSCs as a form of cellular therapy, efficacy in accelerating engraftment of HSCs could not be demonstrated conclusively due to the uncontrolled nature of those pilot studies. Since then larger clinical trials have been conducted but the clinical effectiveness of MSCs in the context of HSC transplantation to enhance engraftment has not been conclusively proven yet [15,16].

Next wave of studies that showed the ability of MSCs to modulate the immune responses, both in vitro and in vivo, generated the impetus for their use as a cellular immunotherapy for a wide range of immune related disorders. Importantly, these immunomodulatory properties of MSCs appear to be independent of the level of HLA matching between the donor and recipient [17]. One common complication of HSC transplantation is graft versus host disease (GVHD), in which the transplanted immune cells attack the immune compromised recipient leading to significant morbidity and mortality. Lee et al. was the first to suggest that cotransplantation of MSCs together with HSCs from a haploidentical donor played a role in prevention of GVHD in a patient with acute myeloid leukemia [18]. However, Le et al. were the first to report the actual successful treatment of a case of very sever and refractory GVHD in a 9-year-old boy who had previously received a HLA-matched, unrelated donor HSC transplant for his leukemia using MSCs generated from the BM of his mother [19]. A subsequent larger study from the same group further verified the safety of MSC administration to patients with steroid-refractory GVHD, irrespective of their HLA status [20], with a significantly improved survival rate in the group that responded to MSC infusion. Indeed, the GVHD could be considered the indication for which MSCs had been studied in the most comprehensive manner as Osiris Therapeutics Inc. (Columbia, MD) has already performed two Phase III, double-blind, placebo-controlled, randomized trials with their proprietary formulation of MSCs (tradename Prochymal). Their MSCs were derived from the BM of a single healthy donor and used in two double blind, controlled randomized studies for the treatment of steroid refractory acute GVHD and as a first-line of therapy for acute GVHD. While subsets of patients with liver or gastrointestinal GVHD showed an improved response rate in the refractory acute GVHD trial, the primary endpoint of prolonged survival in the treated adult patients did not reach statistical significance [10]. However, its use in pediatric patients with severe refractory acute GVHD were shown to be more effective [21] and has led to conditional approval of Prochymal for only pediatric patients with GVHD in Canada and New Zealand [22]. Despite this seemingly negative study, numerous investigators keep studying MSCs (derived from BM or other tissues) for the treatment and/or prevention of GVHD [10]. However, these other clinical trials are still mostly including only a small number of patients. Direct comparison between these smaller, but seemingly more successful, trials are challenging because the clinical protocols could be different, including in their patient population, dose, frequency, and timing of MSC administration. There is always some degree of heterogeneity in the production protocol for MSCs, from the original material used to the passage and type of media used for cell production.

Currently, utilization of MSCs has expanded far beyond transplantation medicine and has attracted attention for numerous tissue repair and regeneration indications in a wide variety of clinical indication. MSCs are being investigated in more than 300 phase I-II clinical trials for a variety of therapeutic applications such as myocardial infarction, heart failure, chronic obstructive pulmonary disease, multiple sclerosis, amyotrophic lateral sclerosis, stroke, spinal cord injury, metachromatic leukodystrophy, Hurler disease Crohn disease, diabetes mellitus, systemic sclerosis, systemic lupus erythematosus, and refractory wounds among numerous others [23–33]. Use of MSCs in transplantation medicine has also extended to the field of solid organ transplantation. Perico et al. were the first to report the safety of use of autologous MSC generated from sternal bone marrow aspirates 4 months prior to patients receiving their living, related donor kidney transplants [34]. Later, Tan et al. reported results from a large, prospective, open-label, randomized clinical trial on end-stage renal disease patients that received a kidney from a living, related donor and showed autologous MSCs reduced the incidence of acute rejection and preserved kidney function after 1 year [35]. It should be also noted that since one of the defining characteristics of MSCs is their capabilities to differentiate into mesodermal tissues (specifically into adipocytes, osteoblasts, and chondrocytes) they have been, very early, and heavily investigated for treatment of localized or systemic musculoskeletal disorders, such as osteogenesis imperfecta [36]. However, it should be also noted that despite huge interest in use of MSCs for musculoskeletal repair only 16% of reported cell therapy protocols for cartilage repair actually used MSCs, while in the majority of them the actual type of cells that were used were chondrocytes [37]. In general, similar to use of MSCs in HSC transplantation, verification of these results in large randomized phase III clinical trials are needed before any definite conclusions about their clinical efficacy could be made.

12.3 CLINICAL CONSIDERATIONS IN USE OF MSCs

12.3.1 Intended Mode of Action: Tissue Regeneration Versus Immune Modulation

Based on the original work of Friedenstein, the clinical application of MSCs was originally focused on their hematopoietic supportive and/or their bone and cartilage forming potential. One of the major motivations for the sudden expansion of the use of MSCs beyond this narrow field was their perceived ability to not only differentiate into cells of mesodermal lineage but into cells of endodermal and ectodermal lineages, such as neurons [38], cardiac muscle cells [39], lung epithelial cells [40], liver cells [41], and pancreatic islet cells [42], named "transdifferentiation." However, by the time that the validity of these original findings were challenged, hundreds of patients had already received MSCs through mainly small clinical trials and, interestingly, many of them reported encouraging results [43–45]. One potential explanation for these promising results, despite the failed verification of basic science behind them, could be the small scale and the uncontrolled nature of those trials. On the other hand, our knowledge about the potential mechanisms of action of MSCs in different disease conditions has also changed considerably over the last decade [46]. So, many of the observed effects could be explained by the newly discovered modes of action that is not related to their multipotentiality, but instead due to their paracrine effects including secretion of cytokines, angiogenic and growth factors, extracellular matrix proteins, and microvesicles such as exosomes. Although MSCs were presumed to preferentially home to BM niches after intravenous infusion [47], later experimental models have shown that MSCs can be detected at low levels in many tissues after their intravenous administration [48]. Indeed, evidence indicates that MSCs preferentially migrate and home to sites of tissue injury and inflammation, and then by secreting trophic factors they promote recovery of damaged tissues, and furthermore induce proliferation of resident progenitors to further promote tissue regeneration [49]. All these newer findings could help to explain why MSCs could be beneficial in such a wide range of systemic diseases.

Originally, it was also assumed that integration of the MSCs into patient tissues and there continued presences were needed to achieve a long lasting therapeutic effect. However, due to expansion of our knowledge about the potential mechanisms of action of these cells such as their secretory and immunomodulatory effects, it is now believed that persistence of the infused MSCs in the recipient tissues is not necessary to yield a therapeutic effect. For example, recent animal studies have shown that while most intravenously infused MSCs are trapped in the lungs they can still induce beneficial whole body systemic effects, including on the heart, via their paracrine effects [50]. These new knowledge about potential mechanism of action of MSCs could influence design of clinical trials on many levels; from the type of disease treated to the route and frequency of MSC administration (ie, direct intracardiac injection versus intravenous infusion for repair of myocardial damage). Furthermore, it could also support the notion that the therapeutic outcomes may be enhanced and or sustained for longer periods by administering repeated doses of MSCs.

MSCs have been shown in vitro and in vivo to be able to suppress the excessive immune and inflammatory responses of lymphocytes (T, B, and NK cells), dendritic cells and macrophages and favor the formation of regulatory T (and other type of regulatory) cells. For T cells, MSCs have been shown to suppress the proliferation of effector T cells in response to various cellular and nonspecific mitogen stimulation [51] and hinder antigen recognition of naive and memory T lymphocytes [52]. MSCs have also been shown to increase the level of regulatory CD4+ CD25 high FoxP3+ T cells at least in part to the production of TGF-beta [53]. In addition to T cells, MSCs have shown to inhibit B cell differentiation and proliferation [54,55]. However, in contradictory reports MSCs have also been shown to induce proliferation and maturation of B cells [56]. MSCs have been shown to inhibit proliferation and differentiation of NK cells [57]. It has also been shown that MSCs induce regulatory phenotype and suppress differentiation of dendritic cells [58,59]. Many of these effects are exerted via secreted factors such transforming growth factor beta (TGF-beta), hepatocyte growth factor (HGF), prostaglandin E2 (PGE2), and indoleamine 2,3-dioxygenase (IDO).

In addition to our new discoveries about the new potential mechanisms of action for MSCs, our understanding of the pathophysiology of many diseases is also changing. Importantly, we now know that the role of inflammatory and immune cells in many human diseases is much greater than previously appreciated. Thus, observation of beneficial effects even when there is very low level of engraftment of MSCs could be potentially explained by their secretory-based antiinflammatory properties. More recent studies suggest that one of the most important antiinflammatory properties of MSCs could be mediated through the macrophages. A variety of in vivo and in vitro studies have shown that a major mediator of MSC effect could be through their modulation of immunophenotype and functional properties of macrophages [60–62], essentially polarizing the immunophenotype of macrophages toward alternatively activated or regulatory macrophages. Because macrophages are the major mediators of inflammation and its subsequent resolution in all the tissues in the body, we propose that the effect of MSCs on macrophages could provide a unified view on the potential mechanism of action of MSCs in such a wide range of clinical conditions, as MSCs could change the repertoire of inflammatory cells in damaged tissues and thus prevent inflammation-induced damage or promote the generation of tissue regenerating macrophages.

12.3.2 Donor Source of MSCs

Many studies have used MSCs derived from tissues of the patients themselves, but evidence of their immune-privileged status and clinical experience indicate that MSCs from HLA-matched donors or even from third party allogeneic donors without any consideration for HLA-matching may be safely used. Autologous MSCs could potentially avoid any concern about tissue incompatibility and rejection, however it could be argued that MSCs generated from the patients might not be the best source, as these MSCs may be also afflicted by the patient's primary disease depending on the disease and tissue source affected. While most studies show that MSCs derived from patients versus normal healthy donors had similar phenotypes such as growth properties, differentiation potential, and surface immunophenotype, there were also intrinsic abnormalities found in the patients' MSCs [63]. Furthermore, MSCs from normal healthy MSC donors may possess therapeutically desirable donor-specific attributes. For example, harvesting MSCs from young allogeneic donors provide another potential advantage in that they have been shown to grow faster and have an increased capacity to differentiate compared to older donors [64]. Importantly the generation of MSCs in sufficient amounts could take several weeks to months, making the use of autologous MSCs prohibitive in clinical scenarios such as acute phases of myocardial infarction or GVHD, where cells are urgently needed. Overall, the use of MSCs from allogeneic donors has been shown repeatedly to be very safe and do not cause infusional toxicity or long term complications. Therefore, the production and use of characterized material isolated from qualified healthy allogenic donors that can be banked for future use is an attractive and practical source of MSCs for clinical purposes.

12.3.3 **Tissue Source for MSCs**

MSCs have now been derived from many adult tissues, such as adipose tissue, pancreas, heart, lung, and dental pulps [65], neonatal tissues such as placenta and amniotic membranes [66], fetal tissues such as liver, lung, and blood [67], and even embryonic stem cells [68]. Thus, not surprisingly, the source of MSCs used in clinical trials has now expanded to include, and the list is growing, adipose tissue, cord blood, and placenta; however, BM has remained the source of cells in more than 50% of clinical trials overall [69]. Nevertheless, due to the ease and availability of harvesting, use of MSCs from adipose tissue has been investigated for GVHD too [70]. However, it should be noted that adipose tissue MSCs has generated highest interest in the fields of plastic and reconstructive surgery [71]. It is important to note that biological material widely used in cosmetic surgery such as fat grafts, lipoaspirates, or stromal vascular fraction (SVF) derived from lipoaspirates, contain a large number of other types of cells in addition to stromal cells and thus, very importantly, should not be considered equal to ex vivo culture expanded MSCs from adipose tissues. Other types of non-BM-MSCs that have been investigated in the clinic include umbilical cord-derived MSCs

(UC-MSCs) for the treatment of systemic lupus erythematosus [72] and placenta-derived MSCs for the treatment of GVHD [73].

Despite phenotypic and immunologic similarities between MSCs from different tissues, it should be noted that there are differences in proliferative and differentiation potential, immunomodulatory activity, transcriptome, and proteome [74,75]. However, whether these differences are therapeutically important continues to be a matter of debate and investigation. As we have theorized before, one proposition is that MSCs derived from the same organ that is targeted for therapy could provide clinically significant advantageous compared to MSCs derived form a heterotypic organ [76].

12.3.4 Cell Dose, Timing, and Frequency

There has been a wide range dosage of cells that have been used therapeutically, but typically, they are between 0.5×10^6 and 5×10^6 MSCs/kg body weight of the recipient. Alternatively, in some studies, mostly for practical purposes, a set amount of cells, for example, 200×10^6 MSCs per patient has been administered. Nevertheless, no clear dose response has been defined yet and the maximum tolerated dose of MSCs without adverse side effects has not yet been established either. The optimal dose of MSC would most likely depend on many factors such as on the disease, age of the patient, route of administration, tissue source of MSCs, and potentially other variables such as the potency of MSCs generated. One clinical study in which patients with de novo acute GVHD were either treated with a low dose (2×10^6 MSCs/kg) versus a high dose (8×10^6 MSCs/kg) showed no difference in respect to safety and efficacy [77]. Interestingly, another study using MSCs to treat cardiac ischemia showed that the lower doses, and autologous, MSCs were actually more effective than the higher doses [78]. Indeed, based on in vitro assays such as mixed lymphocyte cultures, MSCs in high concentrations could inhibit T cell proliferation but at low concentrations may indeed stimulate lymphocyte proliferation [17]; however, the clinical implication of this experimental observation in actual clinical scenarios is not clear.

In addition to dose, another unresolved issue is the timing of MSC administration. For example, should MSCs be administered in the early phase after onset of disease, such as in the acute stages of myocardial infarction, or is it safer and more practical to delay infusion of MSCs till the patient is completely stable. The published data, mostly in preclinical studies, are conflicting; in one murine study, MSCs infused on the day of bone marrow transplantation did not prevent GVHD but when cells were infused on day 2, they significantly reduced mortality [79]. In contrast, two preclinical studies in kidney transplantations showed a better outcome with better kidney function when the MSCs were administered 1 day before transplantation compared to 1 week afterward [80,81].

Another clinical consideration is whether repeated injections of MSCs would provide additional benefit over a single treatment. In general it appears that repeated doses of MSCs at different intervals give a better more long lasting outcome [23,24]. However, interpretation is difficult since the intervals between treatment times are typically not consistent between most studies.

12.3.5 Route of Administration

The reported routes of MSC administration are systemic (intravenous) or locally such as intraarterial, intracoronary, intracardiac, intraperitoneal, intramuscular, and intraarticular, where the injections occur at or near the site of tissue damage. The routes of delivery are typically determined empirically, based on the location of the disease, and to optimize the traffic and accumulation of MSCs at the desired site. However because little is known about the pharmacology of the MSCs injected in the body, the best route for a particular disease such as cardiac disease is unknown [82]. Overall, larger randomized studies that address issues such as dosage, treatment schedule, and route of administration are needed to optimize the therapeutic effects of MSCs in clinical trials. However, conduction of such studies could be cost prohibitive and challenging.

12.3.6 Combination of MSCs with Other Agents Including Immunosuppressive Drugs

It should be noted that most patients who receive MSCs are already receiving standard of care therapy for their disease in addition to many other medications for secondary issues. For example, all transplant patients who are infused with MSCs often receive immunosuppressive medications but can also be on other medications such as antibiotics. So it is important to (1) be aware of the potential effect of these drugs on the viability and/or biological function of the MSCs and (2) conversely recognize that the pharmacokinetics of these concomitant drugs may be influenced by the MSCs. It is also important to understand that MSCs and concomitant drugs might have synergistic and/or antagonists effects on each other as a whole or on their specific properties. A number of studies have investigated the effect of immunosuppressive drugs on biological and especially immunomodulatory properties of MSCs in vitro and also their interactions with MSCs in vivo, sometimes with contradictory results [83,84]. Further, complicating the matter is the fact that use of MSCs in combination with other drugs could make it difficult to discern the exact contribution of each component of the treatment to the observed clinical outcome.

12.3.7 Immunogenicity of the Cells

It is now obvious that the infusion of culture expanded MSCs from HLA disparate donors is safe. Use of MSCs without regard for HLA typing was started based on the assumption, and some animal evidence, that these cells are not immunogenic. However, later studies have shown that MSCs do not always suppress proliferation of allogeneic lymphocytes and actually MSCs under certain experimental conditions could function as antigen presenting cells or even activate immune responses [85,86]. Furthermore, the long-term survival of infused MSCs is now very questionable [87], as indeed many recent studies, both in animal models and in clinical trials, suggest that the extent of MSC engraftment is only minimal and not sufficient to explain the observed clinical effects [88]. Although these observations question the validity of the assumption that MSCs are not immunogenic they have not dampened the enthusiasm for clinical utilization of these cells because we have discovered new mechanisms of action for MSCs based on the new paradigm that MSCs exert their effects without a need for persistent engraftment [89]. Indeed, MSCs do not replace damaged cells but instead contribute to their regeneration via indirect effects such as activation of resident progenitor cells. Low survivability of MSCs could also explain why repeated infusions of MSCs could be necessary to achieve a clinical effect. Ironically, lack of long-term engraftment and their ultimate demise after their administration should alleviate the potential safety concerns for the presumed tumorigenicity and ectopic tissue formation capability of these cells.

12.3.8 Safety Issues

Using MSCs as a cellular therapeutic will have inherent safety risks. These include adverse immune effects, immunogenicity, tumorigenicity, and ectopic tissue formation. Although as of today, several thousand patients have received MSCs for a wide variety of indications using different doses, frequency, and different routes of administration, outcomes of most of those trials have yet to be published in the medical literature. Rare adverse events are likely to be identified only from a publicly available database of a large number of patients after a long-term follow-up. In general, concerns regarding the safety of MSCs have been raised only extremely rarely. For example, Ning et al. reported a higher risk of leukemia relapse in patients who received MSCs for prevention of GVHD [90], but such a risk had not been seen in numerous other similar studies and thus raise the question about the validity of that single report. Also, there is the theoretical risk for the development and accumulation of genetic mutations during long-term culture expansion due to replicative senescence [91,92], and also the possibility of infused MSCs to contribute to the growth of existing tumors or promote their metastasis as shown in some experimental murine models [93,94]. However, it is very reassuring that, as of today, there is still no documented case of tumor formation that could be attributed to MSCs [95].

12.3.9 Relevance of Preclinical Models

The utility of MSCs in clinical medicine will be defined preferably through systematic and well designed, randomized clinical trials. However, prior to embarking on enormously expensive clinical trials most investigators depend on experimental preclinical animal models to identify dose, safety, and efficacy, which serve as a guide for a human clinical trial design. However, major differences in biological properties of MSCs derived from different species and differences in animal models limit the predictive value of such models in preclinical studies. Nevertheless, preclinical animal studies and completed human clinical trials have shown that both autologous and allogenic MSC treatment do not produce any significant adverse effects. Still, much more research is required in order to define the appropriate animal model for each clinical indication. For example, MSCs did not show any beneficial effect in a rodent model of systemic lupus erythematosus (SLE) [96] but showed a promising result using a different murine model of SLE [97]. Based on the animal work, the latter group then pursued use of allogeneic bone marrow derived [98,99] and umbilical cord derived MSCs [100] in small cohorts of patients with SLE and generated promising results. While these results need to be verified in larger clinical trials, this observation points to the fact that preclinical models could provide very different results with major implications for pursuing a clinical trial. What is needed is a continuing comprehensive analysis of preclinical models used to support the clinical trials compared to the outcome of the trial to determine the best and most predictive animal model(s) for every medical application for MSCs. This information should be available as more and more preclinical and their respective clinical trials are conducted, published, analyzed, and scrutinized.

12.3.10 Choice of Potency Assays

Despite extensive use of MSCs in the clinical setting, there is still no standard consensus on any general release criteria to qualify functionality and potency of MSCs before their clinical application. One major obstacle in defining the most appropriate potency assay is the fact that MSCs have been used for such diverse conditions as GVHD, stroke, heart attack, and cartilage repair that one potency assay could not be appropriately reflect the potential mechanism of action intended for the specific clinical indication. Consequently there may be different sets of potency assay each appropriate for a different clinical indication. Indeed, much more work is needed to address this very important issue.

Well established and widely utilized assays of MSC function are the trilineage differentiation assays. These assays qualify MSCs by demonstrating they have the capacity to develop into either osteogenic, chondrogenic, or adipogenic lineages. Although, these assays should be mainly relevant for MSCs that are being used to develop into bone and cartilage, they are often used as potency assays for many other clinical applications. We propose if MSCs are intended to be used for the repair of myocardial infarction the use of the osteogenic assay, for example, is probably irrelevant. MSCs that were used for the treatment of GVHD, presumably because of their effect on T cell suppression, were also used for the treatment of myocardial infarction [23]. However, myocardial infarction in contrast to GVHD is not a T cell-mediated disease, so potency assays to measure the T cell suppressive capability of MSCs should not be used as a potency assay for a myocardial clinical trial. Unfortunately, the lack of relevant functional assays specific for their intended application compels researchers to use assays that are conveniently available. Another widely used assay is the colony-forming unit (CFU) assay, which evaluates the self-renewal capacity of the cells and is based on the ability of single MSC plated at a low density to form a colony. In a recent paper, Deskins compared 10 BM derived MSC lines by testing them with variety of in vitro assays. These lines were then implanted in vivo to evaluate their capacity to engraft and form granulation tissue in a murine wound model. They found that three quantitative assays (growth rate, proliferation, and measurement of cellular ATP) when used together correlated well with biological activity in vivo [101]. Nevertheless, more studies of this nature must be conducted to develop a standardized set of quantifiable assays that are reproducible, and can predict the potency of MSCs necessary for a successful clinical outcome.

12.4 MANUFACTURING CONSIDERATIONS FOR CLINICAL USE

12.4.1 Isolation and Culture Medium for MSCs

Considering that there are now hundreds of clinical trials worldwide using MSCs, it is desirable to know that the MSCs generated for each of these studies have somewhat a similar phenotype. Only if MSCs are produced in a similar way, one can more accurately compare the outcomes between different trials, which frequently report contradictory results. Understandably, in addition to the issues related to donor and tissue source, there is considerable heterogeneity in every step of culture methodology [102,103]. Furthermore, only a few large pharmaceutical supported clinical studies generated MSCs under current good manufacturing practice (cGMP) standards, mainly using proprietary methodologies that are not openly available to the scientific community. In contrast, MSCs in many smaller studies were produced in academic center cell processing laboratories based on the preferred or in-house developed protocols. These variations in culture methodologies, and consequently, the end cellular products generated, are a major issue in comparability of the results between different groups. To ensure the reproducibility of safety and efficacy in clinical trials, the steps of the MSC manufacturing process should be more standardized. However, the manufacturing process for MSCs has multiple steps consisting of procurement, isolation, cultivation, and expansion together with quality control assays. All these factors make it very difficult to identify the important factors, which can impact the clinical outcome and it is currently unrealistic to expect that all investigators will come to a consensus on what is the best production methodology. However, it is also possible that this will might change in the future as the field evolves and more standardized and well-defined technologies emerge.

MSCs are present in the mononuclear cell (MNC) fraction of BM with many variations in their initial handling and processing; these include the use of fresh versus frozen BM, Ficoll versus no Ficoll to collect the MNC fraction, using RBC lysis reagent, passage plating density, and the nature of the surface coating material in the culture flasks [104]. Not surprisingly, even the basal growth medium for MSCs is not standardized, as some investigators favor using α-minimum essential medium while others favor using Dulbecco's modified Eagle's medium [105].

Fetal bovine serum (FBS) has historically been considered a major critical component in MSC culture medium (as a source of growth factors, hormones, and nutrients) to generate high quality MSCs in high numbers [106]. However, FBS is a highly complex mixture of biological components that by nature is not well defined and has variations from batch to batch, some crucial to MSC growth. Moreover, the use of FBS carries the risk of transmitting infectious agents, such as spongiform encephalopathy [107]. Consequently, most FBS used in clinical trials are sourced from countries considered free of Creutzfeldt-Jakob disease. Furthermore, there have been some studies indicating that patients who had received MSCs cultured using serum develop antibodies to immunogenic xenogeneic proteins present in the FBS [108]. Although, many regulatory agencies accept the use of qualified FBS in MSC production many in the cell therapy community are trying to replace FBS with non-FBS supplements in the culture media.

One potential alternative to FBS would be use of xeno-free medium by replacing FBS with human-derived supplements, such as human serum, plasma, or human platelet-derivatives. One of the more commonly used substitute for MSC culture is pooled human platelet lysate (HPL) with several human clinical trials using MSCs generated with HPL [109–111]. Although reportedly, MSCs grown with HPL tend to grow faster than using FBS, however, batch-to-batch differences that occur in FBS also can occur between batches of platelets due to the preparation method, platelet or white blood cell content of the batch and even age of the donor [112]. Other alternatives to FBS and HPL include human AB serum, which according to some reports provides an advantage in expansion of MSCs while preserving their differentiation and immunomodulatory properties [113]. In one clinical study, the investigators collected autologous serum via a plasmapheresis procedure and used it alone or combined with HPL for expansion of allogeneic MSCs prior to their infusion [114]. It should be also noted that investigators have described a variety of protocols by adding a wide range of growth factors and hormones such as basic fibroblast growth factors, PDGF, epidermal growth factor, and TGF-β, and dexamethasone [115]. A chemically defined or serum-free medium would be also an ideal alternative for clinical use. Different serum-free medium specifically formulated for MSCs are available with reported differences in their growth characteristics on MSCs [116]. While using human-derived supplements like HPL or other defined culture components could have theoretical advantages over FBS, it remains to be seen whether their use will provide an actual therapeutic advantage. We propose that what matters the most is production of a high quality and clinically efficacious cellular product even if it is based on a more cumbersome and expensive procedure as the effects of many of these modifications on the regenerative properties and immunomodulatory characteristics of MSCs are not adequately studied.

12.4.2 **Cell Density, Oxygen Tension, and Culture Devices**

Besides the MSC isolation and types of growth media, there have been many other differences in the cultivation methods for MSCs used for preclinical and clinical studies. MSCs grow as adherent cells and after reaching a certain level of confluency are passaged and further expanded. Thus, the number of cells that can be harvested in an ex vivo expansion culture is determined by the surface area of the culture platform and cell density of plated cells. Although MSCs were originally expanded in conventional monolayer cultures for preclinical studies multilayered cell factories and bioreactors, including fully automated bioreactors, have been used to produce large quantities of material for clinical use. Although passage numbers are being used as a representative of proliferative age, population doublings is a more accurate reflection of the proliferation history of cells in culture.

Recently, cell density has emerged as a paramount factor in MSC expansion and their functionality, there is no standard in BM-MNC plating density at the initial stage of expansion, which is usually different from plating density at the subsequent passages [117]. Consequently, the number of cell doublings before each passage in each of these growth conditions can vary considerably and the potential for clonal expansion of rapidly growing subsets can lead to a very different population of MSCs depending on the process. It is also important to note that MSCs reach a senescent state, usually after several passages, and become, in theory, unfit for therapeutic applications; however, maximum or optimal length of culture for therapeutic fitness is not yet well established. Therefore, it should be considered that the phenotype of MSCs produced in small scale for preclinical studies may be different compared to that of large scale for clinical studies. Although the choice of cell density could play a critical role in self-renewing and multipotential differentiation capability of MSCs [118], the ultimate role of cell density in evaluating the therapeutic efficacy of MSCs would be difficult to evaluate.

Similarly, oxygen tension has been investigated as a factor with major influence on proliferation of MSCs. It has been suggested that the lower O_2 concentration (eg, 5%), that is closer to physiologic values could have an impact on the proliferation of stem cells in ex vivo cultures [119]. Reports comparing normoxia conditions (20% O_2) with hypoxic conditions (2–9% O_2) indicated that the hypoxic environment improved the growth rate, chemokine expression, and genetic stability of MSCs during in vitro expansion [120]. These results could foster interest to use MSCs generated under low O_2 culture conditions in the future clinical trials.

MSCs are grown as adherent cells and there are many types of plastic flasks produced by different manufacturers that are suitable for MSC expansion by the adherence-based method. However, a systematic investigation of the isolation and proliferation capacity of MSCs cultured on four different types of culture flasks showed they were not all comparable, although it should be noted that this study did not show differences in the quality and functional capabilities of MSCs generated [121]. It should be also noted that the optimal flask type for initial isolation of MSCs, based on their plastic adherence property, could be different from the type of flask that is optimal for MSC expansion. Since MSCs are usually needed in tens to hundreds of millions, understandably expanding them in typical culture flasks could make it a challenging manual task. To reduce the number of culture flasks needed, multilayered culture flasks has been developed that can be fit into the standard usual cell culture incubators and connected by tubes for convenient medium exchange, and cell harvesting. More recently, automated bioreactors have been developed that could lower the labor and cost even further [122].

12.4.3 Storage and Cryopreservation

Cryopreservation of MSCs has not been specifically optimized for MSCs. Indeed most MSCs in clinical use, like many other types of cells, are commonly frozen in 10% dimethyl sulfoxide with an electrolyte solution (eg, PlasmaLyteA) and a protein source (eg, human serum albumin) using a typical rate-control freezer followed by long-term storage in liquid nitrogen. A comprehensive review by Marquez-Curtis et al. compared many different cryopreservation techniques of MSCs and reported that most different freezing protocols, cryoprotectants, and storage conditions can produce acceptably viable cells [123]. Under these conditions, most MSCs were viable after thawing and their phenotype and functional characteristics preserved. However, one study showed that although no significant decrease in viability of MSCs were observed with time and it remained above 70% even after storage for up to 30 months, the MSCs that were frozen for less than 6 months were significantly more suppressive than both fresh MSC and MSCs stored for more than 6 months. This study also showed that if MSCs were expanded from frozen aliquots of BM-MNC they had lower expansion capacity compared to freshly plated MNC, and MSCs from frozen MNC were also less immunosuppressive. [124]. There are also studies to suggest that MSCs should be allowed to recover following thawing to resume their optimal functionality [125], however in this strategy could be difficult to implement in clinical practice.

12.5 CONCLUSIONS

MSCs have been and continue to be investigated extensively in clinical trials for their regenerative and immunomodulatory properties. However, like many other new therapeutic modalities many issues remain to be addressed, including the best methods for cell production, the optimal frequency, dose, and route of administration, and very importantly the most appropriate clinical indications. The collaborative efforts of basic and clinical researchers are essential to advance our understanding of the mechanism of action and the clinical potential of these cells. Conventional small molecule drug testing in clinical trials are fairly well characterized and the heterogeneity of therapeutic responses is based on the absorption, metabolism, and clearance of the drug in each individual patient. These drugs can fail in the late stage trials because the heterogeneity of the patients' population in regard to the disease could become greater. In contrast, cellular products used for clinical trials often possess enormous heterogeneity of the delivered material, that is, MSCs, which will make interpretation of the results even much more complicated. Unlike small molecules or biologicals, cellular products such as MSCs are living drugs that can undergo changes in each step of their production, including their freeze and thaw process. Theoretically, any subtle changes in MSC culture might impact the biology of the cells; however, delineating the ultimate impact of these changes in culture methodology on the clinical outcome remains very difficult to determine. Other factors besides characterization that need to be considered include the optimal treatment regimen such as dose, frequency, and timing of administration in relation to the stage of the disease, and interaction between MSCs and other medications.

The original clinical trials of MSCs in the setting of HSC transplantation provided the reassurance about the safety of these cells and resulted in other disciplines of medicine to enter this field. The impressive safety record of MSCs and lack of any long-term complications is now well known and is a very good foundation for continued development. However, what is still debated is the extent of clinical benefit provided by these cells. Nevertheless, as long as there are no better therapeutic options for the wide range of degenerative and debilitative diseases that afflicts humans the research on MSCs will continue. Systematic and well-designed clinical trials with full attention to production methodologies could provide some degree of consensus regarding the ultimate place of MSCs in the field of clinical regenerative medicine.

REFERENCES

[1] Friedenstein AJ, Petrakova KV, Kurolesova AI, Frolova GP. Heterotopic of bone marrow. Analysis of precursor cells for osteogenic and hematopoietic tissues. Transplantation 1968;6(2):230–47.

[2] Castro-Malaspina H, Gay RE, Resnick G, Kapoor N, Meyers P, Chiarieri D, et al. Characterization of human bone marrow fibroblast colony-forming cells (CFU-F) and their progeny. Blood 1980;56(2):289–301.

[3] Dexter TM. Stromal cell associated haemopoiesis. J Cell Physiol 1982;1:87–94.

[4] Caplan AI. Mesenchymal stem cells. J Orthop Res 1991;9(5):641–50.

[5] Prockop DJ, Oh JY. Mesenchymal stem/stromal cells (MSCs): role as guardians of inflammation. Mol Ther 2012;20(1):14–20.

[6] Pittenger MF, Mackay AM, Beck SC, Jaiswal RK, Douglas R, Mosca JD, et al. Multilineage potential of adult human mesenchymal stem cells. Science 1999;284(5411):143–7.

[7] Keating A. Mesenchymal stromal cells: new directions. Cell Stem Cell 2012;10(6):709–16.

[8] Dominici M, Le Blanc K, Mueller I, Slaper-Cortenbach I, Marini F, Krause D, et al. Minimal criteria for defining multipotent mesenchymal stromal cells. The International Society for Cellular Therapy position statement. Cytotherapy 2006;8(4):315–7.

[9] Ho AD, Wagner W, Franke W. Heterogeneity of mesenchymal stromal cell preparations. Cytotherapy 2008;10(4):320–30.

[10] Battiwalla M, Hematti P. Mesenchymal stem cells in hematopoietic stem cell transplantation. Cytotherapy 2009;11(5):503–15.

[11] Almeida-Porada G, Flake AW, Glimp HA, Zanjani ED. Cotransplantation of stroma results in enhancement of engraftment and early expression of donor hematopoietic stem cells in utero. Exp Hematol 1999;27(10):1569–75.

[12] Lazarus HM, Haynesworth SE, Gerson SL, Rosenthal NS, Caplan AI. Ex vivo expansion and subsequent infusion of human bone marrow-derived stromal progenitor cells (mesenchymal progenitor cells): implications for therapeutic use. Bone Marrow Transplant 1995;16(4):557–64.

[13] Koc ON, Gerson SL, Cooper BW, Dyhouse SM, Haynesworth SE, Caplan AI, et al. Rapid hematopoietic recovery after coinfusion of autologous-blood stem cells and culture-expanded marrow mesenchymal stem cells in advanced breast cancer patients receiving high-dose chemotherapy. J Clin Oncol 2000;18(2):307–16.

[14] Lazarus HM, Koc ON, Devine SM, Curtin P, Maziarz RT, Holland HK, et al. Cotransplantation of HLA-identical sibling culture-expanded mesenchymal stem cells and hematopoietic stem cells in hematologic malignancy patients. Biol Blood Marrow Transplant 2005;11(5):389–98.

[15] Sanchez-Guijo FM, Lopez-Villar O, Lopez-Anglada L, Villaron EM, Muntion S, Diez-Campelo M, et al. Allogeneic mesenchymal stem cell therapy for refractory cytopenias after hematopoietic stem cell transplantation. Transfusion 2012;52(5):1086–91.

[16] Le Blanc K, Samuelsson H, Gustafsson B, Remberger M, Sundberg B, Arvidson J, et al. Transplantation of mesenchymal stem cells to enhance engraftment of hematopoietic stem cells. Leukemia 2007;21(8):1733–8.

[17] Le Blanc K, Tammik L, Sundberg B, Haynesworth SE, Ringden O. Mesenchymal stem cells inhibit and stimulate mixed lymphocyte cultures and mitogenic responses independently of the major histocompatibility complex. Scand J Immunol 2003;57(1):11–20.

[18] Lee ST, Jang JH, Cheong JW, Kim JS, Maemg HY, Hahn JS, et al. Treatment of high-risk acute myelogenous leukaemia by myeloablative chemoradiotherapy followed by co-infusion of T cell-depleted haematopoietic stem cells and culture-expanded marrow mesenchymal stem cells from a related donor with one fully mismatched human leucocyte antigen haplotype. Br J Haematol 2002;118(4):1128–31.

[19] Le Blanc K, Rasmusson I, Sundberg B, Gotherstrom C, Hassan M, Uzunel M, et al. Treatment of severe acute graft-versus-host disease with third party haploidentical mesenchymal stem cells. Lancet 2004;363(9419):1439–41.

[20] Le Blanc K, Frassoni F, Ball L, Locatelli F, Roelofs H, Lewis I, et al. Mesenchymal stem cells for treatment of steroid-resistant, severe, acute graft-versus-host disease: a phase II study. Lancet 2008;371(9624):1579–86.

[21] Prasad VK, Lucas KG, Kleiner GI, Talano JA, Jacobsohn D, Broadwater G, et al. Efficacy and safety of ex vivo cultured adult human mesenchymal stem cells (Prochymal) in pediatric patients with severe refractory acute graft-versus-host disease in a compassionate use study. Biol Blood Marrow Transplant 2011;17(4):534–41.

[22] Prasad VK, Lucas KG, Kleiner GI, Talano JA, Jacobsohn D, Broadwater G, et al. Efficacy and safety of ex-vivo cultured adult human mesenchymal stem cells (Prochymal(TM)) in pediatric patients with severe refractory acute graft-versus-host disease in a compassionate use study. Biol Blood Marrow Transplant 2010;.

[23] Taupin P. OTI-010 osiris therapeutics/JCR pharmaceuticals. Curr Opin Investig Drugs 2006;7(5):473–81.

[24] Koc ON, Day J, Nieder M, Gerson SL, Lazarus HM, Krivit W. Allogeneic mesenchymal stem cell infusion for treatment of metachromatic leukodystrophy (MLD) and Hurler syndrome (MPS-IH). Bone Marrow Transplant 2002;30(4):215–22.

[25] Chen SL, Fang WW, Ye F, Liu YH, Qian J, Shan SJ, et al. Effect on left ventricular function of intracoronary transplantation of autologous bone marrow mesenchymal stem cell in patients with acute myocardial infarction. Am J Cardiol 2004;94(1):92–5.

[26] Sueblinvong V, Weiss DJ. Cell therapy approaches for lung diseases: current status. Curr Opin Pharmacol 2009;9(3):268–73.

[27] Mazzini L, Mareschi K, Ferrero I, Vassallo E, Oliveri G, Boccaletti R, et al. Autologous mesenchymal stem cells: clinical applications in amyotrophic lateral sclerosis. Neurol Res 2006;28(5):523–6.

[28] Bang OY, Lee JS, Lee PH, Lee G. Autologous mesenchymal stem cell transplantation in stroke patients. Ann Neurol 2005;57(6):874–82.

[29] Yoshikawa T, Mitsuno H, Nonaka I, Sen Y, Kawanishi K, Inada Y, et al. Wound therapy by marrow mesenchymal cell transplantation. Plast Reconstr Surg 2008;121(3):860–77.

[30] Abdi R, Fiorina P, Adra CN, Atkinson M, Sayegh MH. Immunomodulation by mesenchymal stem cells: a potential therapeutic strategy for type 1 diabetes. Diabetes 2008;57(7):1759–67.

[31] Christopeit M, Schendel M, Foll J, Muller LP, Keysser G, Behre G. Marked improvement of severe progressive systemic sclerosis after transplantation of mesenchymal stem cells from an allogeneic haploidentical-related donor mediated by ligation of CD137L. Leukemia 2008;22(5):1062–4.

[32] Tyndall A, Uccelli A. Multipotent mesenchymal stromal cells for autoimmune diseases: teaching new dogs old tricks. Bone Marrow Transplant 2009;43(11):821–8.

[33] Karussis D, Karageorgiou C, Vaknin-Dembinsky A, Gowda-Kurkalli B, Gomori JM, Kassis I, et al. Safety and immunological effects of mesenchymal stem cell transplantation in patients with multiple sclerosis and amyotrophic lateral sclerosis. Arch Neurol 2010;67(10):1187–94.

[34] Perico N, Casiraghi F, Introna M, Gotti E, Todeschini M, Cavinato RA, et al. Autologous mesenchymal stromal cells and kidney transplantation: a pilot study of safety and clinical feasibility. Clin J Am Soc Nephrol 2011;6(2):412–22.

[35] Tan J, Wu W, Xu X, Liao L, Zheng F, Messinger S, et al. Induction therapy with autologous mesenchymal stem cells in living-related kidney transplants: a randomized controlled trial. JAMA 2012;307(11):1169–77.

[36] Horwitz EM, Prockop DJ, Fitzpatrick LA, Koo WW, Gordon PL, Neel M, et al. Transplantability and therapeutic effects of bone marrow-derived mesenchymal cells in children with osteogenesis imperfecta. Nat Med 1999;5(3):309–13.

[37] Martin I, Baldomero H, Bocelli-Tyndall C, Emmert MY, Hoerstrup SP, Ireland H, et al. Tissue Eng Part A. The survey on cellular and engineered tissue therapies in Europe in 2011 2014;20(3-4):842–53.

[38] Woodbury D, Schwarz EJ, Prockop DJ, Black IB. Adult rat and human bone marrow stromal cells differentiate into neurons. J Neurosci Res 2000;61(4):364–70.

[39] Makino S, Fukuda K, Miyoshi S, Konishi F, Kodama H, Pan J, et al. Cardiomyocytes can be generated from marrow stromal cells in vitro. J Clin Invest 1999;103(5):697–705.

[40] Wang G, Bunnell BA, Painter RG, Quiniones BC, Tom S, Lanson NA Jr, et al. Adult stem cells from bone marrow stroma differentiate into airway epithelial cells: potential therapy for cystic fibrosis. Proc Natl Acad Sci USA 2005;102(1):186–91.

[41] Sato Y, Araki H, Kato J, Nakamura K, Kawano Y, Kobune M, et al. Human mesenchymal stem cells xenografted directly to rat liver are differentiated into human hepatocytes without fusion. Blood 2005;106(2):756–63.

[42] Tang DQ, Cao LZ, Burkhardt BR, Xia CQ, Litherland SA, Atkinson MA, et al. In vivo and in vitro characterization of insulin-producing cells obtained from murine bone marrow. Diabetes 2004;53(7):1721–32.

[43] Phinney DG, Prockop DJ. Concise review: mesenchymal stem/multipotent stromal cells: the state of transdifferentiation and modes of tissue repair--current views. Stem Cells 2007;25(11):2896–902.

[44] Prockop DJ. "Stemness" does not explain the repair of many tissues by mesenchymal stem/multipotent stromal cells (MSCs). Clin Pharmacol Ther 2007;82(3):241–3.

[45] Lu P, Blesch A, Tuszynski MH. Induction of bone marrow stromal cells to neurons: differentiation, transdifferentiation, or artifact? J Neurosci Res 2004;77(2):174–91.

[46] Tolar J, Le Blanc K, Keating A, Blazar BR. Concise review: hitting the right spot with mesenchymal stromal cells. Stem Cells 2010;28(8):1446–55.

[47] Devine SM, Bartholomew AM, Mahmud N, Nelson M, Patil S, Hardy W, et al. Mesenchymal stem cells are capable of homing to the bone marrow of non-human primates following systemic infusion. Exp Hematol 2001;29(2):244–55.

[48] Devine SM, Cobbs C, Jennings M, Bartholomew A, Hoffman R. Mesenchymal stem cells distribute to a wide range of tissues following systemic infusion into nonhuman primates. Blood 2003;101(8):2999–3001.

[49] Dar A, Kollet O, Lapidot T. Mutual, reciprocal SDF-1/CXCR4 interactions between hematopoietic and bone marrow stromal cells regulate human stem cell migration and development in NOD/SCID chimeric mice. Exp Hematol 2006;34(8):967–75.

[50] Lee RH, Pulin AA, Seo MJ, Kota DJ, Ylostalo J, Larson BL, et al. Intravenous hMSCs improve myocardial infarction in mice because cells embolized in lung are activated to secrete the anti-inflammatory protein TSG-6. Cell Stem Cell 2009;5(1):54–63.

[51] Di Nicola M, Carlo-Stella C, Magni M, Milanesi M, Longoni PD, Matteucci P, et al. Human bone marrow stromal cells suppress T-lymphocyte proliferation induced by cellular or nonspecific mitogenic stimuli. Blood 2002;99(10):3838–43.

[52] Krampera M, Glennie S, Dyson J, Scott D, Laylor R, Simpson E, et al. Bone marrow mesenchymal stem cells inhibit the response of naive and memory antigen-specific T cells to their cognate peptide. Blood 2003;101(9):3722–9.

[53] English K, Ryan JM, Tobin L, Murphy MJ, Barry FP, Mahon BP. Cell contact, prostaglandin E(2) and transforming growth factor beta 1 play non-redundant roles in human mesenchymal stem cell induction of CD4 + CD25(High) forkhead box P3+ regulatory T cells. Clin Exp Immunol 2009;156(1):149–60.

[54] Rasmusson I, Le Blanc K, Sundberg B, Ringden O. Mesenchymal stem cells stimulate antibody secretion in human B cells. Scand J Immunol 2007;65(4):336–43.
[55] Tabera S, Perez-Simon JA, Diez-Campelo M, Sanchez-Abarca LI, Blanco B, Lopez A, et al. The effect of mesenchymal stem cells on the viability, proliferation and differentiation of B-lymphocytes. Haematologica 2008;93(9):1301–9.
[56] Traggiai E, Volpi S, Schena F, Gattorno M, Ferlito F, Moretta L, et al. Bone marrow-derived mesenchymal stem cells induce both polyclonal expansion and differentiation of B cells isolated from healthy donors and systemic lupus erythematosus patients. Stem Cells 2008;26(2):562–9.
[57] Spaggiari GM, Capobianco A, Becchetti S, Mingari MC, Moretta L. Mesenchymal stem cell-natural killer cell interactions: evidence that activated NK cells are capable of killing MSCs, whereas MSCs can inhibit IL-2-induced NK-cell proliferation. Blood 2006;107(4):1484–90.
[58] Nauta AJ, Kruisselbrink AB, Lurvink E, Willemze R, Fibbe WE. Mesenchymal stem cells inhibit generation and function of both CD34 + -derived and monocyte-derived dendritic cells. J Immunol 2006;177(4):2080–7.
[59] Ramasamy R, Fazekasova H, Lam EW, Soeiro I, Lombardi G, Dazzi F. Mesenchymal stem cells inhibit dendritic cell differentiation and function by preventing entry into the cell cycle. Transplantation 2007;83(1):71–6.
[60] Nemeth K, Leelahavanichkul A, Yuen PS, Mayer B, Parmelee A, Doi K, et al. Bone marrow stromal cells attenuate sepsis via prostaglandin E(2)-dependent reprogramming of host macrophages to increase their interleukin-10 production. Nat Med 2009;15(1):42–9.
[61] Kim J, Hematti P. Mesenchymal stem cell-educated macrophages: a novel type of alternatively activated macrophages. Exp Hematol 2009;37(12):1445–53.
[62] Zhang QZ, Su WR, Shi SH, Wilder-Smith P, Xiang AP, Wong A, et al. Human gingiva-derived mesenchymal stem cells elicit polarization of m2 macrophages and enhance cutaneous wound healing. Stem Cells 2010;28(10):1856–68.
[63] Kastrinaki MC, Sidiropoulos P, Roche S, Ringe J, Lehmann S, Kritikos H, et al. Functional, molecular and proteomic characterisation of bone marrow mesenchymal stem cells in rheumatoid arthritis. Ann Rheum Dis 2008;67(6):741–9.
[64] Zhou S, Greenberger JS, Epperly MW, Goff JP, Adler C, Leboff MS, et al. Age-related intrinsic changes in human bone-marrow-derived mesenchymal stem cells and their differentiation to osteoblasts. Aging Cell 2008;7(3):335–43.
[65] da Silva Meirelles L, Chagastelles PC, Nardi NB. Mesenchymal stem cells reside in virtually all post-natal organs and tissues. J Cell Sci 2006;119(Pt. 11):2204–13.
[66] In 't Anker PS, Scherjon SA, Kleijburg-van der Keur C, de Groot-Swings GM, Claas FH, Fibbe WE, et al. Isolation of mesenchymal stem cells of fetal or maternal origin from human placenta. Stem Cells 2004;22(7):1338–45.
[67] in 't Anker PS, Noort WA, Scherjon SA, Kleijburg-van der Keur C, Kruisselbrink AB, van Bezooijen RL, et al. Mesenchymal stem cells in human second-trimester bone marrow, liver, lung, and spleen exhibit a similar immunophenotype but a heterogeneous multilineage differentiation potential. Haematologica 2003;88(8):845–52.
[68] Hematti P. Human embryonic stem cell-derived mesenchymal progenitors: an overview. Methods Mol Biol 2011;690:163–74.
[69] Sharma RR, Pollock K, Hubel A, McKenna D. Mesenchymal stem or stromal cells: a review of clinical applications and manufacturing practices. Transfusion 2014;54(5):1418–37.

[70] Fang B, Song Y, Liao L, Zhang Y, Zhao RC. Favorable response to human adipose tissue-derived mesenchymal stem cells in steroid-refractory acute graft-versus-host disease. Transplant Proc 2007;39(10):3358–62.

[71] Hanson SE, Bentz ML, Hematti P. Mesenchymal stem cell therapy for nonhealing cutaneous wounds. Plast Reconstr Surg 2010;125(2):510–6.

[72] Gu Z, Akiyama K, Ma X, Zhang H, Feng X, Yao G, et al. Transplantation of umbilical cord mesenchymal stem cells alleviates lupus nephritis in MRL/lpr mice. Lupus 2010;19(13):1502–14.

[73] Brooke G, Rossetti T, Pelekanos R, Ilic N, Murray P, Hancock S, et al. Manufacturing of human placenta-derived mesenchymal stem cells for clinical trials. Br J Haematol 2009;144(4):571–9.

[74] Bongso A, Fong CY. The therapeutic potential, challenges and future clinical directions of stem cells from the Wharton's jelly of the human umbilical cord. Stem Cell Rev 2013;9(2):226–40.

[75] Strioga M, Viswanathan S, Darinskas A, Slaby O, Michalek J. Same or not the same? Comparison of adipose tissue-derived versus bone marrow-derived mesenchymal stem and stromal cells. Stem Cells Dev 2012;21(14):2724–52.

[76] Hematti P. Role of mesenchymal stromal cells in solid organ transplantation. Transplant Rev 2008;22(4):262–73.

[77] Kebriaei P, Isola L, Bahceci E, Holland K, Rowley S, McGuirk J, et al. Adult human mesenchymal stem cells added to corticosteroid therapy for the treatment of acute graft-versus-host disease. Biol Blood Marrow Transplant 2009;15(7):804–11.

[78] Hare JM, Fishman JE, Gerstenblith G, DiFede Velazquez DL, Zambrano JP, Suncion VY, et al. Comparison of allogeneic vs autologous bone marrow-derived mesenchymal stem cells delivered by transendocardial injection in patients with ischemic cardiomyopathy: the POSEIDON randomized trial. JAMA 2012;308(22):2369–79.

[79] Polchert D, Sobinsky J, Douglas G, Kidd M, Moadsiri A, Reina E, et al. IFN-gamma activation of mesenchymal stem cells for treatment and prevention of graft versus host disease. Eur J Immunol 2008;38(6):1745–55.

[80] Casiraghi F, Azzollini N, Todeschini M, Cavinato RA, Cassis P, Solini S, et al. Localization of mesenchymal stromal cells dictates their immune or proinflammatory effects in kidney transplantation. Am J Transplant 2012;12(9):2373–83.

[81] Perico N, Casiraghi F, Gotti E, Introna M, Todeschini M, Cavinato RA, et al. Mesenchymal stromal cells and kidney transplantation: pretransplant infusion protects from graft dysfunction while fostering immunoregulation. Transpl Int 2013;26(9):867–78.

[82] Sanganalmath SK, Bolli R. Cell therapy for heart failure: a comprehensive overview of experimental and clinical studies, current challenges, and future directions. Cir Res 2013;113(6):810–34.

[83] Eggenhofer E, Renner P, Soeder Y, Popp FC, Hoogduijn MJ, Geissler EK, et al. Features of synergism between mesenchymal stem cells and immunosuppressive drugs in a murine heart transplantation model. Transpl Immunol 2011;25(2–3):141–7.

[84] Hoogduijn MJ, Crop MJ, Korevaar SS, Peeters AM, Eijken M, Maat LP, et al. Susceptibility of human mesenchymal stem cells to tacrolimus, mycophenolic acid, and rapamycin. Transplantation 2008;86(9):1283–91.

[85] Chan JL, Tang KC, Patel AP, Bonilla LM, Pierobon N, Ponzio NM, et al. Antigen-presenting property of mesenchymal stem cells occurs during a narrow window at low levels of interferon-gamma. Blood 2006;107(12):4817–24.

[86] Stagg J, Pommey S, Eliopoulos N, Galipeau J. Interferon-gamma-stimulated marrow stromal cells: a new type of nonhematopoietic antigen-presenting cell. Blood 2006;107(6):2570–7.

[87] Uccelli A, Moretta L, Pistoia V. Mesenchymal stem cells in health and disease. Nat Rev 2008;8(9):726–36.

[88] von Bahr L, Batsis I, Moll G, Hagg M, Szakos A, Sundberg B, et al. Analysis of tissues following mesenchymal stromal cell therapy in humans indicate limited long-term engraftment and no ectopic tissue formation. Stem Cells 2012;30(7):1575–8.

[89] Prockop DJ, Kota DJ, Bazhanov N, Reger RL. Evolving paradigms for repair of tissues by adult stem/progenitor cells (MSCs). J Cell Mol Med 2010;14(9): 2190–9.

[90] Ning H, Yang F, Jiang M, Hu L, Feng K, Zhang J, et al. The correlation between cotransplantation of mesenchymal stem cells and higher recurrence rate in hematologic malignancy patients: outcome of a pilot clinical study. Leukemia 2008;22(3):593–9.

[91] Rubio D, Garcia-Castro J, Martin MC, de la Fuente R, Cigudosa JC, Lloyd AC, et al. Spontaneous human adult stem cell transformation. Cancer Res 2005;65(8): 3035–9.

[92] Wang Y, Huso DL, Harrington J, Kellner J, Jeong DK, Turney J, et al. Outgrowth of a transformed cell population derived from normal human BM mesenchymal stem cell culture. Cytotherapy 2005;7(6):509–19.

[93] Djouad F, Plence P, Bony C, Tropel P, Apparailly F, Sany J, et al. Immunosuppressive effect of mesenchymal stem cells favors tumor growth in allogeneic animals. Blood 2003;102(10):3837–44.

[94] Karnoub AE, Dash AB, Vo AP, Sullivan A, Brooks MW, Bell GW, et al. Mesenchymal stem cells within tumour stroma promote breast cancer metastasis. Nature 2007;449(7162):557–63.

[95] Prockop DJ, Keating A. Relearning the lessons of genomic stability of human cells during expansion in culture: implications for clinical research. Stem Cells 2012;30(6):1051–2.

[96] Youd M, Blickarz C, Woodworth L, Touzjian T, Edling A, Tedstone J, et al. Allogeneic mesenchymal stem cells do not protect NZBxNZW F1 mice from developing lupus disease. Clinical Exp Immunol 2010;161(1):176–86.

[97] Zhou K, Zhang H, Jin O, Feng X, Yao G, Hou Y, et al. Transplantation of human bone marrow mesenchymal stem cell ameliorates the autoimmune pathogenesis in MRL/lpr mice. Cell Mol Immunol 2008;5(6):417–24.

[98] Sun L, Akiyama K, Zhang H, Yamaza T, Hou Y, Zhao S, et al. Mesenchymal stem cell transplantation reverses multiorgan dysfunction in systemic lupus erythematosus mice and humans. Stem Cells 2009;27(6):1421–32.

[99] Liang J, Zhang H, Hua B, Wang H, Lu L, Shi S, et al. Allogenic mesenchymal stem cells transplantation in refractory systemic lupus erythematosus: a pilot clinical study. Ann Rheum Dis 2010;69(8):1423–9.

[100] Sun L, Wang D, Liang J, Zhang H, Feng X, Wang H, et al. Umbilical cord mesenchymal stem cell transplantation in severe and refractory systemic lupus erythematosus. Arthritis Rheum 2010;62(8):2467–75.

[101] Deskins DL, Bastakoty D, Saraswati S, Shinar A, Holt GE, Young PP. Human mesenchymal stromal cells: identifying assays to predict potency for therapeutic selection. Stem Cells Transl Med 2013;2(2):151–8.

[102] Haack-Sorensen M, Bindslev L, Mortensen S, Friis T, Kastrup J. The influence of freezing and storage on the characteristics and functions of human mesenchymal stromal cells isolated for clinical use. Cytotherapy 2007;9(4):328–37.

[103] Neuhuber B, Swanger SA, Howard L, Mackay A, Fischer I. Effects of plating density and culture time on bone marrow stromal cell characteristics. Exp Hematol 2008;36(9):1176–85.

[104] Dal Pozzo S, Urbani S, Mazzanti B, Luciani P, Deledda C, Lombardini L, et al. High recovery of mesenchymal progenitor cells with non-density gradient separation of human bone marrow. Cytotherapy 2010;12(5):579–86.

[105] Sotiropoulou PA, Perez SA, Salagianni M, Baxevanis CN, Papamichail M. Cell culture medium composition and translational adult bone marrow-derived stem cell research. Stem Cells 2006;24(5):1409–10.

[106] Sensebe L. Clinical grade production of mesenchymal stem cells. Biomed Mater Eng 2008;18(Suppl. 1):S3–S10.

[107] Asher DM. Bovine sera used in the manufacture of biologicals: current concerns and policies of the U.S. Food and Drug Administration regarding the transmissible spongiform encephalopathies. Dev Biol Stand 1999;99:41–4.

[108] Sundin M, Ringden O, Sundberg B, Nava S, Gotherstrom C, Le Blanc K. No alloantibodies against mesenchymal stromal cells, but presence of anti-fetal calf serum antibodies, after transplantation in allogeneic hematopoietic stem cell recipients. Haematologica 2007;92(9):1208–15.

[109] von Bonin M, Stolzel F, Goedecke A, Richter K, Wuschek N, Holig K, et al. Treatment of refractory acute GVHD with third-party MSC expanded in platelet lysate-containing medium. Bone Marrow Transplant 2009;43(3):245–51.

[110] Lucchini G, Introna M, Dander E, Rovelli A, Balduzzi A, Bonanomi S, et al. Platelet-lysate-expanded mesenchymal stromal cells as a salvage therapy for severe resistant graft-versus-host disease in a pediatric population. Biol Blood Marrow Transplant 2010;16(9):1293–301.

[111] Sanchez-Guijo F, Caballero-Velazquez T, Lopez-Villar O, Redondo A, Parody R, Martinez C, et al. Sequential third-party mesenchymal stromal cell therapy for refractory acute graft-versus-host disease. Biol Blood Marrow Transplant 2014;20(10):1580–5.

[112] Horn P, Bokermann G, Cholewa D, Bork S, Walenda T, Koch C, et al. Impact of individual platelet lysates on isolation and growth of human mesenchymal stromal cells. Cytotherapy 2010;12(7):888–98.

[113] Le Blanc K, Samuelsson H, Lonnies L, Sundin M, Ringden O. Generation of immunosuppressive mesenchymal stem cells in allogeneic human serum. Transplantation 2007;84(8):1055–9.

[114] Perez-Simon JA, Lopez-Villar O, Andreu EJ, Rifon J, Muntion S, Diez Campelo M, et al. Mesenchymal stem cells expanded in vitro with human serum for the treatment of acute and chronic graft-versus-host disease: results of a phase I/II clinical trial. Haematologica 2011;96(7):1072–6.

[115] Kratchmarova I, Blagoev B, Haack-Sorensen M, Kassem M, Mann M. Mechanism of divergent growth factor effects in mesenchymal stem cell differentiation. Science 2005;308(5727):1472–7.

[116] Gottipamula S, Muttigi MS, Kolkundkar U, Seetharam RN. Serum-free media for the production of human mesenchymal stromal cells: a review. Cell Prolif 2013;46(6):608–27.

[117] Hanley PJ, Mei Z, da Graca Cabreira-Hansen M, Klis M, Li W, Zhao Y, et al. Manufacturing mesenchymal stromal cells for phase I clinical trials. Cytotherapy 2013;15(4):416–22.

[118] Colter DC, Sekiya I, Prockop DJ. Identification of a subpopulation of rapidly self-renewing and multipotential adult stem cells in colonies of human marrow stromal cells. Proc Natl Acad Sci USA 2001;98(14):7841–5.

[119] Ma T, Grayson WL, Frohlich M, Vunjak-Novakovic G. Hypoxia and stem cell-based engineering of mesenchymal tissues. Biotechnol Prog 2009;25(1):32–42.

[120] Das R, Jahr H, van Osch GJ, Farrell E. The role of hypoxia in bone marrow-derived mesenchymal stem cells: considerations for regenerative medicine approaches. Tissue Eng Part B Rev 2010;16(2):159–68.

[121] Sotiropoulou PA, Perez SA, Salagianni M, Baxevanis CN, Papamichail M. Characterization of the optimal culture conditions for clinical scale production of human mesenchymal stem cells. Stem Cells 2006;24(2):462–71.

[122] Hanley PJ, Mei Z, Durett AG, da Graca Cabreira-Harrison M, Klis M, Li W, et al. Efficient manufacturing of therapeutic mesenchymal stromal cells with the use of the quantum cell expansion system. Cytotherapy 2014;16(8):1048–58.

[123] Marquez-Curtis LA, Janowska-Wieczorek A, McGann LE, Elliott JA. Cryopreservation of mesenchymal stromal cells derived from various tissues. Cryobiology 2015;181–97.

[124] Samuelsson H, Ringden O, Lonnies H, Le Blanc K. Optimizing in vitro conditions for immunomodulation and expansion of mesenchymal stromal cells. Cytotherapy 2009;11(2):129–36.

[125] Francois M, Copland IB, Yuan S, Romieu-Mourez R, Waller EK, Galipeau J. Cryopreserved mesenchymal stromal cells display impaired immunosuppressive properties as a result of heat-shock response and impaired interferon-gamma licensing. Cytotherapy 2012;14(2):147–52.

Index

A

B

G

J

K

L

M

Q

R